The Body in Pain

Etude de Géricault d'après Eugène Delacroix 1818-19

The Body in Pain

THE MAKING AND UNMAKING
OF THE WORLD

Elaine Scarry

New York   Oxford

OXFORD UNIVERSITY PRESS

1985

BJ
1409
.S35
1985

Oxford University Press

Oxford   New York   Toronto
Delhi   Bombay   Calcutta   Madras   Karachi
Kuala Lumpur   Singapore   Hong Kong   Tokyo
Nairobi   Dar es Salaam   Cape Town
Melbourne   Auckland

and associated companies in
Beirut   Berlin   Ibadan   Mexico City   Nicosia

Published by Oxford University Press, Inc.,
200 Madison Avenue, New York, New York 10016

Library of Congress Cataloging-in-Publication Data
Scarry, Elaine.
The body in pain.
Includes index.
1. Pain.   2. War.   3. Torture.
I. Title.
BJ1409.S35   1985   128   85-15585
ISBN 0-19-503601-8

Printing (last digit): 9 8 7 6 5 4 3 2 1
Printed in the United States of America

ACKNOWLEDGMENTS

THE WORK for this book has been sustained by the generosity of many institutions and people, and I wish to thank them here.

A 1977 NEH Summer Grant and a 1977–78 fellowship from the Institute for
Human Values in Medicine made possible the initial research into the aesthetic,
medical, and political literatures, and funded travel both to the International
Secretariat of Amnesty International in London and to McGill University in
Montreal. To Amnesty International I am deeply grateful for allowing me access
to the published and unpublished materials in their research department, for
granting me permission to quote from those materials, and for their day-by-day
assistance during the weeks when I worked in their midst; their generosity was
as unfailing as it was unsurprising. I would also like to thank Dr. Ronald Melzack
of McGill University for both the substance and spirit of his conversation during
the summer of 1977, as well as for his advice at several later moments.

Attention to the legal contexts of pain first became possible when a 1979
University of Pennsylvania Summer Grant enabled me to devote an extended
period to reading the trial transcripts of personal injury cases. Because such
transcripts are not publically available, I am indebted to two Philadelphia firms—
LaBrum and Doak; and Beasley, Hewson, Casey, Colleran, Erbstein, and Thistle—for their hospitality throughout those months. I was able in the summer of
1980 to return to the problems posed by the legal materials, thanks to the research
provisions that Harvard Law School so generously extends to its Visiting Scholars.

I am fortunate to have been part of two working groups that brought together
people from the humanities, social sciences, medicine, and law. From 1979–
81, the Research Group on Suffering at the Hastings Center (Institute of Society,
Ethics and Life Sciences) met periodically to discuss both theoretical and practical
problems of healing; from all the participants in this seminar I learned a great
deal. A 1979–81 grant from the National Humanities Center provided an un-
interrupted year of writing, as well as the intellectual camaraderie of a large

group of people. My thanks to Jan Paxton, Madolene Stone, Dick Eaton, Hal
Berman, and many others for their lively ideas, friendship, and laughter; to
librarian Alan Tuttle for his research assistance; to Quentin Anderson, Joe Beatty,
and Emory Elliott for their tough-minded and provocative readings of the first
chapter; and to moral philosopher David Falk for his reading of several chapters
and for many hours of fruitful discussion, both in that year and the years that
followed.

By one path or another, sections of the manuscript reached many people,
some whom I knew personally and others whom I did not. Their readings were
often scrupulous and imaginative; and the quality of their comments helped to
create the intellectual pressure necessary to complete the as yet unwritten por-
tions. Catherine Gallagher, Elizabeth Hardwick, Steven Marcus, Joseph Scarry,
and Stephen Toulmin all played a larger part in the final writing of the book
than they themselves perhaps realize.

A 1982–83 University of Pennsylvania sabbatical leave and a 1983 University
of Pennsylvania Research Council grant for manuscript preparation made possible
the final stages of work on the book. The continual re-emergence of the name
"University of Pennsylvania" accurately suggests the ongoing support provided
by my colleagues both in English and other fields. Research leaves were taken
during the chairmanships of Stuart Curran and Robert Lucid; to them, as well
as to Daniel Hoffman, Roland Frye, Elizabeth Flower, and Jean Alter, my special
thanks for their encouragement and assistance. The opportunity to present parts
of the manuscript at Penn, as well as at Berkeley, Cornell, and the Hastings
Center, was of great value to me.

One of the subjects of this book is the passage of what is only imagined into
a material form, and the book has enacted its own content by itself gradually
acquiring a material form. Many people at Oxford participated in the physical
construction of this book; I am especially fortunate to have had William Sisler
as editor and Rosemary Wellner as manuscript editor. At the moment when this
book was first passing into typescript, Barbara Schulman devoted many generous
hours to proofreading. Eva Scarry has read the manuscript at every stage, pa-
tiently tracing the vagaries of handwriting into type, type into proof, and proof
into print, as though it were a pleasure to do so, even where the subject matter
most distressed or the arguments disturbed. I am grateful to have received permission to use Géricault's *Etude de Géricault d'après Eugène Delacroix 1818–
19* from its owner, a private collector in Switzerland. My thanks also to Michael
Fried, who first showed me Géricault's extraordinary drawings from the *Raft of
the Medusa* period.

The steady support of several people throughout the long writing of this book
has been decisive. Dr. Eric Cassell's responses to the manuscript have been as
important to me as his own writings on behalf of his patients have been inspiring.
Jack Davis has entered into the book's arguments with the unsparing intellectual

rigor familiar to all who know him. Allen Grossman's knowledge, moral fervor, and capacity for intellectual friendship are perhaps not unlimited, but he has made it difficult to identify the limits.

Work on this project, as on any project, has often seemed lonely and long. The people listed in these pages have conspired to assure that whenever I looked up from that work I would find a sturdy and bountiful world. No one has done more to construct that bounty than Philip Fisher, who seemed to have built a new desk each time I began a new chapter, and in this and many other ways created the surface on which the work could be done. For his unceasing habits of argument and invention, for the pressure of his belief and the energy of his disbelief, my deep thanks.

Philadelphia   E.S.
June 1985

CONTENTS

The Body in Pain

INTRODUCTION

A LTHOUGH this book has only a single subject, that subject can itself be divided into three different subjects: *first*, the difficulty of expressing physical pain; *second*, the political and perceptual complications that arise as a result of that difficulty; and *third*, the nature of both material and verbal expressibility or, more simply, the nature of human creation.

It might be best to picture these three subjects as three concentric circles, for when we enter into the innermost space of the first, we quickly discover that we are (whether or not this is what we intended) already standing within the wider circumference of the second, and no sooner do we make that discovery than we learn we have all along been standing in the midst of the third. To be at the center of any one of them is to be, simultaneously, at the center of all three.

Physical pain has no voice, but when it at last finds a voice, it begins to tell a story, and the story that it tells is about the inseparability of these three subjects, their embeddedness in one another. Although it is the task of this book to record that story—and hence to make visible the larger structures of entailment—it may be useful here at the opening to speak briefly of each subject in isolation.

The Inexpressibility of Physical Pain

When one hears about another person's physical pain, the events happening within the interior of that person's body may seem to have the remote character of some deep subterranean fact, belonging to an invisible geography that, however portentous, has no reality because it has not yet manifested itself on the visible surface of the earth. Or alternatively, it may seem as distant as the interstellar events referred to by scientists who speak to us mysteriously of not yet detectable intergalactic screams[1] or of "very distant Seyfert galaxies, a class

3

of objects within which violent events of unknown nature occur from time to time.''[2]

Vaguely alarming yet unreal, laden with consequence yet evaporating before the mind because not available to sensory confirmation, unseeable classes of objects such as subterranean plates, Seyfert galaxies, and the pains occurring in other people's bodies flicker before the mind, then disappear.

Physical pain happens, of course, not several miles below our feet or many miles above our heads but within the bodies of persons who inhabit the world through which we each day make our way, and who may at any moment be separated from us by only a space of several inches. The very temptation to invoke analogies to remote cosmologies (and there is a long tradition of such analogies) is itself a sign of pain's triumph, for it achieves its aversiveness in part by bringing about, even within the radius of several feet, this absolute split between one's sense of one's own reality and the reality of other persons.

Thus when one speaks about ''one's own physical pain'' and about ''another person's physical pain,'' one might almost appear to be speaking about two wholly distinct orders of events. For the person whose pain it is, it is ''effort-lessly'' grasped (that is, even with the most heroic effort it cannot *not* be grasped); while for the person outside the sufferer's body, what is ''effortless'' is *not* grasping it (it is easy to remain wholly unaware of its existence; even with effort, one may remain in doubt about its existence or may retain the astonishing freedom of denying its existence; and, finally, if with the best effort of sustained attention one successfully apprehends it, the aversiveness of the ''it'' one apprehends will only be a shadowy fraction of the actual ''it''). So, for the person in pain, so incontestably and unnegotiably present is it that ''having pain'' may come to be thought of as the most vibrant example of what it is to ''have certainty,'' while for the other person it is so elusive that ''hearing about pain'' may exist as the primary model of what it is ''to have doubt.'' Thus pain comes unsharably into our midst as at once that which cannot be denied and that which cannot be confirmed.

Whatever pain achieves, it achieves in part through its unsharability, and it ensures this unsharability through its resistance to language. ''English,'' writes Virginia Woolf, ''which can express the thoughts of Hamlet and the tragedy of Lear has no words for the shiver or the headache. . . . The merest schoolgirl when she falls in love has Shakespeare or Keats to speak her mind for her, but let a sufferer try to describe a pain in his head to a doctor and language at once runs dry.''[3] True of the headache, Woolf's account is of course more radically true of the severe and prolonged pain that may accompany cancer or burns or phantom limb or stroke, as well as of the severe and prolonged pain that may occur unaccompanied by any nameable disease. Physical pain does not simply resist language but actively destroys it, bringing about an immediate reversion to a state anterior to language, to the sounds and cries a human being makes before language is learned.

Though Woolf frames her observation in terms of one particular language, the essential problem she describes, not limited to English, is characteristic of all languages. This is not to say that one encounters *no* variations in the expressibility of pain as one moves across different languages. The existence of culturally stipulated responses to pain—for example, the tendency of one population to vocalize cries; the tendency of another to suppress them—is well documented in anthropological research. So, too, a particular constellation of sounds or words that make it possible to register alterations in the felt-experience of pain in one language may have no equivalent in a second language: thus Sophocles's agonized Philoctetes utters a cascade of *changing* cries and shrieks that in the original Greek are accommodated by an array of formal words (some of them twelve syllables long), but that at least one translator found could only be rendered in English by the uniform syllable "Ah" followed by variations in punctuation (Ah! Ah!!!!). But even if one were to enumerate many additional examples, such cultural differences, taken collectively, would themselves constitute only a very narrow margin of variation and would thus in the end work to expose and confirm the universal sameness of the central problem, a problem that originates much less in the inflexibility of any one language or in the shyness of any one culture than in the utter rigidity of pain itself: its resistance to language is not simply one of its incidental or accidental attributes but is essential to what it is.

Why pain should so centrally entail, require, this shattering of language will only gradually become apparent over the course of many pages; but an approximation of the explanation may be partially apprehended by noticing the exceptional character of pain when compared to all our other interior states. Contemporary philosophers have habituated us to the recognition that our interior states of consciousness are regularly accompanied by objects in the external world, that we do not simply "have feelings" but have feelings *for* somebody or something, that love is love of x, fear is fear of y, ambivalence is ambivalence about z. If one were to move through all the emotional, perceptual, and somatic states that take an object—hatred for, seeing of, being hungry for—the list would become a very long one and, though it would alternate between states we are thankful for and those we dislike, it would be throughout its entirety a consistent affirmation of the human being's capacity to move out beyond the boundaries of his or her own body into the external, sharable world.[4] This list and its implicit affirmation would, however, be suddenly interrupted when, moving through the human interior, one at last reached physical pain, for physical pain—unlike any other state of consciousness—has no referential content. It is not *of* or *for* anything. It is precisely because it takes no object that it, more than any other phenomenon, resists objectification in language.

Often, a state of consciousness other than pain will, if deprived of its object, begin to approach the neighborhood of physical pain; conversely, when physical pain is transformed into an objectified state, it (or at least some of its aversiveness) is eliminated. A great deal, then, is at stake in the attempt to invent linguistic structures that will reach and accommodate this area of experience normally so inaccessible to language; the human attempt to reverse the de-objectifying work of pain by forcing *pain itself* into avenues of objectification is a project laden with practical and ethical consequence.

Who are the authors of this attempted reversal, the creators or near-creators of a language for pain? Because the words of five different groups of women and men have been regularly consulted in the preliminary thinking for this book, it will be helpful to name them here, though they together constitute only a very partial list of all those who have entered into the long history of this struggle.

First, of course, are individuals who have themselves been in great pain and whose words are later available either because they themselves remember them, because a friend remembers them, or because they have been recorded and memorialized in, for example, a written case history. Though the total number of words may be meager, though they may be hurled into the air unattached to any framing sentence, something can be learned from these verbal fragments not only about pain but about the human capacity for word-making. To witness the moment when pain causes a reversion to the pre-language of cries and groans is to witness the destruction of language; but conversely, to be present when a person moves up out of that pre-language and projects the facts of sentience into speech is almost to have been permitted to be present at the birth of language itself.

Because the person in pain is ordinarily so bereft of the resources of speech, it is not surprising that the language for pain should sometimes be brought into being by those who are not themselves in pain but who speak *on behalf of* those who are. Though there are very great impediments to expressing another's sentient distress, so are there also very great reasons why one might want to do so, and thus there come to be avenues by which this most radically private of experiences begins to enter the realm of public discourse. Here are four such avenues.

Perhaps the most obvious is medicine, for the success of the physician's work will often depend on the acuity with which he or she can hear the fragmentary language of pain, coax it into clarity, and interpret it. The hesitation built into the previous sentence—"*perhaps* the most obvious"—acknowledges the fact that many people's experience of the medical community would bear out the opposite conclusion, the conclusion that physicians do not trust (hence, hear) the human voice, that they in effect perceive the voice of the patient as an "unreliable narrator" of bodily events, a voice which must be bypassed as quickly as possible so that they can get around and behind it to the physical events themselves. But if the only external sign of the felt-experience of pain (for which there is no alteration in the blood count, no shadow on the X ray, no pattern on the CAT scan) is the patient's verbal report (however itself inadequate), then to bypass the voice is to bypass the bodily event, to bypass the patient, to bypass the person in pain. Thus the reality of a patient's X-rayable cancer may be believed-in but the accompanying pain disbelieved and the pain medication underprescribed. Medical contexts, like all other contexts of human experience, provide instances of the alarming phenomenon noted earlier: to have great pain is to have certainty; to hear that another person has pain is to have doubt. (The doubt of other persons, here as elsewhere, amplifies the suffering of those already in pain.)

Medical contexts, however, also provide many counterinstances; and though this has historically long been the case (for there have always been individual physicians whose daily work was premised on both a deep affection for the human body and a profound respect for the human voice), it has become especially true of the present era of medicine, which has begun to focus increasing attention on the nature and treatment of pain.

The extent to which medical research on the physical problem of pain is simultaneously bound up with the problem of language creation is best illustrated by what may at first appear to be only a coincidence: the person who discovered what is now considered the most compelling and potentially accurate theoretical model of the physiology of pain is also the person who invented a diagnostic tool that enables patients to articulate the individual character of their pain with greater precision than was previously possible. Ronald Melzack, who has with his colleague Patrick Wall authored the widely respected and celebrated "Gate-Control Theory of Pain," has also with his colleague W. S. Torgerson developed the "McGill Pain Questionnaire" that, less well-known, is itself quietly celebrated in the day-by-day world of the hospital and pain clinic.

The invention of the diagnostic questionnaire was in part occasioned by Melzack's recognition that the conventional medical vocabulary ("moderate pain," "severe pain") described only one limited aspect of pain, its intensity; and that describing pain only in terms of this solitary dimension was equivalent to describing the complex realm of visual experience exclusively in terms of light flux.[5] Thus he and Torgerson, after gathering the apparently random words most often spoken by patients, began to arrange those words into coherent groups which, by making visible the consistency interior to any one set of words, worked to bestow visibility on the characteristics of pain. When heard in isolation, any one adjective such as "*throbbing* pain" or "*burning* pain" may appear to convey very little precise information beyond the general fact that the speaker is in distress. But when "throbbing" is placed in the company of certain other commonly occurring words ("flickering," "quivering," "pulsing," "throbbing," and "beating"), it is clear that all five of them express, with varying degrees of intensity, a rhythmic on-off sensation, and thus it is also clear that one coherent dimension of the felt-experience of pain is this "temporal dimension." Similarly, when "burning" is placed in the context of three other words ("hot pain," "burning pain," "scalding pain," "searing pain"), it is apparent that these words, though once more differing importantly in their intensity, are alike in registering the existence of a "thermal dimension" to pain. Again, the words "pinching," "pressing," "gnawing," "cramping," and "crushing," together express what Melzack and Torgerson have designated as "constrictive pressure." Out of these categories larger categories are formed; for the "temporal," "thermal," and "constrictive" groups are among those that together express the *sensory* content of pain, while certain other word groupings express pain's *affective* content, and still others its evaluative or *cognitive* content.

Although the precise sensitivity of this diagnostic tool will only be fully determined after additional years of testing and use, it is already certain that the questionnaire enables patients to generate descriptions more easily.[6] It has also become evident that the particular array of words chosen by the patient may help to indicate the presence or absence of a particular disease as well as the most effective means of diminishing the pain. The choice of the three words "searing," "pulsing," and "shooting," for example, tells the physician that the patient's pain is characterized by the thermal, temporal, and spatial dimensions. Because this particular triad of dimensions is more characteristic of some diseases than of others, the physician knows whether arthritis or instead cancer or instead nerve damage should be suspected as a possible accompaniment. Again, because pain characterized by this particular triad of dimensions has begun to be shown (in the initial years of the questionnaire's use) to be more susceptible to some forms of therapy or medication than others, the physician knows how best to begin the longed-for healing process.

It would be inaccurate to suggest that either the medical problem of pain or the problem of expressing pain in medical contexts has been solved. But through the mediating structures of this diagnostic questionnaire, language ("as if," T. S. Eliot might say, "a magic lantern threw the nerves in patterns on a screen") has begun to become capable of providing an external image of interior events. Melzack and Torgerson have not discovered new words but have instead uncovered a structure residing in the narrow, already-existing vocabulary, the vocabulary originated by patients themselves. Thus necessary to the invention of this diagnostic tool was Melzack's assumption that the human voice, far from being untrustworthy, is capable of accurately exposing even the most resistant aspects of material reality. The depth of his belief in the referential powers of the human voice only becomes visible, however, when one recognizes that he has found in language not only the record of the felt-experience of pain, the signs of accompanying disease, and the invitation to appropriate treatment (as are all suggested by the McGill Questionnaire) but has found there even the secrets of the neurological and physiological pathways themselves; for, according to his own account,[7] it was while listening to the language of his patients that he first intuited what in its later formulation became known to us as the "Gate-Control Theory of Pain."

This same trust in language also characterizes the work occurring in several nonmedical contexts; and so, in addition to medical case histories and diagnostic questionnaires, there come to be other verbal documents—the publications of Amnesty International, the transcripts of personal injury trials, the poems and narratives of individual artists—that also record the passage of pain into speech. Each of these three enables pain to enter into a realm of shared discourse that is wider, more social, than that which characterizes the relatively intimate conversation of patient and physician. Because this public realm is of central concern in this book, each of the three will be extensively drawn on, at times appearing in the foreground and at other times in the background of the arguments being made.

Amnesty International's ability to bring about the cessation of torture depends centrally on its ability to communicate the reality of physical pain to those who are not themselves in pain. When, for example, one receives a letter from Amnesty in the mail, the words of that letter must somehow convey to the reader the aversiveness being experienced inside the body of someone whose country may be far away, whose name can barely be pronounced, and whose ordinary life is unknown except that it is known that that ordinary life has ceased to exist. The language of the letter must also resist and overcome the inherent pressures toward tonal instability: that language must at once be characterized by the greatest possible tact (for the most intimate realm of another human being's body is the implicit or explicit subject) and by the greatest possible immediacy (for the most crucial fact about pain is *its presentness* and the most crucial fact about torture is that it *is happening*). Tact and immediacy ordinarily work against one another; thus the difficulty of sustaining either tone is compounded by the necessity of sustaining both simultaneously.

The goal of the letter is not simply to make the reader a passive recipient of information about torture but to encourage his or her active assistance in eliminating torture. The "reader of the letter" may now, for example, become the "writer of a letter": that is, the person may begin to use his own language (though he may also draw on the language provided by Amnesty International, as Amnesty International in its formulations in turn has drawn on the language of former political prisoners) to address appropriate government officials or others who may have the authority to stop the torture. As even this brief description suggests, embedded in Amnesty's work, as in medical work, is the assumption that the act of verbally expressing pain is a necessary prelude to the collective task of diminishing pain. It is also true that here, as in medicine, the human voice must aspire to become a precise reflection of material reality: Amnesty's ability to stop torture depends on its international authority, and its international authority depends on its reputation for consistent accuracy; the words "someone is being tortured'' cannot be, and are never, pronounced unless it *is* the case that someone is being tortured.[8]

A fourth arena in which physical pain begins to enter language is the courtroom, for sometimes when a person has been very seriously injured, a civil suit follows; and the concept of compensation extends not only to the visible bodily injury but to the invisible experience of physical suffering. Viewed from a distance, such litigation may seem to lack the moral clarity of the work occurring at Amnesty International or in medical contexts. Here, for example, it is not immediately apparent in exactly what way the verbal act of expressing pain (which may result in a monetary award to the plaintiff) helps to eliminate the physical fact of the pain. Furthermore, built into the very structure of the case is a dispute about the correspondence between language and material reality: the accuracy of the descriptions of suffering given by the plaintiff's lawyer may be contested by the defendant's lawyer (though in instances involving extreme hurt, this tends not to be so). Why a civilization that invents medical institutions and international organizations like Amnesty International should also invent legal "remedies" for bodily suffering will eventually become clear. For the moment it is enough simply to notice that, whatever else is true, such litigation provides a situation that once again requires that the impediments to expressing pain be overcome. Under the pressure of this requirement, the lawyer, too, becomes an inventor of language, one who speaks *on behalf of* another person (the plaintiff) and attempts to communicate the reality of that person's physical pain to people who are not themselves in pain (the jurors).

A fifth and final source is art, and thus we come full circle back to Virginia Woolf's complaint about the absence (or what should more accurately be designated the "near-absence") of literary representations of pain. Alarmed and dismayed by his or her own failure of language, the person in pain might find it reassuring to learn that even the artist—whose lifework and everyday habit are to refine and extend the reflexes of speech—ordinarily falls silent before pain. The isolated instances in which this is not so, however, provide a much more compelling (because usable) form of reassurance—fictional analogues, perhaps whole paragraphs of words, that can be borrowed when the real-life crisis of silence comes.

Here and there in the vast expanse of literary texts, one comes upon an isolated play, an exceptional film, an extraordinary novel that is not just incidentally but centrally and uninterruptedly about the nature of bodily pain. In Sophocles's *Philoctetes*, the fate of an entire civilization is suspended in order to allow the ambassadors of that civilization to stop and take account of the nature of the human body, the wound in that body, the pain in that wound. Bergman's *Cries and Whispers* opens with a woman's diary entry, "It is Monday morning and I am in pain," and becomes throughout its duration (a duration that required that its cinematographer photograph two hundred different background shades of red) a sustained attempt to lift the interior facts of bodily sentience out of the inarticulate pre-language of "cries and whispers" into the realm of shared objectification.

More often, though still with great rarity, the subject may enter briefly into a small corner of a literary text, and such passages, whether a single line or a scene, may work to expose its attributes, even if the writer has merely shouted at pain, has resorted to name-calling ("the useless, unjust, incomprehensible, inept abomination that is physical pain," writes Huysmans[9]), or has instead bestowed on it a single name: "I have given a name to my pain and call it 'dog,' " announces Nietzsche in a brilliantly magisterial pretense of having at last gained the upper hand; "It is just as faithful, just as obtrusive and shameless, just as entertaining, just as clever as any other dog—and I can scold it and vent my bad mood on it, as others do with their dogs, servants, and wives."[10] In the isolation of pain, even the most uncompromising advocate of individualism might suddenly prefer a realm populated by companions, however imaginary and safely subordinate.

The rarity with which physical pain is represented in literature is most striking when seen within the framing fact of how consistently art confers visibility on other forms of distress (the thoughts of Hamlet, the tragedy of Lear, the heartache of Woolf's "merest schoolgirl"). *Psychological* suffering, though often difficult for any one person to express, *does* have referential content, *is* susceptible to verbal objectification, and is so habitually depicted in art that, as Thomas Mann's Settembrini reminds us, there is virtually no piece of literature that is *not* about suffering, no piece of literature that does not stand by ready to assist us. The issue of "assistance" is not, of course, a self-evident one: there is always the danger that a fictional character's suffering (whether physical or psychological) will divert our attention away from the living sister or uncle who can be helped by our compassion in a way that the fictional character cannot be; there is also the danger that because artists so successfully express suffering, they may themselves collectively come to be thought of as the most authentic class of sufferers, and thus may inadvertently appropriate concern away from others in radical need of assistance.

These possibilities, however, only call our attention back to the general question about the relation between expressing pain and eliminating pain that has arisen in each of the contexts of verbalization described earlier. The importance of this question will become more apparent once we move to our second subject.

The Political Consequences of Pain's Inexpressibility

Though the overt subject of the preceding discussion was the difficulty of expressing physical pain, at every moment lingering nearby was another subject, the political complications that arise as a result of that difficulty. How intricately the problem of pain is bound up with the problem of power can be briefly indicated by returning to the four central observations that surfaced earlier and seeing how laden with political consequence each of the four is.

First, we noticed that it often happens that two people can be in a room together, the one in pain, the other either partially or wholly unaware of the first person's pain. But the implicit question that is being asked here, "How is it that one person can be in the presence of another person in pain and not know it?," leads inevitably to a second question that will be dealt with extensively in this book, "How is it that one person can be in the presence of another person in pain and not know it—not know it to the point where he himself inflicts it, and goes on inflicting it?"

Second, it was observed that ordinarily there is no language for pain, that it (more than any other phenomenon) resists verbal objectification. But the relative ease or difficulty with which any given phenomenon can be *verbally represented* also influences the ease or difficulty with which that phenomenon comes to be *politically represented*. If, for example, it were easier to express intellectual aspiration than bodily hunger, one would expect to find that the problem of education had a greater degree of social recognition than the problem of malnutrition or famine; or again, if property (as well as the ways in which property can be jeopardized) were easier to describe than bodily disability (as well as the ways in which a disabled person can be jeopardized), then one would not be astonished to discover that a society had developed sophisticated procedures for protecting "property rights" long before it had succeeded in formulating the concept of "the rights of the handicapped." It is not simply accurate but tautological to observe that given any two phenomena, the one that is more visible will receive more attention. But the sentient fact of physical pain is not simply somewhat less easy to express than some second event, not simply somewhat less visible than some second event, but so nearly impossible to express, so flatly invisible, that the problem goes beyond the possibility that almost any other phenomenon occupying the same environment will distract attention from it. Indeed, even where it is virtually the only content in a given environment, it will be possible to describe that environment as though the pain were not there. Thus, for example, torture comes to be described—not only by regimes that torture but sometimes by people who stand outside those regimes—as a form of *information*-gathering or (in its even more remarkable formulation) *intelligence*-gathering; and uncovering the perceptual processes that permit this misdescription will be the first step in the extended structural analysis of torture to which Chapter 1 is devoted. Similarly (though by no means identically), while the central activity of war is injuring and the central goal in war is to out-injure the opponent, the fact of injuring tends to be absent from strategic and political descriptions of war: thus Chapter 2 will open with a review of writings by Clausewitz, Liddell Hart, Churchill, Sokolovskiy and other theorists of war in order to make visible the particular paths by which it disappears. The act of misdescribing torture or war, though in some instances intentional and in others unintentional, is in either case partially made possible by the inherent difficulty of accurately describing any event whose central content is bodily pain or injury.

The third central point that emerged earlier was an extension of the second: though there is ordinarily no language for pain, under the pressure of the desire to eliminate pain, an at least fragmentary means of verbalization is available both to those who are themselves in pain and to those who wish to speak on behalf of others. As physical pain is monolithically consistent in its assault on language, so the verbal strategies for overcoming that assault are very small in number and reappear consistently as one looks at the words of patient, physician, Amnesty worker, lawyer, artist: these verbal strategies revolve around the verbal sign of the weapon or what will eventually be called here the language of "agency." But we will also see that this verbal sign is so inherently unstable that when not carefully controlled (as it is in the contexts just cited) it can have different effects and can even be intentionally enlisted for the opposite purposes, invoked not to coax pain into visibility but to push it into further invisibility, invoked not to assist in the elimination of pain but to assist in its infliction, invoked not to extend culture (as happens in medicine, law, and art) but to dismantle that culture. The fact that the language of agency has on the one hand a radically benign potential and on the other hand a radically sadistic one does not lead to the conclusion that the two are inseparable, nor to the conclusion that those who use it in the first way are somehow implicated in the actions of those who use it in the second way. On the contrary: the two uses are not simply distinct but mutually exclusive; in fact we will see that one of the central tasks of civilization is to stabilize this most elementary sign.

The fourth major point that surfaced in the opening discussion was the recognition of the way pain enters into our midst as at once something that cannot be denied and something that cannot be confirmed (thus it comes to be cited in philosophic discourse as an example of conviction, or alternatively as an example of scepticism). To have pain is to have *certainty*; to hear about pain is to have *doubt*. But we will see that the relation between pain and belief is even more problematic than has so far been suggested. If the felt-attributes of pain are (through one means of verbal objectification or another) lifted into the visible world, *and if the referent for these now objectified attributes is understood to be the human body*, then the sentient fact of the person's suffering will become knowable to a second person. It is also possible, however, for the felt-attributes of pain to be lifted into the visible world but now attached to *a referent other than the human body*. That is, the felt-characteristics of pain—one of which is its compelling vibrancy or its incontestable reality or simply its "certainty"— can be appropriated away from the body and presented as the attributes of something else (something which by itself lacks those attributes, something which does not in itself appear vibrant, real, or certain). This process will throughout the argument of this book be called *"analogical verification"* or *"analogical substantiation."* It will gradually become apparent that at particular moments when there is within a society a crisis of belief—that is, when some central idea or ideology or cultural construct has ceased to elicit a population's belief either because it is manifestly fictitious or because it has for some reason been divested of ordinary forms of substantiation—the sheer material factualness of the human body will be borrowed to lend that cultural construct the aura of "realness" and "certainty." Part One, the first half of this book, will show how centrally those periods during which there is a breakdown in the framing assumptions of civilization depend on this process. Chapter 1 unfolds the nature of analogical verification as it occurs in torture, and Chapter 2 makes visible the crucial place it has in the structural logic of war. Part Two returns to the subject and shows that it is part of the original and ongoing project of civilization to diminish the reliance on (and to find substitutes for) this process of substantiation, and that this project comes in the west to be associated with an increased pressure toward material culture, or material self-expression.

As has become evident even in this brief review of the four initial assertions of this book (and as will become much more evident), the difficulty of articulating physical pain permits political and perceptual complications of the most serious kind. The failure to express pain—whether the failure to objectify its attributes or instead the failure, once those attributes are objectified, to refer them to their original site in the human body—will always work to allow its appropriation and conflation with debased forms of power; conversely, the successful expression of pain will always work to expose and make impossible that appropriation and conflation.

The paths by which these political and perceptual complications arise will be traced descriptively and with little reliance on formal terminology (in part because there is no pre-existing terminology for much of what will become visible here). Very occasionally it becomes necessary to introduce into the argument a somewhat formal term in order to make it clear that a particular phenomenon encountered in an earlier chapter is now being reencountered in a later one. For example, Chapter 1 provides an extended description of the process by which the attributes of pain can be severed from the pain itself and conferred on a political construct, but does so without insisting on any single name for this process. When, however, versions of this same phenomenon reappear in Chapter 2, sections IV and V, and Chapter 4, sections I–III, it is useful to have (almost as a kind of shorthand) the name "analogical verification" so that the similarities as well as the critically important differences between the various versions can be talked about and assessed. There are four or five other terms (e.g., referential instability, intentional object) that at certain points become similarly necessary; but in each case the phenomenon to which the term refers will already have been set forth, and thus the particular way it is being used (which may or may not overlap with the use of this vocabulary in other frameworks), as well as the particular array of questions that attend it, will be clear. The one exception is the phrase "language of agency" which arises at such an early moment in the book that a preliminary description and illustration may be helpful here.

Because the existing vocabulary for pain contains only a small handful of adjectives, one passes through direct descriptions very quickly and (as V. C. Medvei noted in his 1948 treatise on pain[11]) almost immediately encounters an "as if" structure: it feels as if . . . ; it is as though. . . . On the other side of the ellipse there reappear again and again (regardless of whether the immediate context of the vocalization is medical or literary or legal) two and only two metaphors, and they are metaphors whose inner workings are very problematic. The first specifies an external agent of the pain, a weapon that is pictured as producing the pain; and the second specifies bodily damage that is pictured as accompanying the pain. Thus a person may say, "It feels as though a hammer is coming down on my spine" even where there is no hammer; or "It feels as if my arm is broken at each joint and the jagged ends are sticking through the skin" even where the bones of the arms are intact and the surface of the skin is unbroken. Physical pain is not identical with (and often exists without) either agency or damage, but these things are referential; consequently we often call on them to convey the experience of the pain itself.

In order to avoid confusion here, it should be noted that it is of course true that in any given instance of pain, there may actually be present a weapon (the hammer may really be there) or wound (the bones may really be coming through the skin); and the weapon or wound may immediately convey to anyone present the sentient distress of the person hurt; in fact, so suggestive will they be of the sensation of hurt that the person, if not actually in pain, may find it difficult to assure the companion that he or she is not in pain. In medical case histories of people whose pain began with an accident, the sentences describing the accident (the moment when the hammer fell from the ladder onto the person's spine) may more successfully convey the sheer fact of the patient's agony than those sentences that attempt to describe the person's pain directly, even though the impact of the hammer (lasting one second) and the pain (lasting one year) are obviously not the same (and the patient, if asked whether she has the feeling of "hammering" pain might correct us and say no, it is knife-like). The central point here is that insofar as an actual agent (a nail sticking into the bottom of the foot) and an imagined agent (a person's statement, "It feels as if there's a nail sticking into the bottom of my foot") both convey something of the felt-experience of pain to someone outside the sufferer's body, they both do so for the same reason: in neither case is the nail identical with the sentient experience of pain; and yet because it has shape, length, and color, because it either exists (in the first case) or can be pictured as existing (in the second case) at the external boundary of the body, it begins to externalize, objectify, and make sharable what is originally an interior and unsharable experience.

Both weapon (whether actual or imagined) and wound (whether actual or imagined) may be used associatively to express pain. To some extent the inner workings of the two metaphors, as well as the perceptual complications that attend their use, overlap because the second (bodily damage) sometimes occurs as a version of the first (agency). The feeling of pain entails the feeling of being acted upon, and the person may either express this in terms of the world acting on him ("It feels like a knife . . . ") or in terms of his own body acting on him ("It feels like the bones are cutting through . . . "). Thus, though the phrase "language of agency" refers primarily to the image of the weapon, its meaning also extends to the image of the wound. Ordinarily, however, the metaphor of bodily damage also entails a wholly distinct set of perceptual complications; and these complications, as well as the ways in which they get sorted out by culture, will require a separate treatment and be dealt with in a later work.

As an actual physical fact, a weapon is an object that goes into the body and produces pain; as a perceptual fact, it can lift pain and its attributes out of the body and make them visible. The mental habit of *recognizing* pain *in* the weapon (despite the fact that an inanimate object cannot "have pain" or any other sentient experience) is both an ancient and an enduring one. Thus Homer speaks of an arrow "freighted with dark pains," as though the heavy hurt the arrow will cause is already visibly contained in and carried by the object—is palpably there as its weight and cargo.[12] Margery Kempe, the fourteenth-century mystic, speaks of a "boisterous nail," as though not only the pain that can be produced by the nail but the noises and cries in turn produced by the person in pain are already audible in the nail itself.[13] It is in the spirit of the same observation that Wittgenstein asks whether we ought not to be able to speak of the stone that causes hurt as having "pain patches" on it.[14] And the implications of the observation are extended in Joseph Beuys's small sculpture of a knife blade bound in gauze, exhibited at the Guggenheim in 1979 and entitled, "When you cut your finger, bandage the knife."

The point here is not just that pain can be apprehended in the image of the weapon (or wound) but that it almost cannot be apprehended without it: few people would have difficulty understanding Michael Walzer's troubled statement, "I cannot conceptualize infinite pain without thinking of whips and scorpions, hot irons and other people";[15] and the fact that the very word "pain" has its etymological home in "*poena*" or "punishment" reminds us that even the elementary act of naming this most interior of events entails an immediate mental somersault out of the body into the external social circumstances that can be pictured as having caused the hurt.

Given the expressive potential of the language of agency, it is not surprising that it reappears continually in the words of those working to objectify and eliminate pain. Many of the elementary adjectives listed on the McGill Pain Questionnaire (e.g., burning, stabbing, drilling, pinching, gnawing) are embedded forms of this language since, as Melzack's own account makes clear, a patient may characterize the pain in her arm as "burning" or instead say "It feels as if my arm's on fire," may characterize the pain behind his eyes as "drilling" or instead say "It feels as though a drill. . . . " (Some forms of pain therapy explicitly invite the patient to conceptualize a weapon or object inside the body and then mentally push it out—a process that has precedents in much older remedies that often entailed a shaman or doctor mimetically "pulling" the pain out of the body with some appropriately shaped object.) Medical researchers also use agency language in their descriptions and maps of physiological mechanisms: the term "*trigger* points" (used to indicate the bodily points where pain usually originates or the paths along which it spreads) is an instance.

Those working within the nonmedical contexts described earlier—Amnesty International, the law, art—also show this same awareness of the expressive potential of the sign of the weapon: thus Amnesty International realized they would be able to enlist the help of men and women in many walks of life when a 1963 newspaper image of a torture weapon elicited from the public an immediate outcry against the human hurt visibly suggested by the object;[16] the sign of the weapon is repeatedly introduced into that section of the closing argument in a personal injury trial that is explicitly devoted to describing the plaintiff's "pain and suffering"; and Odysseus's original adeptness at wholly ignoring Philoctetes's pain is subverted by the intervening sign of the weapon, for he eventually "sees" Philoctetes's pain only because circumstances arise that require him to attend to—what else?—Philoctetes's bow.

This brief array of examples illustrates the benign potential of the language of agency, its invocation by those who wish to express their own pain (Melzack's patients), to express someone else's pain (Amnesty, Sophocles), or to imagine other people's pain (Walzer); and a detailed examination of any one of these uses would confirm the critically important point stressed earlier, that in order to express pain one must *both* objectify its felt-characteristics *and* hold steadily visible the referent for those characteristics. That is, the image of the weapon only enables us to see *the attributes* of pain if it is clear that the attributes we are seeing are the attributes *of pain* (and not of something else). The deeply problematic character of this language, its inherent instability, arises precisely because it permits a break in the identification of the referent and thus a misidentification of the thing to which the attributes belong. While the advantage of the sign is its proximity to the body, its disadvantage is the ease with which it can then be spatially separated from the body.

Given the fact that actual weapons ordinarily hurt rather than heal persons, it would be surprising if the iconography of weapons ordinarily worked to assist those in pain, and of course it does not. When, for example, the language of agency enters political discourse, its use is often very far removed from the one just summarized, as the following unpleasant examples suggest: it is said that Richard Nixon's favorite saying whenever he had triumphed over and therefore discomforted a journalist was, "That really flicks the scab off";[17] George Wallace once spoke of wanting to give his political enemy a "barbed-wire enema" (and when publically called on to apologize for his statement, he appeared to believe its scatology rather than its cruelty was the problem);[18] the language of agency is again recognizable in Lyndon Johnson's verbal habit during the Vietnam period of describing a military or political victory as "nailing the coon skin to the wall"; and, in a startling confusion of the large with the small, Ronald Reagan complained of the Soviet reaction to the American decision to produce a neutron bomb by saying, "[The Russians] are squealing like they're sitting on a sharp nail."[19] It would be possible to debate for a long time the significance or insignificance of this language (for though clearly not innocent, the precise extent to which it is harmful is unclear). But what is both self-evident and undebatable is this: whatever these sentences express, what they do not express is physical pain. In none of the four does the sign of the weapon work to bestow visibility on the aversiveness of physical suffering, and in none of the four does the speaker invoke that sign in order to direct sympathetic attention to the felt-experience of the person pictured as acted upon by the object.

It will eventually become apparent that the particular perceptual confusion sponsored by the language of agency is the conflation of pain with power. For now, it is enough to notice that the mere appearance of the sign of a weapon in a spoken sentence, a written paragraph, or a visual image (e.g., the litany of weapons in the writings of Sade; their occasional presence in a fashion photograph or painting) does not mean that there has been any attempt to present pain and, on the contrary, often means that the nature of pain has just been pushed into deeper obscurity.

The negative use of the language of agency, only fleetingly suggested here, will in the opening chapter be shown not as it occurs in isolated sentences but as it enters into the structure of larger events where it achieves the full extremity of its sadistic potential. In torture, it is in part the obsessive display of agency that permits one person's body to be translated into another person's voice, that allows real human pain to be converted into a regime's fiction of power. The sign of the weapon will again be attended to in the second chapter, for the perceptual confusion sponsored by the sign increases the difficulty of accurately identifying the function of injuring in war (and thus increases also the difficulty of identifying the precise character of the action that could plausibly be used in its place).

As the language of agency has a central place in torture and war—the two events in which the ordinary assumptions of culture are suspended—so, conversely, the basic structures of culture are centrally devoted to stabilizing this sign. The discussion of civilization's ongoing modifications of "agency" in the second half of the book (Part Two: Chapters 3–5) is sometimes framed explicitly in terms of changes in verbal or visual *iconography* (e.g., the sign of the cross; the signs on the flags of nations), at other times is framed in terms of the restructuring not of the icon or image but of the *actual object* (e.g., the modifications in the form of the weapon that allow it to become transformed into a tool or into an artifact), and at still other times is framed in terms of the human *actions* associated with such objects (e.g., the elaborate mental labor of dissociating "wounding" from "creating" in the Hebrew scriptures). The idiom of this last sentence—"tool," "artifact," "restructure," "creating"—calls attention to the fact that there is a third subject in this book that has so far not been introduced.

The Nature of Human Creation

We have seen that physical pain is difficult to express, and that this inexpressibility has political consequences; but we will also see that those political consequences—by making overt precisely what is at stake in "inexpressibility"—begin to expose by inversion the essential character of "expressibility," whether verbal or material. Thus as our first subject led to a second, so the second leads just as inevitably to a third: the nature of human creation.

What it is to "uncreate" and what it is to create eventually become central preoccupations of this book, as the overarching two-part division—*Unmaking* and *Making*—suggests. The way in which the material in the first half necessitates attention to the problem of creation addressed in the second can be briefly indicated here by first identifying in skeletal outline the central argument about torture and war, and then identifying what within the argument carries with it the requirement that "making" itself become better understood.

Based on the verbal accounts of people who were political prisoners during the 1970s, Chapter 1 shows that torture has a structure that is as narrow and consistent as its geographical incidence is widespread. That structure entails the simultaneous and inseparable occurrence of three events which if described sequentially would occur in the following order: first, the infliction of physical pain; second, the objectification of the eight central attributes of pain; and third, the translation of those attributes into the insignia of the regime.

The work of the opening chapter is only to identify and make manifest this three-part structure. But an example of the objectification and appropriation of *any one* of pain's attributes also begins to make it clear why the book at a much later point turns of necessity to the subject of creating. Physical pain—to invoke what is at this moment its single most familiar attribute—is language-destroying. Torture inflicts bodily pain that is itself language-destroying, but torture also mimes (objectifies in the external environment) this language-destroying capacity in its interrogation, the purpose of which is not to elicit needed information but visibly to deconstruct the prisoner's voice. The word "deconstruct" rather than "destroy" is used in the previous sentence because to say the interrogation "visibly destroys" the prisoner's voice only implies that the *outcome* of the event is the shattering of the person's voice (and if this alone were the goal, there would be no need for a verbal interrogation since the inflicted pain alone accomplishes this outcome). The prolonged interrogation, however, also graphically objectifies the step-by-step backward movement along the path by which language comes into being and which is here being reversed or uncreated or deconstructed. We will see that this same mime of uncreating reappears consistently throughout all the random details of torture—not only in relation to verbal constructs (e.g., sentences, names) but also in relation to material artifacts (e.g., a chair, a cup) and mental objects (i.e., the objects of consciousness). Thus it eventually becomes clear that this is not simply a repeated element within the large framing event but is the framing event itself. In other words, as the overall three-part structure of action emerges before our eyes, we gradually begin to recognize what it is we are looking at; and what we are looking at *is* the structure of unmaking. The import of this will be returned to after summarizing Chapter 2.

War, too, has a structure. As might be expected when one moves from what is essentially a two-person event to one involving hundreds of thousands of persons, and again when one moves from an event premised on one-directional injuring to one premised on two-directional or reciprocal injuring, the structure of war is more complex and the identification of that structure requires complicated sets of arguments and sub-arguments. The chapter is, however, divided into five sections and the overarching argument is carried through those divisions.

Sections I ("War Is Injuring") and II ("War Is a Contest") set two premises in place in order to ask the question, "What differentiates injuring from any other activity on which a contest can be based in order to arrive at a winner and a loser?" This is a critically important question since if injuring has only the solitary function of allowing one side to out-injure the other and thus of designating one of the disputants the winner, almost any other human activity could by now have been substituted in its place: thus injuring must have a second function. Section III shows that the single answer that has been given to this question (explicitly by Clausewitz and implicitly by twentieth-century political and military theorists)—namely, that war carries the power of its own enforcement—cannot possibly be true (Clausewitz worried about the probable falseness of the explanation, though his counterparts in this century seem less aware of its falsity). Section IV, the longest and most important phase of the argument, provides a different answer, showing the way the compelling reality of the injured bodies is being used at the end of war to lend the aura of material reality to the winning construct (as well as to the concept of winning itself) until there is time for the world participants to provide more legitimate means of substantiation. Section V compares the use of the human body in torture and in war in order to account for the moral distance that separates their respective procedures of analogical verification. The basis of the distinction is "consent": in war, the persons whose bodies are used in the confirmation process have given their consent over this most radical use of the human body while in torture no such consent is exercised. The chapter ends by showing that nuclear war more closely approximates the model of torture than the model of conventional war because it is a structural impossibility that the populations whose bodies are used in the confirmation process can have exercised any consent over this use of their bodies.

As was true of the torture chapter, the central purpose of the chapter on war is to identify the nature of the monolithic event under discussion and not to expose the interior shape of "unmaking." But, again as in the torture chapter, we will see that that central purpose cannot be accomplished without also undertaking the second task because "the structure of war" and "the structure of unmaking" are not two subjects but one. If as the intricacies of this conflation emerge before us they have the defect of sometimes seeming astonishing, they will also at every moment have the virtue of confirming the obvious; for that torture and war are acts of destruction (and hence somehow the opposite of creation), that they entail the suspension of civilization (and are somehow the opposite of that civilization), are things we have always known and things one immediately apprehends even when viewing these two events from a great distance; the only thing that could not have been anticipated from a distance but that is forced upon us as self-evident once we enter the interior of these two events is that they are, in the most literal and concrete way possible, an appropriation, aping, and reversing of the action of creating itself. Once the structures of torture and war have been exposed and compared, it becomes clear that the human action of making entails two distinct phases—making-up (mental imagining) and making-real (endowing the mental object with a material or verbal form)—and that the appropriation and deconstruction of making occur sometimes at the first and sometimes at the second of these two sites.

Part Two clarifies the structure of creating that emerges only in inverted outline in Part One. Chapter 3 attends specifically to "mental imagining" (or what was a moment ago called the phase of "making-up"), and Chapters 4 and 5 together examine the action of creating verbal and material artifacts (or what was a moment ago called the phase of "making-real"). Because the central arguments of these chapters themselves depend on the detailed and substantive observations about bodily pain and injury that occur in the earlier chapters, they will be summarized and introduced at the transition from Part One to Part Two, as we move onto the path that eventually carries us past Pegasus and telegraphs, past altars, light bulbs, and coats, past blankets, product liability trials, and songs, as well as medical vaccines and sacred texts.

The vocabulary of "creating," "inventing," "making," "imagining," is not in the twentieth century a morally resonant one: "imagining," for example, is usually described as an ethically neutral or amoral phenomenon; the phrase "material making" is similarly flat in its connotations, and is even (because of its conflation with "materialism") sometimes pronounced with a derisive inflection. But an unspoken question begins to arise in Part One which might be formulated in the following way: given that the deconstruction of creation is present in the structure of one event which is widely recognized as close to being an absolute of immorality (torture), and given that the deconstruction of creation is again present in the structure of a second event regarded as morally problematic by everyone and as radically immoral by some (war), is it not peculiar that the very thing being deconstructed—creation—does not in its intact form have a moral claim on us that is as high as the others' is low, that the action of creating is not, for example, held to be bound up with justice in the way those other events are bound up with injustice, that it (the mental, verbal, or material process of making the world) is not held to be centrally entailed in the elimination of pain as the unmaking of the world is held to be entailed in pain's infliction? The morality of creating cannot, of course, be inferred from the immorality of uncreating, and will instead be shown on its own terms. That we ordinarily perceive it as empty of ethical content is, it will be argued, itself a signal to us of how faulty and fragmentary our understanding of creation is, not only in this respect but in many others. It is not the valorization of making but its accurate description that is crucial, for *if* it is in fact laden with ethical consequence, then it may be that a firm understanding of *what it is* will in turn enable us to recognize more quickly what is happening not only in large-scale emergencies like torture or war but in other long-standing dilemmas, such as the inequity of material distribution.

In the long run, we will see that the story of *physical pain* becomes as well a story about the expansive nature of human *sentience*, the felt-fact of aliveness that is often sheerly happy, just as the story of *expressing* physical pain eventually opens into the wider frame of *invention*. The elemental "as if" of the person in pain ("It feels as if . . . ," "It is as though . . . ") will lead out into the array of counterfactual revisions entailed in making.

This book is about the way other persons become visible to us, or cease to be visible to us. It is about the way we make ourselves (and the originally interior facts of sentience) available to one another through verbal and material artifacts, as it is also about the way the derealization of artifacts may assist in taking away another person's visibility. The title of the book, *The Body in Pain*, designates as the book's subject the most contracted of spaces, the small circle of living matter; and the subtitle designates as its subject the most expansive territory, *The Making and Unmaking of the World.* But the two go together, for what is quite literally at stake in the body in pain is the making and unmaking of the world.

PART ONE

Unmaking

1

THE STRUCTURE OF TORTURE
The Conversion of Real Pain into
the Fiction of Power

NOWHERE is the sadistic potential of a language built on agency so visible as in torture. While torture contains language, specific human words and sounds, it is itself a language, an objectification, an acting out. Real pain, agonizing pain, is inflicted on a person; but torture, which contains specific acts of inflicting pain, is also itself a demonstration and magnification of the felt-experience of pain. In the very processes it uses to produce pain within the body of the prisoner, it bestows visibility on the structure and enormity of what is usually private and incommunicable, contained within the boundaries of the sufferer's body. It then goes on to deny, to falsify, the reality of the very thing it has itself objectified by a perceptual shift which converts the vision of suffering into the wholly illusory but, to the torturers and the regime they represent, wholly convincing spectacle of power. The physical pain is so incontestably real that it seems to confer its quality of "incontestable reality" on that power that has brought it into being. It is, of course, precisely because the reality of that power is so highly contestable, the regime so unstable, that torture is being used.[1]

What assists the conversion of absolute pain into the fiction of absolute power is an obsessive, self-conscious display of agency. On the simplest level, the agent displayed is the weapon. Testimony given by torture victims from many different countries almost inevitably includes descriptions of being made to stare at the weapon with which they were about to be hurt: prisoners of the Greek Junta (1967–71), for example, were made to contemplate a wall arrangement of whips, canes, clubs, and rods, were made to examine the size of the torturer's fist and the monogrammed ring which "he wore and which made his blows more painful," or were compelled to look at a bull's pizzle coated with the dried blood of a fellow prisoner.[2] But whatever the regime's primary weapon, it is only one of many weapons and its display is only one of many endlessly multiplied acts of display: torture is a process which not only converts but announces the conversion of every conceivable aspect of the event and the environmnent

27

into an agent of pain. It is not accidental that in the torturers' idiom the room in which the brutality occurs was called the "production room" in the Philippines,[3] the "cinema room" in South Vietnam,[4] and the "blue lit stage" in Chile[5]: built on these repeated acts of display and having as its purpose the production of a fantastic illusion of power, torture is a grotesque piece of compensatory drama.

Moral stupidity, then, here as in its less savage and obscene forms, has an unconscious structure. Torture is in its largest outlines the invariable and simultaneous occurrence of three phenomena which, if isolated into separate and sequential steps, would occur in the following order. First, pain is inflicted on a person in ever-intensifying ways. Second, the pain, continually amplified within the person's body, is also amplified in the sense that it is objectified, made visible to those outside the person's body. Third, the objectified pain is denied as pain and read as power, a translation made possible by the obsessive mediation of agency. The workings of these three phenomena will very gradually emerge during the following description of the place of body and voice in torture.

I. Pain and Interrogation

Torture consists of a primary physical act, the infliction of pain, and a primary verbal act, the interrogation. The first rarely occurs without the second. As is true of the present period, most historical episodes of torture, such as the Inquisition, have inevitably included the element of interrogation: the pain is traditionally accompanied by "the Question." Ancient history, too, confirms the insistent coupling; strangers caught by the *yaksha* cults in India, for example, were sacrificed after being subjected to a series of riddles.[6] The connection between the physical act and the verbal act, between body and voice, is often misstated or misunderstood. Although the information sought in an interrogation is almost never credited with being a *just* motive for torture, it is repeatedly credited with being the motive for torture. But for every instance in which someone with critical information is interrogated, there are hundreds interrogated who could know nothing of remote importance to the stability or self-image of the regime.[7] Just as within a precarious regime the motive for arrest is often a fiction (the eggseller's eggs were too small—Greece[8]), and just as the motive for punishing those imprisoned is often a fiction (the men, although locked in their cells, watched and applauded the television report that a military plane had crashed—Chile[9]), so what masquerades as the motive for torture is a fiction.

The idea that the need for information is the motive for the physical cruelty arises from the tone and form of the questioning rather than from its content: the questions, no matter how contemptuously irrelevant their content, are announced, delivered, *as though* they motivated the cruelty, *as if* the answers to them were crucial. Few other moments of human speech so conflate the modes of the interrogatory, the declarative, the imperative, as well as the emphatic form of each of these three, the exclamatory. Each mode implies a radically different relation of speaker to listener, and so the rapid slipping and colliding of these voices and the relations they imply—the independence of the declarative, the uncertain dependence of the interrogatory, the dominance of the imperative, each as though unaccompanied and unqualified by the others, raised to its most absolute in the urgency of the exclamatory—suggest a level of instability so extreme that the questioner might seem involved in the outcome to the very extent of his being. In fact, when this kind of conflation occurs in private, nonpolitical human speech, and where it is unaccompanied by physical brutality, when, say, a jealous lover or a terrified parent asks questions and asserts the answers in a way that rocks between utterly self-sufficient conviction and a pleading need of the listener's crediting or confirmation, the person may well be involved in the response to the very extent of his or her being. But as the content and context of the torturer's questions make clear, the fact that something is asked *as if* the content of the answer matters does not mean that it matters. It is crucial to see that the interrogation does not stand outside an episode of torture as its motive or justification: it is internal to the structure of torture, exists there because of its intimate connections to and interactions with the physical pain.

Pain and interrogation inevitably occur together in part because the torturer and the prisoner each experience them as opposites. The very question that, within the political pretense, matters so much to the torturer that it occasions his grotesque brutality will matter so little to the prisoner experiencing the brutality that he will give the answer. For the torturers, the sheer and simple fact of human agony is made invisible, and the moral fact of inflicting that agony is made neutral by the feigned urgency and significance of the question. For the prisoner, the sheer, simple, overwhelming fact of his agony will make neutral and invisible the significance of any question as well as the significance of the world to which the question refers. Intense pain is world-destroying. In compelling confession, the torturers compel the prisoner to record and objectify the fact that intense pain is world-destroying. It is for this reason that while the content of the prisoner's answer is only sometimes important to the regime, the form of the answer, the fact of his answering, is always crucial.

There is not only among torturers but even among people appalled by acts of torture and sympathetic to those hurt, a covert disdain for confession. This disdain is one of many manifestations of how inaccessible the reality of physical pain is to anyone not immediately experiencing it.[10] The nature of confession is falsified by an idiom built on the word "betrayal": in confession, one betrays oneself and all those aspects of the world—friend, family, country, cause—that the self is made up of. The inappropriateness of this idiom is immediately apparent in any nonpolitical context. It is a commonplace that at the moment when a dentist's drill hits and holds an exposed nerve, a person sees stars. What is meant by "seeing stars" is that the contents of consciousness are, during those moments, obliterated, that the name of one's child, the memory of a friend's face, are all absent. But the nature of this "absence" is not illuminated by the word "betrayal." One cannot betray or be false to something that has ceased to exist and, in the most literal way possible, the created world of thought and feeling, all the psychological and mental content that constitutes both one's self and one's world, and that gives rise to and is in turn made possible by language, ceases to exist.

For the most part, neither actual nor fictional accounts of torture focus on the process of disintegrating perception brought about by great pain and objectified in confession. Sartre's short story "The Wall" is a partial description of the process. The story begins shortly before a prisoner of the fascist government in Spain learns that he has been sentenced to be executed and it ends shortly after he learns that his death sentence has been, at least temporarily, reversed. For a moment his sudden reprieve seems senseless: when the regime had offered him his life in return for information about the hiding place of a friend, he had ensured his own death by naming a false location. In the final sentences of the story, he learns that his friend has been captured in the spot he named.

From this description one might assume that "The Wall" closes with a sense of crushing irony. The ending is almost a paradigm of ironic structures in its quick series of reversals (the expectation of execution gives way to sudden reprieve; the relief of reprieve gives way to a recognition of its cost), in its coupling of opposites (the seriousness of betrayal is brought about by the banality of gratuitous coincidence), and in its insistent exposure of the limits of awareness (what the prisoner understood himself to be concealing he was instead exposing; the realm in which his will could operate, which he assessed to be and accepted as being very very small, was in fact far smaller). But the ending is almost wholly without ironic impact: although anyone while reading will probably recognize as though from a great distance that the final events have a rhythm that would ordinarily make the mind wince, contract, even for a moment shut down, the actual response will probably be closer to a shrug. The utter flatness of the closing is indicative of—is in fact a way of allowing us to see our own participation in—the state of consciousness the story describes, a state that has at its center the single, overwhelming discrepancy between an increasingly palpable body and an increasingly substanceless world, a discrepancy that makes all the lesser discrepancies we normally identify as "ironic" seem as remote and full of dissolution as the world to which they belong.[11]

Pablo Ibbieta's experience in "The Wall" is close to if not identical with that of a person subjected to great pain. He is not tortured: he comments quickly at one point that if tortured he would certainly give the information asked. But he is sentenced to be executed, and is then suddenly released, and so in fact undergoes what has been in the recent past a form of torture common in places like Chile, Brazil, Greece, and the Philippines—the "mock execution" or, as it was called in the Philippines, "the process of dying." Of course, no particular form of torture is required to make visible the kinship between pain and death, both of which are radical and absolute, found only at the boundaries they themselves create. That pain is so frequently used as a symbolic substitute for death in the initiation rites of many tribes[12] is surely attributable to an intuitive human recognition that pain is the equivalent in felt-experience of what is unfeelable in death. Each only happens because of the body. In each, the contents of consciousness are destroyed. The two are the most intense forms of negation, the purest expressions of the anti-human, of annihilation, of total aversiveness, though one is an absence and the other a felt presence, one occurring in the cessation of sentience, the other expressing itself in grotesque overload. Regardless, then, of the context in which it occurs, physical pain always mimes death and the infliction of physical pain is always a mock execution.

The moment in which Pablo Ibbieta learns he is to be shot the next morning by a firing squad is the moment once described by George Eliot as that in which a person's general awareness "All men must die" is displaced by the specific awareness, "I must die—and soon." Whenever death can be designated as "soon" the dying has already begun. Ibbieta is dying not because he has yet experienced the damage that will end his life but because he has begun to experience the body that will end his life, the body that can be killed, and which when killed will carry away the conditions that allow him to exist. Throughout the night that leads up to the morning of the scheduled execution, his body alternates somewhat aimlessly between two forms of calling attention to itself: at times it expresses itself in a heightened sensitivity, in burning cheeks[13] or in a disturbingly acute sense of smell (8); at other times it asserts its presence through a greatly diminished sensitivity, no longer signalling him when it sweats (6) or urinates (12). For neither Ibbieta nor Sartre is there anything intriguingly paradoxical or even puzzling about this duality: the two simply belong together, exist uncommented upon side by side. Ibbieta's body is full of "pains . . . like a crowd of tiny scars" (8). The body is its pains, a shrill sentience that hurts and is hugely alarmed by its hurt; and the body is its scars, thick and forgetful, unmindful of its hurt, unmindful of anything, mute and insensate. The body, this intensely—and sometimes, as in pain, obscenely—alive tissue is also the thing that allows Ibbieta, or anyone, to be one day dead.

Ibbieta perceives the body as an "enormous vermin" to which he is tied (12), a colossus to which he is bound but with which he feels no kinship. In its huge, heavy presence, the rest of the world grows light, as though all else has been upended and emptied of its contents. What was full is now an outline, a sketch, a caricature. Spain and anarchy, dramatic realities a few days earlier, are now without immediacy and meaning. Nor is his sense of country and cause revitalized by his face-to-face confrontation with his enemy interrogators: he describes their gestures, dress, opinions, and acts as looking ludicrously small, "shocking and burlesqued" (15); rather than feeling angered at their ideology or their brutality, he feels embarrassed by their seriousness, their silly and sententious ignorance of their own mortality. This loss of country and conviction is only one of many signs of the new weightlessness of world and self. The woman he loved, the woman who a single day earlier had so pressed upon his consciousness that words of affectionate description had spilled forth from him even when there was no appropriate listener present, is now so irrecoverably remote that he rejects the chance to speak the few words of a farewell message (11). Even the physical objects in his prison cell, the most immediate and concrete objects of consciousness, have been emptied of their content, have each become a mere sketch: bench, lamp, and coal pile, Ibbieta comments, had "a funny look: they were more obliterated, less dense than usual" (12).

The narration as a whole also has the quality of a sketch: the experience it describes is utterly clear in its outline but all the emotional edges have been eliminated, absorbed into a surface as uniform and undifferentiated as Ibbieta's world, Ibbieta's wall. For the reader, the fear of the matter has been replaced by the simple fact of the matter, except perhaps at the periphery of vision where one is still able to apprehend, at least for a few seconds at a time, not merely the empty outline but the speed and force and direction of the events described, the crisis of world collapse that is taking place. The objects of consciousness from the most expansive to the most intimate, from those that exist in the space at the very limits of vision to those that exist in the space immediately outside the boundaries of the body, from the Big Dipper down to Spain and in through the realm of personal memories to the most abiding objects of love and belief and arriving finally at the bench beneath him and the coal pile at his feet—all in one patient rush are swept through and annihilated. It is in part the horrible momentum of this world contraction that is mirrored in the sudden agonized grimace of a person overcome by great pain or by the recognition of imminent death.

The process of perception Sartre describes, obviously not dependent on a political context, belongs anywhere where death is near and so belongs to aging. Sometimes assisted by younger human beings, the body works to obliterate the world and self of the old person. Something of this world dissolution is already at work even in the tendency of those in late middle age, no longer working, to see their former jobs, their life actions, their choices as wrong or trivial, jobs, actions, and choices that are probably no more insignificant than Ibbieta's Spain but that seem insignificant by virture of the same process, though these people may only be at the beginning of what he has almost finished. As the body breaks down, it becomes increasingly the object of attention, usurping the place of all other objects, so that finally, in very very old and sick people, the world may exist only in a circle two feet out from themselves; the exclusive content of perception and speech may become what was eaten, the problems of excreting, the progress of pains, the comfort or discomfort of a particular chair or bed. Stravinsky once described aging as: "the ever-shrinking perimeter of pleasure." This constantly diminishing world ground is almost a given in representations of old age. As Ibbieta's bench dissolves beneath him, so the ground beneath the old grows insubstantial, ceases to belong to them.[14] Sophocles's Oedipus, forbidden from entering his homeland, Thebes, is also violator and trespasser of the ground at Colonus; Shakespeare's Lear, having at last after long humiliation consented to enter the small but sharable space of a cage, stands instead alone on the narrow edge of a country and a cliff; Beckett's Winnie, the most literal victim of Stravinsky's ever-shrinking perimenter, is caught by a piece of ground that has snapped shut around her waist and that soon will close on the smaller circle of her neck. Each of these plays, though dense with other meanings, is in part the dramatization of the struggle to stay alive, to stay a little, to maintain one's extension out into the world whether that world, that self-extension, resides in a full-sized retinue or in a handbag full of familiar objects, in a young city's need of an elder's blessing or in, most simply, your beautiful child.[15] For each of the three, the voice becomes a final source of self-extension; so long as one is speaking, the self extends out beyond the boundaries of the body, occupies a space much larger than the body.[16] It is not accidental that a substantial part of the power of each play is its verbal virtuosity, that these old people talk so much, that for each the *tour de force* is less a display of style than a mode of survival whether it is Oedipus's highly charged alternation between the ritualized past and future of confession and oath, or Lear's commands, pleas, shrieks, his howling noise, or Winnie's brave cluckings. Their ceaseless talk articulates their unspoken understanding that only in silence do the edges of the self become coterminous with the edges of the body it will die with.

As in dying and death, so in serious pain the claims of the body utterly nullify the claims of the world. The annihilating power of pain is visible in the simple fact of experience observed by Karl Marx, "There is only one antidote to mental suffering, and that is physical pain,"[17] a pronouncement whose premises are only slightly distorted in Oscar Wilde's "God spare me physical pain and I'll take care of the moral pain myself."[18] As though in anticipation of a century that would produce out of its own physical well-being (or the well-being of its most articulate class) an endless fascination with the details of psychic distress and dislocation, the nineteenth century periodically reminded itself and its heirs of the privileges implicit in madness, whether that reminder took the form of aphorism, as in Marx and Wilde, or was expanded into narrative, as in George Eliot's Arthur Donnithorne who notices petulantly that physical pain might lift him out of his own self-absorbed boredom long enough to help him avoid damaging himself, a very young woman, and the hierarchical norms of their community,[19] or as in Emile Zola's Monsieur Hennebeau who recognizes that his obsessive sorrow about his unfulfilled marriage would be instantly eliminated if, like the coal miners, his "empty belly were twisted with pains that made his brain reel."[20] Physical pain is able to obliterate psychological pain because it obliterates all psychological content, painful, pleasurable, and neutral. Our recognition of its power to end madness is one of the ways in which, knowingly or unknowingly, we acknowledge its power to end all aspects of self and world.

Another manifestation of this power is its continual reappearance in religious experience. The self-flagellation of the religious ascetic, for example, is not (as is often asserted) an act of denying the body, eliminating its claims from attention, but a way of so emphasizing the body that the contents of the world are cancelled and the path is clear for the entry of an unworldly, contentless force. It is in part this world-ridding, path-clearing logic that explains the obsessive presence of pain in the rituals of large, widely shared religions as well as in the imagery of intensely private visions, that partly explains why the crucifixion of Christ is at the center of Christianity, why so many primitive forms of worship climax in pain ceremonies, why Brontë's *Wuthering Heights* is built on the principle first announced in Lockwood's dream that the pilgrim's staff is also a cudgel, why even Huysmans's famous dandy recognizes in his sieges of great pain a susceptibility to religious conversion, why in the brilliant ravings of Artaud some ultimate and essential principle of reality can be compelled down from the heavens onto a theatre stage by the mime of cruelty, why, though it occurs in widely different contexts and cultures, the metaphysical is insistently coupled with the physical with the equally insistent exclusion of the middle term, world.

The position of the person who is tortured is in many ways, of course, radically different from that of the person who experiences pain in a religious context, or that of an old person facing death, or that of the person who is hurt in a dentist's office. One simple and essential difference is duration: although a dentist's drill may in fact be the torturer's instrument, it will not land on a nerve for the eternity of a few seconds but for the eternity of the uncountable number of seconds that make up the period of torture, a period that may be seventeen hours on a single day or four hours a day on each of twenty-nine days. A second difference is control: the person tortured does not will his entry into and withdrawal out of the pain as the religious communicant enters and leaves the pain of a Good Friday meditation, or as the patient enters and leaves the pain of the healing therapy. A third difference is purpose: the path of worldly objects is swept clean not, as in religion, to make room for the approach of some divinely intuited force nor, as in medicine and dentistry, to repair the ground for the return of the world itself; there is in torture not even a fragment of a benign explanation as there is in old age where the absence of the world from oneself can be understood as an experienceable inversion of the eventual but unexperienceable absence of oneself from the world. Perhaps only in the prolonged and searing pain caused by accident or by disease or by the breakdown of the pain pathway itself is there the same brutal senselessness as in torture. But these other non-political contexts are called upon because they make immediately self-evident a central fact about pain that, although emphatically present in torture, is also obscured there by the idiom of "betrayal." It is the intense pain that destroys a person's self and world, a destruction experienced spatially as either the contraction of the universe down to the immediate vicinity of the body or as the body swelling to fill the entire universe. Intense pain is also language-destroying: as the content of one's world disintegrates, so the content of one's language disintegrates; as the self disintegrates, so that which would express and project the self is robbed of its source and its subject.

World, self, and voice are lost, or nearly lost, through the intense pain of torture and not through the confession as is wrongly suggested by its connotations of betrayal. The prisoner's confession merely objectifies the fact of their being almost lost, makes their invisible absence, or nearby absence, visible to the torturers. To assent to words that through the thick agony of the body can be only dimly heard, or to reach aimlessly for the name of a person or a place that has barely enough cohesion to hold its shape as a word and none to bond it to its worldly referent, is a way of saying, yes, all is almost gone now, there is almost nothing left now, even this voice, the sounds I am making, no longer form my words but the words of another.

Torture, then, to return for a moment to the starting point, consists of a primary physical act, the infliction of pain, and a primary verbal act, the interrogation. The verbal act, in turn, consists of two parts, "the question" and "the answer," each with conventional connotations that wholly falsify it. "The question" is mistakenly understood to be "the motive"; "the answer" is mistakenly understood to be "the betrayal." The first mistake credits the torturer, providing him with a justification, his cruelty with an explanation. The second discredits the prisoner, making him rather than the torturer, his voice rather than his pain, the cause of his loss of self and world. These two misinterpretations are obviously neither accidental nor unrelated. The one is an absolution of responsibility; the other is a conferring of responsibility; the two together turn the moral reality of torture upside down. Almost anyone looking at the *physical* act of torture would be immediately appalled and repulsed by the torturers. It is difficult to think of a human situation in which the lines of moral responsibility are more starkly or simply drawn, in which there is a more compelling reason to ally one's sympathies with the one person and to repel the claims of the other. Yet as soon as the focus of attention shifts to the *verbal* aspect of torture, those lines have begun to waver and change their shape in the direction of accommodating and crediting the torturers.[21] This inversion, this interruption and redirecting of a basic moral reflex, is indicative of the kind of interactions occurring between body and voice in torture and suggests why the infliction of acute physical pain is inevitably accompanied by the interrogation.

However near the prisoner the torturer stands, the distance between their physical realities is colossal, for the prisoner is in overwhelming physical pain while the torturer is utterly without pain; he is free of any pain originating in his own body; he is also free of the pain originating in the agonized body so near him. He is so without any human recognition of or identification with the pain that he is not only able to bear its presence but able to bring it continually into the present, inflict it, sustain it, minute after minute, hour after hour. Although the distance separating the two is probably the greatest distance that can separate two human beings, it is an invisible distance since the physical realities it lies between are each invisible. The prisoner experiences an annihilating negation so hugely felt throughout his own body that it overflows into the spaces before his eyes and in his ears and mouth; yet one which is unfelt, unsensed by anybody else. The torturer experiences the absence of this annihilating negation. These physical realities, an annihilating negation and an absence of negation, are therefore translated into verbal realities in order to make the invisible distance visible, in order to make what is taking place in terms of pain take place in terms of power, in order to shift what is occurring exclusively in the mode of sentience into the mode of self-extension and world. The torturer's questions—asked, shouted, insisted upon, pleaded for—objectify the fact that he has a world, announce in their feigned urgency the critical importance of that world, a world whose asserted magnitude is confirmed by the cruelty it is able to motivate and justify. Part of what makes his world so huge is its continual juxtaposition with the small and shredded world objectified in the prisoner's answers, answers that articulate and comment on the disintegration of all objects to which he might have been bonded in loyalty or love or good sense or long familiarity. It is only the prisoner's steadily shrinking ground that wins for the torturer his swelling sense of territory. The question and the answer are a prolonged comparative display, an unfurling of world maps.

This display of worlds can alternatively be understood as a display of selves or a display of voices, for the three are close to being a single phenomenon. The vocabulary of "motive" and "betrayal," for example, is itself an indication of a perceived difference in selfhood: to credit the torturer with having a motive is, among other things, to credit him with having psychic content, the very thing the prisoner's confession acknowledges the absence of and which the idiom of "betrayal" accuses him of willfully abandoning. The question and answer also objectify the fact that while the prisoner has almost no voice—his confession is a halfway point in the disintegration of language, an audible objectification of the proximity of silence—the torturer and the regime have doubled their voice since the prisoner is now speaking their words.

The interrogation is, therefore, crucial to a regime. Within the physical events of torture, the torturer "has" nothing: he has only an absence, the absence of pain. In order to experience his distance from the prisoner in terms of "having," their physical difference is translated into a verbal difference: the absence of pain is a presence of world; the presence of pain is the absence of world. Across this set of inversions pain becomes power. The direct equation, "the larger the prisoner's pain, the larger the torturer's world" is mediated by the middle term, "the prisoner's absence of world": the larger the prisoner's pain (the smaller the prisoner's world and therefore, by comparison) the larger the torturer's world. This set of inversions at once objectifies and falsifies the pain, objectifies one crucial aspect of pain in order to falsify all other aspects. The obliteration of the contents of consciousness, the elimination of world ground, which is a condition brought about by the pain and therefore one that once objectified (as it is in confession) should act as a sign of the pain, a call for help, an announcement of a radical occasion for attention and assistance, instead acts to discredit the claims of pain, to repel attention, to ensure that the pain will be unseen and unattended to. That not only the torturers but the world at large should tend to identify the confession as a "betrayal" makes very overt the fact that his absence of world earns the person in pain not compassion but contempt. This phenomenon in which the claims of pain are eclipsed by the very loss of world it has brought about is a crucial step in the overall process of perception that allows one person's physical pain to be understood as another person's power.

When one human being "recognizes" the incontestable legitimacy of another human being's existence, he or she is locating the other's essential reality in one of two places—either in the complex fact of sentience or in the objects of sentience, in the fact of consciousness or in the objects of consciousness. In normal and benign contexts, the two occur together and imply one another: we respect the objects of sentience, the worldly forms of self-extension, precisely because they lead one in to the fact of another's sentience; Gloucester's earldom, Winnie's handbag, Ibbieta's Spain, Lear's feather are like luminous breadcrumbs leading home, traces in the external world of the overwhelming fact at the center. But the two can also become utterly split off from one another. When this happens, the very habit of seeing in the one the proximity of the other encourages the mistake of seeing in the absence of the one the absence of the other, seeing in the loss of Gloucester's earldom the loss of Gloucester's sentience, an act of perceptual brutality that is a private and silent form of putting out his eyes.

A political situation is almost by definition one in which the two locations of selfhood are in a skewed relation to one another or have wholly split apart and have begun to work, or be worked, against one another. Torture is the most extreme instance of this situation, for one person gains more and more world-ground not in spite of but because of the other's sentience: the overall equation it works to bring about, "the larger the prisoner's pain, the larger the torturer's power" can be restated, "the more sentient the prisoner, the more numerous and extensive the torturer's objects of sentience."[22] The middle steps in the equation can also be rewritten in this language: to say "the torturer inflicts pain in order to produce a confession" is to say "the torturer uses the prisoner's sentience to obliterate the objects of the prisoner's sentience" or "the torturer uses the prisoner's aliveness to crush the things that he lives for." And, finally, the entire process is self-amplifying, for as the prisoner's sentience destroys his world, so now his absence of world, as described earlier, destroys the claims of sentience: the confession which displays the fact that he has nothing he lives for now obscures the fact that he is violently alive. Over and over, in each stage and step, the torturer's mime of expanding world-ground depends on a demonstration of the prisoner's absence of world. The confession is one crucial demonstration of this absent world, but there are others.

II. The Objectification of the Prisoner's World Dissolution

The disintegration of the contents of consciousness, the contraction and ultimate dissolution of the prisoner's world, acknowledged and objectified in confession, is also objectified in the physical objects the torturer uses as weapons, in the torturer's actions, and in the torturer's language. His weapons, his acts, and his words, all necessarily drawn from the world, all make visible the nature of his engagement with the world, though the scale differs in the three. The more physical each is, the smaller the world it represents. As physical objects, the weapons occupy the basic unit of shelter, the room, and so represent the world in its most contracted and distilled form. The room is barred, sealed, guarded; little of the larger world is allowed to enter; but in his physical acts and in his words, the torturer alludes to and allows to enter aspects of the world in a more expansive state, civilization in its more spacious form. In each of the three realms, as will be shown, the torturer dramatizes the disintegration of the world, the obliteration of consciousness that is happening within the prisoner himself. Brutal, savage, and barbaric, torture (even if unconsciously) self-consciously and explicitly announces its own nature as an undoing of civilization, acts out the uncreating of the created contents of consciousness.

In normal contexts, the room, the simplest form of shelter, expresses the most benign potential of human life. It is, on the one hand, an enlargement of the body: it keeps warm and safe the individual it houses in the same way the body encloses and protects the individual within; like the body, its walls put boundaries around the self preventing undifferentiated contact with the world, yet in its windows and doors, crude versions of the senses, it enables the self to move out into the world and allows that world to enter. But while the room is a magnification of the body, it is simultaneously a miniaturization of the world, of civilization. Although its walls, for example, mimic the body's attempt to secure for the individual a stable internal space—stabilizing the temperature so that the body spends less time in this act; stabilizing the nearness of others so that the body can suspend its rigid and watchful postures; acting in these and other ways like the body so that the body can act less like a wall—the walls are also, throughout all this, independent objects, objects which stand apart from and free of the body, objects which realize the human being's impulse to project himself out into a space beyond the boundaries of the body in acts of making, either physical or verbal, that once multiplied, collected, and shared are called civilization.

There is nothing contradictory about the fact that the shelter is at once so graphic an image of the body and so emphatic an instance of civilization: only because it is the first can it be the second. It is only when the body is comfortable, when it has ceased to be an obsessive object of perception and concern, that consciousness develops other objects, that for any individual the external world (in part already existing and in part about to be formed) comes into being and begins to grow. Both in the details of its outer structure and in its furniture (from "furnir" meaning "to further" or "to forward," to project oneself outward) the room accommodates and thereby eliminates from human attention the human body: the simple triad of floor, chair, and bed (or simpler still, floor, stool, and mat) makes spatially and therefore steadily visible the collection of postures and positions the body moves in and out of, objectifies the three locations within the body that most frequently hold the body's weight, objectifies its need continually to shift within itself the locus of its weight, objectifies, finally, its need to become wholly forgetful of its weight, to move weightlessly into a larger mindfulness. As the elemental room is multiplied into a house of rooms and the house into a city of houses, the body is carried forward into each successive intensification of civilization. In western culture, whole rooms within a house attend to single facts about the body, the kitchen and eating, the bathroom and excreting, the bedroom and sleeping; so, too, entire cities become attentive to single facts about the body, as movement is visible in the car industry in Detroit, or eyesight and memory in the film and copying of Rochester. It is, though, back in the inward and enclosing space of the single room and its domestic content that the outward unfolding (so appropriately called "the flowering") of civilization originates. One very beautiful honoring of this fact is the monument of a clothespin—a tiny domestic object transformed into something forty feet high, upright, arching, and magnificent—that Philadelphia has placed in the lap of its City Hall. The shielding, "holding" gestures of the domestic are overtly present in the ordinary clothespin: one piece of wood is held to another piece of wood by a metal arm and the three together now act as one to enable other things to be coupled to each other and to itself. In Oldenburg's "Clothespin," these successive acts of inanimate holding find their origin and destination in the inclusive sentience of the human hug, for the monument holds within itself the gracious and self-confident embrace of two lovers. The scale of this clothespin, its complete ease in the presence of City Hall, its recognition that the enduring and monumental reside in the daily, its discovery of the broad pleasures of magnanimous intelligence in the narrow reflexes of punning wit, its identification of pressing with expressing and holding in with reaching out—all these are translations of and tributes to the central, overwhelming characteristic of the domestic, that its protective, narrowing act is the location of the human being's most expansive potential.

In torture, the world is reduced to a single room or set of rooms. Called "guest rooms" in Greece[23] and "safe houses" in the Philippines,[24] the torture rooms are often given names that acknowledge and call attention to the generous, civilizing impulse normally present in the human shelter. They call attention to this impulse only as prelude to announcing its annihilation. The torture room is not just the setting in which the torture occurs; it is not just the space that happens to house the various instruments used for beating and burning and producing electric shock. It is itself literally converted into another weapon, into an agent of pain. All aspects of the basic structure—walls, ceiling, windows, doors—undergo this conversion. Basques tortured by the Spanish describe "el cerrojo," the rapid and repeated bolting and unbolting of the door in order to keep them at all times in immediate anticipation of further torture, as one of the most terrifying and damaging acts.[25] Found among PIDES' paraphernalia in Portugal were manuscripts of gibberish which, according to the men and women brutalized there, were read at the doors of prisoners deprived of sleep for days.[26] Solzhenitsyn describes how in Russia guards were trained to slam the door in as jarring a way as possible or to close it in equally unnerving silence.[27] Former prisoners in the Philippines report having had their heads repeatedly banged into the wall.[28] Israeli soldiers held in Syria describe being suspended from the ceiling in a tire that was swung as they were beaten, or having one's genitals tied by a string to a door handle and having the string beaten.[29] According to the testimony of Greeks tortured under the Colonels' Regime, the act of looking out a window was made the occasion for beatings; prisoners were taken to the window and threatened that they would be "de-fenestrated"; they were made to stand against the wall and recite obscenities or push against the wall while repeatedly reciting the line, "Make way wall that I may pass"; they were subjected to the Greek equivalent of "el cerrojo"; the door was left open to make audible conversations threatening to the prisoner, his friends, or family.[30]

Just as all aspects of the concrete structure are inevitably assimilated into the process of torture, so too the contents of the room, its furnishings, are converted into weapons: the most common instance of this is the bathtub that figures prominently in the reports from numerous countries, but it is only one among many. Men and women tortured during the period of martial law in the Philippines, for example, described being tied or handcuffed in a constricted position for hours, days, and in some cases months to a chair, to a cot, to a filing cabinet, to a bed;[31] they describe being beaten with "family-sized soft drink bottles" or having a hand crushed with a chair, of having their heads "repeatedly banged on the edges of a refrigerator door" or "repeatedly pounded against the edges of a filing cabinet."[32] The room, both in its structure and its content, is converted into a weapon, deconverted, undone. Made to participate in the annihilation of the prisoners, made to demonstrate that everything is a weapon, the objects themselves, and with them the fact of civilization, are annihilated: there is no wall, no window, no door, no bathtub, no refrigerator, no chair, no bed.

Beside the overwhelming fact that a human being is being severely hurt, the exact nature of the weapon or the miming of the deconstruction of civilization is at most secondary. But it is also crucial to see that the two are here forced into being expressions and amplifications of one another: the de-objectifying of the objects, the unmaking of the made, is a process externalizing the way in which the person's pain causes his world to disintegrate; and, at the same time, the disintegration of the world is here, in the most literal way possible, made painful, made the direct cause of the pain. That is, in the conversion of a refrigerator into a bludgeon, the refrigerator disappears; its disappearance objectifies the disappearance of the world (sky, country, bench) experienced by a person in great pain; and it is the very fact of its disappearance, its transition from a refrigerator into a bludgeon, that inflicts the pain. The domestic act of protecting becomes an act of hurting and in hurting, the object becomes what it is not, an expression of individual contraction, of the retreat into the most self-absorbed and self-experiencing of human feelings, when it is the very essence of these objects to express the most expansive potential of the human being, his ability to project himself out of his private, isolating needs into a concrete, objectified, and therefore sharable world. The appearance of these common domestic objects in torture reports of the 1970s is no more gratuitous and accidental than the fact that so much of our awareness of Germany in the 1940s is attached to the words "ovens," "showers," "lampshades," and "soap."

The prisoner's physical world is limited to the room and its contents; no other concrete embodiments of civilization pass through the doors. But two of civilization's institutions, though not physically present, are constantly alluded to in the action of torture, and so hover behind and arch over the physical reality of the sealed room. Like the domestic objects, these institutions are unmade by being made weapons. The first is, of course, the trial. In its basic outlines, torture is the inversion of the trial, a reversal of cause and effect. While the one studies evidence that may lead to punishment, the other uses punishment to generate the evidence. The slogans of the South Vietnamese torturers announce what is there and elsewhere always visible in the process itself: "If they are not guilty, beat them until they are," "If you are not a Vietcong, we will beat you until you admit you are; and if you admit you are, we will beat you until you no longer dare to be one.''[33] The second institution ubiquitously present by inversion is medicine. Solzhenitsyn's *The Gulag Archipelago* describes the process by which in Russia the doctor becomes the torturer's "right hand man."[34] In Greece, a man referred to as "Dr. Kofas" was a major focus throughout the trial of the torturers who had served the Colonels' Regime.[35] Reports of torture from prisoners in the Philippines include references to "unwanted dental treatment."[36] In Chile, absurdly negligent medical treatment is said by many to have worsened the patients' condition and there are reports of prisoners being overdosed with lethal drugs.[37] In Portugal, doctors studied photographs of maimed prisoners as well as the prisoners themselves in order to further the design of the torture procedure.[38] In Brazil, there were forms of torture called "the mad dentist" and "the operating table."[39] Syrian prisoners of Israel claim to have undergone possibly unnecessary amputations, to have had wounds cleaned with petrol, and to have had absurdly large amounts of blood taken.[40] In Uruguay, doctors assisted in the administration of drugs causing hallucinations and acute sensations of pain and asphyxiation; those who refused to assist the torturers disappeared at such a rate that Uruguay's medical and health care programs entered a state of crisis.[41] It is unnecessary to catalogue the instances endlessly. Whether medicine merely provides the equipment[42] or the name for a form of torture, whether the doctor ever was a doctor or has only assumed a role,[43] whether he designs the form of torture used, inflicts the brutality himself, assists the process by healing the person so he can be again tortured, or legitimizes the process by the masquerade of aid, the institution of medicine like that of justice is deconstructed, unmade by being made at once an actual agent of the pain and a demonstration of the effects of pain on human consciousness. While other institutions are alluded to in the process of torture, their appearance is sporadic and usually the result of the accidental location of the torture rooms or headquarters, the sports stadium in Chile, the police station in Paraguay, the traffic control office in Greece, and, in an earlier decade, the sweets factory in Algeria.[44] But it is in the nature of torture that the two ubiquitously present should be medicine and law, health and justice, for they are the institutional elaborations of body and state. These two were also the institutions most consistently inverted in the concentration camps, though they were slightly differently defined in accordance with Germany's position as a modern, industrialized mass society: the "body" occurring not in medicine but in its variant, the scientific laboratory; the "state" occurring not in the process of law, the trial, but in the process of production, the factory.

As the torturer uses the immediate physical setting in a direct deconstruction of the smallest unit of civilization, and as his actions allude to and subvert larger units of civilization, two of its primary institutional forms, so his words reach out, body forth, and destroy more distant and more numerous manifestations of civilization. Amid his insistent questions and exclamations, his jeers, gibberish, obscenities, his incomprehensible laughter, his monosyllables, his grunts—for just as a person in pain reverts to sounds prior to language, the cries and screams of human hurt, so the person inflicting pain reverts to a pre-language, uncaring noises remembered in the accounts given by former political prisoners and sometimes included in fictional accounts of brutality such as Zola's portrait of Bijard in *L'Assommoir*—there are words, random words, names for torture, names for the prisoner's body, and this idiom continually moves out to the realm of the man-made, the world of technology and artifice. The twofold denial of the human, both the particular human being being hurt and the collective human present in the products of civilization, is more easily apprehended if one first recalls the overwhelming experience to which any one of these names is being attached.

> In the centre of the room there was an ordinary bathtub, rather large. From a hole in the wall hung a plastic pipe from which water was flowing to fill the bath. On the opposite wall stood out two iron rings a little smaller than a riding stirrup. On the left side there was a large, red-coloured light like a semaphore. On the other side a cross had been scratched into the wall and painted, about 30 by 15 centimetres, more or less. The man in charge made me look at the cross; running his index finger along the groove he said to me: "This is where a tailor died last December and you run the same risk if you don't tell us what you know...."
>
> As I did not reply to them, they made me look at the red-coloured light, which they lit up at once. Within five minutes I was dazzled: I could see only a large round menacing cloud about two metres big before my eyes, with semi-darkness roundabout.
>
> They made me sit down on the edge of the trough at its highest part, having first tied my feet with ropes and my hands behind my back. I was stripped of my clothes.
>
> Suddenly they grabbed me by the shoulders and pushed me to the bottom of the trough. I held my breath a while making desperate efforts to get my head out of the water and take in some air. I managed to free my head but they submerged me again, and when my efforts to get out became violent, the heaviest members of the group trampled on the top part of my body. I could no longer bear the lack of air and began to swallow water through my mouth, nose and ears.
>
> My ears started to hum as the water made its way in. They seemed to be blowing up like a balloon. Then came a sharp whistling, very loud at first, which has not yet completely gone and which I hear when there is complete silence. The more I swallowed water the more my struggles to breathe also increased and they all pressed me down to the bottom of the trough—my head, chest and hands....
>
> I must have swallowed 8–10 litres of water. When they took me out and laid me on the ground, one of them trod heavily on my stomach: water poured out from my mouth and nose, spurting like a jet from a hose.
>
> After a second session of immersion—having first drained my stomach—they made me sit on a chair. As I was now weak from exhaustion they did not use the ring in the wall but whipped me with a plaited rope....[45]

To attach any name, any word to the willful infliction of this bodily agony is to make language and civilization participate in their own destruction; the specific names chosen merely make this subversion more overt. The form of torture described here, used in the 1970s in the Philippines, Vietnam, Uruguay, Brazil, and Paraguay is in the torturers' idiom almost everywhere referred to as "the submarine" when water or soapy water or dirty water is used, is in some places called "the Portugese submarine" when the water is electrified, and is usually called "the dry submarine" if the person is held in a plastic bag or immersed in feces. The nomenclature for torture is typically drawn from three spheres of civilization. First, as in the above instance, the prolonged, acute distress of the body is in its contortions claimed to be mimetic of a particular invention or technological feat: the person's pain will be called "the telephone" in Brazil, "the plane ride" in Vietnam, "the motorola" in Greece, and "the San Juanica Bridge" in the Philippines.[46] The second sphere is the realm of cultural events, cermonies, and games: there is "the dance" in Argentina, "the birthday party" in the Philippines, and "hors d'oeuvres," "tea party," and "tea party with toast" in Greece.[47] Though the primary act of eating or moving, along with all other primary acts of the body, will at some point be brought into the torture process, it is naturally not the acts of eating or moving themselves but the self-consciously civilized elaborations of these acts, the dinner party or the dance, that the torturer's words reach out for. The third realm is nature or nature civilized; it enters less frequently than the first two and, when it does, is usually limited to that part of nature that is dainty, diminutive, or mythologized, easily assimilated into the human framework. The "tiger cages" of Vietnam are exceptional. More typical are "the little hare" of Greece, "the parrot's perch" of Brazil and Uruguay, and "the dragon's chair" of Brazil.[48] In all these cases the designation of an intensely painful form of bodily contortion with a word usually reserved for an instance of civilization produces a circle of negation: there is no human being in excruciating pain; that's only a telephone; there is no telephone; that is merely a means of destroying a human being who is not a human being, who is only a telephone, who is not a telephone but merely a means of destroying a telephone. The double negation of a human being and a symptom of civilization combine to bring about a third area of negation, the negation of the torturer's recognition of what is happening, a negation that will in turn allow the first two to continue. The torturer's idiom not only indicates but helps bring about the process of perception in which all human reality is made, no matter how screamingly present, invisible, inaudible.

Through the torturer's language, his actions, and the physical setting, the world is brought to the prisoner in three rings: the random technological and cultural embodiments of civilization overarch the two primary social institutions of medicine and law, which in turn overarch the basic unit of shelter, the room. Just as the prisoner's confession makes visible the contraction and closing in of his universe, so the torturer reenacts this world collapse. Civilization is brought to the prisoner and in his presence annihilated in the very process by which it is being made to annihilate him. Civilization itself in its language and its literature records the path that torture in its unconscious miming of the deconstruction of civilization follows in reverse: the protective, healing, expansive acts implicit in "host" and "hostel" and "hospitable" and "hospital" all converge back in "hospes," which in turn moves back to the root "hos" meaning house, shelter, or refuge; but once back at "hos," its generosity can be undone by an alternative movement forward into "hostis," the source of "hostility" and "hostage" and "host"—not the host that willfully abandons the ground of his power in acts of reciprocity and equality but the "host" deprived of all ground, the host of the eucharist, the sacrificial victim. Even fictional representations of torture like Kafka's "In the Penal Colony," where the lethal apparatus is an enlarged and elaborate sewing machine, record the fact that the unmaking of civilization inevitably requires a return to and mutilation of the domestic, the ground of all making. This world unmaking, this uncreating of the created world, which is an external objectification of the psychic experience of the person in pain, becomes itself the cause of the pain. Although the world is, as in Sartre's short story, reduced to the crushingly blank and uniform wall, it is not, as in Sartre's story, merely the harsh undifferentiated surface against which the execution occurs. It is itself the executioner's weapon; it is the world, the wall, that executes. Whatever its political naiveté or its melodramatic intentions, Poe's "The Pit and the Pendulum" discovers in its final moments the single distilled form of torture that in many ways represents all forms of torture, the walls collapsing in on the human center to crush it alive.

III. The Transformation of Body into Voice

The appropriation of the world into the torturer's arsenal of weapon is a crucial step in the overall process of torture for, as was suggested at the opening of this chapter, it is by the obsessive mediation of agency that the prisoner's pain will be perverted into the fraudulent assertion of power,[49] that the objectified pain is denied as pain and read as power. At first, the weapon or agent seems to be only the torturers and whatever piece of apparatus serves as their primary tool, but now the environment too has joined them, been compelled to participate in and amplify their swelling sense of causality. Absolutely everything but the prisoner himself stands present as a weapon, and ultimately he, too, is assimilated into the perceptual strategies of agency. The process by which this final assimilation takes place will be visible in a return to and review of the relation between the pain and the interrogation, the connections between physical and verbal acts, for the translation of pain into power is ultimately a transformation of body into voice, a transformation arising in part out of the dissonance of the two, in part out of the consonance of the two.

A large part of the mime of power emerges out of the opposition of body and voice. The torturer experiences his own body and voice as opposites; the prisoner experiences his own body and voice as opposites; the prisoner's experience of the two is an inversion of the torturer's. Hence there are four sets of oppositions. The pain is hugely present to the prisoner and absent to the torturer; the question is, within the political fiction, hugely significant to the torturer and insignificant to the prisoner; for the prisoner, the body and its pain are overwhelmingly present and voice, world, and self are absent; for the torturer, voice, world, and self are overwhelmingly present and the body and pain are absent. These multiple sets of oppositions at every moment announce and magnify the distance between torturer and prisoner and thereby dramatize the former's power, for power is in its fraudulent as in its legitimate forms always based on distance from the body.

But the very consistency of these oppositions between body and voice means that the two also mirror one another. Assigned identical positions, they reflect and amplify one another. Just as the pain is a physical measure of the colossal discrepancy between the person tortured and the torturer (for whatever their spatial proximity, there are no two experiences farther apart than suffering and inflicting pain), so the interrogation is the verbal objectification of that colossal discrepancy. In his desperate insistence that his questions be answered, the torturer luxuriates in the privilege or absurdity of having a world that the other has ceased to have. Nowhere does language come so close to being the concrete agent of physical pain as here where it not only occurs in such close proximity to the raising of the rod or the turning on of the electricity, but also parallels and thereby doubles the display of distance. Just as the words of the one have become a weapon, so the words of the other are an expression of pain, in many cases telling the torturer nothing except how badly the prisoner hurts. The question, whatever its content, is an act of wounding; the answer, whatever its content, is a scream. This identification of the physical and verbal acts is consciously or unconsciously acknowledged in the language of the torturers themselves. The leading generals in charge of torture under the Colonels' Regime in Greece repeatedly spoke to their prisoners in images dramatizing the connection between two dreaded forms of exposure, open wounds and confession: "Hazijisis punched me in the chest one day and said, 'Here, you're going to tell all. You will open out like a rose.' "[50] Of the many forms of brutality used under this regime, the most persistent, other than falanga, was the delivery of repeated blows to the prisoner's sternum, often causing him to vomit blood. This type of body damage provided the torturers with their idiom for confession: "Here you'll spew it all up"; "You are Mitsii the taxi-driver whom we've been hunting for. Now you're in our hands. We know everything. This is ESA. You'll vomit blood and tell us yourself."[51]

There is a second equally crucial and equally cruel bond between physical pain and interrogation that further explains their inevitable appearance together. Just as the interrogation, like the pain, is a way of wounding, so the pain, like the interrogation, is a vehicle of self-betrayal. Torture systematically prevents the prisoner from being the agent of anything and simultaneously pretends that he is the agent of some things. Despite the fact that in reality he has been deprived of all control over, and therefore all responsibility for, his world, his words, and his body, he is to understand his confession as it will be understood by others, as an act of self-betrayal. In forcing him to confess or, as often happens, to sign an unread confession, the torturers are producing a mime in which the one annihilated shifts to being the agent of his own annihilation. But this mime, though itself a lie, mimes something real and already present in the physical pain; it is a visible counterpart to an invisible but intensely felt aspect of pain. Regardless of the setting in which he suffers (home, hospital, or torture room), and regardless of the cause of his suffering (disease, burns, torture, or malfunctioning of the pain network itself), the person in great pain experiences his own body as the agent of his agony. The ceaseless, self-announcing signal of the body in pain, at once so empty and undifferentiated and so full of blaring adversity, contains not only the feeling ''my body hurts'' but the feeling ''my body hurts me.'' This part of the pain, like almost all others, is usually invisible to anyone outside the boundaries of the sufferer's body, though it sometimes becomes visible when a young child or an animal in the first moments of acute distress takes maddening flight, fleeing from its own body as though it were a part of the environment that could be left behind. If self-hatred, self-alienation, and self-betrayal (as well as the hatred of, alienation from, and betrayal of all that is contained in the self—friends, family, ideas, ideology) were translated out of the psychological realm where it has content and is accessible to language into the unspeakable and contentless realm of physical sensation it would be intense pain.

This unseen sense of self-betrayal in pain, objectified in forced confession, is also objectified in forced exercises that make the prisoner's body an *active* agent, an actual cause of his pain. He may be put in a contorted posture in a cramped space for months or for years as happened in Vietnam; he may be made to walk ceaselessly on bended knees as in Spain; squat until he collapses, and carry a heavy stone while being beaten as in the Philippines; he may be made to stand upright in his cell each day for eleven hours as in Argentina; he may be made to throw his head back as far as possible and, as in Greece where it was called ''making knots,'' repeatedly swallow his own saliva.[52] Part of his sense of his body as agent comes at the moment when his failure to sustain the prescribed posture or exercise brings from the torturer another form of punishment; but, for the prisoner and those present, the most emphatic and direct exhibition of self-agency comes from the exercise itself. Standing rigidly for eleven hours can produce as violent muscle and spine pain as can injury from elaborate equipment and apparatus, though any of us outside this situation, used to adjusting our body positions every few moments before even mild discomfort is felt, may not

immediately recognize this. W. K. Livingston, a leading researcher of the physiology of pain, describes his own incomprehension when a colleague who suffered nauseating pain each day for years as a result of amputation, attributed
much of his agony to a feeling of tenseness:

> I once asked him why the sense of tenseness in the hand was so frequently emphasized
> among his complaints. He asked me to clench my fingers over my thumb, flex my
> wrist, and raise the arm into a hammer lock position and hold it there. He kept me
> in this position as long as I could stand it. At the end of five minutes I was perspiring
> freely, my hand and arm felt unbearably cramped, and I quit. But you can take
> your hand down, he said.[53]

Only when a person throws his head back and swallows three times does he
begin to apprehend what is involved in one hundred and three or three hundred
and three swallows, what atrocities one's own body, muscle, and bone structure
can inflict on oneself. The political prisoner is, of course, reminded of this at
every moment. Each source of strength and delight, each means of moving out
into the world or moving the world in to oneself, becomes a means of turning
the body back in on itself, forcing the body to feed on the body: the eyes are
only access points for scorching light, the ears for brutal noises; eating, the act
at once so incredible and so simple in which the world is literally taken into the
body, is replaced by rituals of starvation involving either no food or food that
nauseates; taste and smell, two whole sensory modes that have emerged to watch
over the entry of the world into the body, are systematically abused with burns
and cuts to the inside of nose and mouth, and with bug-infested or putrefying
substances; normal needs like excretion and special wants like sexuality are made
ongoing sources of outrage and repulsion. Even the most small and benign of
bodily acts becomes a form of agency. In *The First Circle*, Solzhenitsyn describes
how prisoners, while sleeping, were forced to keep their hands outside the
blanket, and he writes, "It was a diabolical rule. It is a natural, deep-rooted,
unnoticed human habit to hide one's hands while asleep, to hold them against
one's body."[54] The prisoner's body—in its physical strengths, in its sensory
powers, in its needs and wants, in its ways of self-delight, and finally even, as
here, in its small and moving gestures of friendship toward itself—is, like the
prisoner's voice, made a weapon against him, made to betray him on behalf of
the enemy, made to be the enemy.

But the relation between body and voice that for the prisoner begins in opposition (the pain is so real that "the question" is unreal, insignificant) and that
goes on to become an identification (the question, like the pain, is a way of
wounding; the pain, like the question, is a vehicle of self-betrayal) ultimately
ends in opposition once more. For what the process of torture does is to split
the human being into two, to make emphatic the ever present but, except in the
extremity of sickness and death, only latent distinction between a self and a body,

between a "me" and "my body." The "self" or "me," which is experienced on the one hand as more private, more essentially at the center, and on the other hand as participating across the bridge of the body in the world, is "embodied" in the voice, in language. The goal of the torturer is to make the one, the body, emphatically and crushingly *present* by destroying it, and to make the other, the voice, *absent* by destroying it. It is in part this combination that makes torture, like any experience of great physical pain, mimetic of death; for in death the body is emphatically present while that more elusive part represented by the voice is so alarmingly absent that heavens are created to explain its whereabouts.

Through his ability to project words and sounds out into his environment, a human being inhabits, humanizes, and makes his own a space much larger than that occupied by his body alone. This space, always contracted under repressive regimes, is in torture almost wholly eliminated. The "it" in "Get it out of him" refers not just to a piece of information but to the capacity for speech itself. The written or tape-recorded confession that can be carried away on a piece of paper or on a tape is only the most concrete exhibition of the torturer's attempt to induce sounds so that they can then be broken off from their speaker so that they can then be taken off and made the property of the regime. The torturer tries to make his own not only the words of the prisoner's confession but all his words and sounds. One form of stress imposed on Portuguese prisoners was making them speak in a constant loud volume to other prisoners.[55] In Greece, a similar rule was extended to nonverbal forms of sound: "[The officer] was not satisfied with my answer and hit me again. . . . Here the guard ordered me to walk so that my steps would be heard. He said I was to walk on the double."[56] He will, while being hurt, be made to speak, to sing, and, of course, to scream— and even those screams, the sounds anterior to language that a human being reverts to when overwhelmed by pain, will in turn be broken off and made the property of the torturers in one of two ways. They will, first of all, be used as the occasion for, be made the agent of, another act of punishment. As the torturer displays his control of the other's voice by first inducing screams, he now displays that same control by stopping them: a pillow or a pistol or an iron ball or a soiled rag or a paper packet of excrement is shoved into the person's mouth, or a loud motor is placed next to his head, or electricity is used to contract his jaws.[57] Secondly, in many countries these screams are, like the words of the confession, tape-recorded and then played where they can be heard by fellow prisoners, close friends, and relatives.[58] Again and again the descriptions given by those who have been imprisoned and tortured are full of cries, phrases, fragments of speech whose source cannot be identified—someone was sobbing, someone was screaming, someone called out, "Stop it, You'll kill him," who was it, is he responding to my being hurt, can he see me, or is it his own hurt, are they too being brutalized, do those screams come from someone now being tortured, are they the tape recording of someone previously tortured, is it my husband's voice, my child's, are these so compelling sounds the sounds of real human hurt or are they sounds made up to mock and torment me?

In this closed world where conversation is displaced by interrogation, where human speech is broken off in confession and disintegrates into human cries, where even those cries can be broken off to become one more weapon against the person himself or against a friend, in this world of broken and severed voices, it is not surprising that the most powerful and healing moment is often that in which a human voice, though still severed, floating free, somehow reaches the person whose sole reality had become his own unthinkable isolation, his deep corporeal engulfment. The prisoner who, alone in long solitary confinement and repeatedly tortured, found within a loaf of bread a matchbox containing a small piece of paper that had written on it the single, whispered word "Corragio!", "Take courage";[59] the Uruguayan man arranging for some tangible signal that his words had reached their destination, "My darling, if you receive this letter put a half a bar of Boa soap in the next parcel";[60] the imprisoned Chilean women who on Christmas Eve sang with all their might to their men in a separate camp the song they had written, "Take heart, Jose, my love" and who, through the abusive shouts of guards ordering silence, heard "faintly on the wind . . . the answering song of the men"[61]—these acts and their multiplication in the extensive and ongoing attempts of Amnesty International to restore to each person tortured his or her voice, to use language to let pain give an accurate account of itself, to present regimes that torture with a deluge of letters and telegrams, a deluge of voices speaking on behalf of, voices speaking in the voice of, the person silenced, these acts that return to the prisoner his most elemental political ground as well as his psychic content and density are finally almost physiological in their power of alteration. As torture consists of acts that magnify the way in which pain destroys a person's world, self, and voice, so these other acts that restore the voice become not only a denunciation of the pain but almost a diminution of the pain, a partial reversal of the process of torture itself. An act of human contact and concern, whether occurring here or in private contexts of sympathy, provides the hurt person with worldly self-extension: in acknowledging and expressing another person's pain, or in articulating one of his nonbodily concerns while he is unable to, one human being who is well and free willingly turns himself into an image of the other's psychic or sentient claims, an image existing in the space outside the sufferer's body, projected out into the world and held there intact by that person's powers until the sufferer himself regains his own powers of self-extension. By holding that world in place, or by giving the pain a place in the world, sympathy lessens the power of sickness and pain, counteracts the force with which a person in great pain or sickness can be swallowed alive by the body.

To acknowledge the radical subjectivity of pain is to acknowledge the simple and absolute incompatibility of pain and the world. The survival of each depends on its separation from the other. To bring them together, to bring pain into the world by objectifying it in language, is to destroy one of them: either, as in the case of Amnesty International and parallel efforts in other areas, the pain is objectified, articulated, brought into the world in such a way that the pain itself is diminished and destroyed; or alternatively, as in torture and parallel forms of sadism, the pain is at once objectified and falsified, articulated but made to refer to something else and in the process, the world, or some dramatized surrogate of the world, is destroyed. As the opening of this chapter asserted and as the previous description has tried to show, torture is a form of savagery and stupidity (words invoked here as literal designations rather than as dismissive labels) that has a structure. This structure may be in part premeditated, seems for the most part unconscious, and is in either case based on the nature of pain, the nature of power, the interaction between the two, and the interaction between the ultimate source of each—the body, the locus of pain, and the voice, the locus of power. It involves the invariable and simultaneous occurrence of three phenomena which, for the sake of description and summary, can be isolated into three separate and sequential steps.

IV. Three Simultaneous Phenomena in the Structure of Torture

(1) the infliction of pain

(2) the objectification of the subjective attributes of pain

(3) the translation of the objectified attributes of pain into the insignia of power

The first of the three steps is the infliction of great physical pain on a human being. Although this is the most heinous part of the process, it alone would never accomplish the torturer's goal. One aspect of great pain—as acknowledged by those who have suffered it in diverse political and private contexts, and as asserted by those who have studied it from the perspective of psychology, philosophy, and physiology, and, finally, as becomes obvious to common sense alone—is that it is to the individual experiencing it overwhelmingly present, more emphatically real than any other human experience, and yet is almost invisible to anyone else, unfelt, and unknown. Even prolonged, agonized human screams, which press on the hearer's consciousness in something of the same way pain presses on the consciousness of the person hurt, convey only a limited dimension of the sufferer's experience. It may be for this reason that images of the human scream recur fairly often in the visual arts, which for the most part avoid depictions of auditory experience. The very failure to convey the sound makes these representations arresting and accurate; the open mouth with no sound reaching anyone in the sketches, paintings, or film stills of Grünewald, Stanzione, Munch, Bacon, Bergman, or Eisenstein, a human being so utterly consumed in the act of making a sound that cannot be heard, coincides with the way in which pain engulfs the one in pain but remains unsensed by anyone else. For the torturer, it is not enough that the prisoner experience pain. Its reality, although already incontestable to the sufferer, must be made equally incontestable to those outside the sufferer. Pain is therefore made visible in the multiple and elaborate processes that evolve in producing it.

In, then, the second step of torture, the subjective characteristics of pain are objectified. Although the prisoner's internal experience may be close to or identical with that of a person suffering severe pain from burns or a stroke or cancer or phantom limb, it is, unlike this other person's, simultaneously being externalized. The following attributes belong equally to the felt-experience of patient and prisoner, although it is only in the second context (or in some other area of objectification) that they become graspable from the outside.

—The first, the most essential, aspect of pain is its sheer aversiveness. While other sensations have content that may be positive, neutral, or negative, the very content of pain is itself negation. If to the person in pain it does not feel averse, and if it does not in turn elicit in that person aversive feelings toward it, it is not in either philosophical discussions or psychological definitions of it called pain.[62] Pain is a pure physical experience of negation, an immediate sensory rendering of "against," of something being against one, and of something one must be against. Even though it occurs within oneself, it is at once identified as "not oneself," "not me," as something so alien that it must right now be gotten rid of. This internal physical experience is in torture accompanied by its external political equivalent, the presence in the space outside the body of a self-proclaimed "enemy," someone who in becoming the enemy becomes the human embodiment of aversiveness; he ceases to have any psychological characteristics or content other than that he is, like physical pain, "not me," "against me." Although there are many averse political contexts—an occupied town or a prison, for example—where the "againstness" exists in an implicit and silent state of readiness, exists not now but only as an always closeby future, it is the very nature of torture to in each present moment identify, announce, act out in brutality, accusation, and challenge the state of its own otherness, the state of being against, the fact of being the enemy.

—A second and third aspect of pain, closely related to the first, are the double experience of agency. While pain is in part a profound sensory rendering of "against," it is also a rendering of the "something" that is against, a something at once internal and external. Even when there is an actual weapon present, the sufferer may be dominated by a sense of internal agency: it has often been observed that when a knife or a nail or pin enters the body, one feels not the knife, nail or pin but one's own body, one's own body hurting one. Conversely, in the utter absence of any actual external cause, there often arises a vivid sense of external agency, a sense apparent in our elementary, everyday vocabulary for pain: knifelike pains, stabbing, boring, searing pains. In physical pain, then, suicide and murder converge, for one feels acted upon, annihilated, by inside and outside alike. The sense of self-agency, visible in many dimensions of torture, is primarily dramatized there in the ritualized self-betrayal of confession and forced exercise. The sense of external agency is objectified in the systematic assimilation of shelter and civilization into the torturer's collection of weapons. But inside and outside and the two forms of agency ultimately give way to and merge with one another: confession and exercises are a form of external as well as internal agency since one's own body and voice now no longer belong to oneself; and the conversion of the physical and cultural setting into torture instruments is internal as well as external since it acts as an image of the impact of pain on human consciousness.

—This dissolution of the boundary between inside and outside gives rise to a fourth aspect of the felt experience of physical pain, an almost obscene conflation of private and public. It brings with it all the solitude of absolute privacy with none of its safety, all the self-exposure of the utterly public with none of its possibility for camaraderie or shared experience. Artistic objectifications of pain often concentrate on this combination of isolation and exposure. Ingmar Bergman's films repeatedly couple physical pain with intense moments of humiliation. In the opening sequence of *Sawdust and Tinsel*, for example, a cuckolded clown, alone amid the jeers of watching soldiers, carries the naked and impossibly heavy body of his faithless, cared for wife up a steep hill of rocks that slice his bare feet. The terrain of Sophocles's *Philoctetes*—the background against which we watch and hear the agonized writhing of the wounded hero, writhing which so repelled his shipmates that they long ago abandoned him here—is a small island of jagged rocks at once utterly cut off from homeland and humanity and utterly open to the elements. The solitary figure in the typical canvas of Francis Bacon is made emphatically alone by his position on a dais, by an arbitrary geometric box inserted over him, and by his naked presence against a uniform (and in its uniformity, almost absolute) orange-red background; yet while he is intensely separate from the viewer (a separation Bacon wanted to heighten further by having the canvasses covered with glass) he is simultaneously mercilessly exposed to us, not merely because he is undressed, unshielded by any material or clothing, but because his melting body is turned inside out, revealing the most inward and secret parts of him. This combination, not usually as in these artistic objectifications visible to an outsider but always present in the felt-experience of pain, is part of the ongoing external action and activity of torture, for the prisoner is forced to attend to the most intimate and interior facts of his body (pain, hunger, nausea, sexuality, excretion) at a time when there is no benign privacy, for he is under continual surveillance, and there is no benign public, for there is no human contact, but instead only an ugly inverting of the two.

—A fifth dimension of physical pain is its ability to destroy language, the power of verbal objectification, a major source of our self-extension, a vehicle through which the pain could be lifted out into the world and eliminated. Before destroying language, it first monopolizes language, becomes its only subject: complaint, in many ways the nonpolitical equivalent of confession, becomes the exclusive mode of speech. Eventually the pain so deepens that the coherence of complaint is displaced by the sounds anterior to learned language. The tendency of pain not simply to resist expression but to destroy the capacity for speech is in torture reenacted in overt, exaggerated form. Even where the torturers do not permanently eliminate the voice through mutilation or murder, they mime the work of pain by temporarily breaking off the voice, making it their own, making it speak their words, making it cry out when they want it to cry, be silent when they want its silence, turning it on and off, using its sound to abuse the one whose voice it is as well as other prisoners. The derisive connotations of "betrayal" surrounding confession also reveal in heightened form the process by which in nonpolitical contexts a person's complaint-filled, deteriorating, or absent language obscures and discredits his needs at the very moment when they are most acute. Even in 1976 and 1977 when the American news media for the first time began to devote space and sympathetic attention to the problems of physical pain, it was not unusual to see sometimes in local papers articles with headlines such as "A pain is a pain if you complain" and "Chronic pain can make you one."[63]

—A sixth element of physical pain, one that overlaps but is not quite coterminous with the previous element, is its obliteration of the contents of consciousness. Pain annihilates not only the objects of complex thought and emotion but also the objects of the most elemental acts of perception. It may begin by destroying some intricate and demanding allegiance, but it may end (as is implied in the expression "blinding pain") by destroying one's ability simply to see. In torture, this world dissolution, acknowledged in confession, is mimed in the conversion into weapons and resulting cancellation of all parts of the room as well as all parts of the larger world that can be bodied forth in the torturer's action and speech.

—A seventh aspect of pain, built on the first six, is its totality. Pain begins by being "not oneself" and ends by having eliminated all that is "not itself." At first occurring only as an appalling but limited internal fact, it eventually occupies the entire body and spills out into the realm beyond the body, takes over all that is inside and outside, makes the two obscenely indistinguishable, and systematically destroys anything like language or world extension that is alien to itself and threatening to its claims. Terrifying for its narrowness, it nevertheless exhausts and displaces all else until it seems to become the single broad and omnipresent fact of existence. From no matter what perspective pain is approached, its totality is again and again faced. Even neurological and physiological descriptions repeatedly acknowledge the breadth of its presence. Its mastery of the body, for example, is suggested by the failure of many surgical attempts to remove pain pathways because the body quickly, effortlessly, and endlessly generates new pathways.[64] Of its location in the brain, Melzack writes:

It is traditionally assumed that pain sensation and response are subserved by a "pain centre" in the brain. The concept of a pain centre, however, is totally inadequate to account for the complexity of pain. Indeed, the concept is pure fiction, unless virtually the whole brain is considered to be the pain centre, because the thalamus, hypothalamus, brainstem reticular formation, limbic system, parietal cortex, and frontal cortex are all implicated in pain perception. Other brain areas are obviously involved in the emotional and motor features of pain.[65]

This same totality is equally descriptive of felt-experience. Although other sensations sometimes have the power to diminish pain by distracting the person's attention, in prolonged and acute pain the body often begins to interpret all sensations as pain. S. W. Mitchell, a Civil War surgeon, a minor though prolific novelist, and a major figure in medical research and observation of wounds and wound pain, writes,

Perhaps few persons who are not physicians can realize the influence which long-continued and unendurable pain may have upon both body and mind. The older books are full of cases in which, after lancet wounds, the most terrible pain and local spasms resulted. When these had lasted for days or weeks, the whole surface became hyperaesthetic, and the senses grew to be only avenues for fresh and increasing tortures, until every vibration, every change of light, and even . . . the effort to read brought on new agony.[66]

Torture aspires to the totality of pain. Antonin Artaud once described the way in which a pain "as it intensifies and deepens, multiplies its resources and means of access at every level of the sensibility."[67] So the torturers, like pain itself, continually multiply their resources and means of access until the room and everything in it becomes a giant externalized map of the prisoner's feelings. Almost as obsessively narrow and repetitive as the pain on which it models itself, torture can be more easily seen because it has dimension and depth, a space that can be walked around in though not walked out of. Here there is nothing audible or visible, there is nothing that can be touched, or tasted, or smelled that is not the palpable manifestation of the prisoner's pain.

—The eighth and, for now, final element carries us back to where we were immediately before starting this enumeration of objectified elements of pain, namely, that one of its most frightening aspects is its resistance to objectification. Though indisputably real to the sufferer, it is, unless accompanied by visible body damage or a disease label, unreal to others. This profound ontological split is a doubling of pain's annihilating power: the lack of acknowledgment and recognition (which if present could act as a form of self-extension) becomes a second form of negation and rejection, the social equivalent of the physical aversiveness. This terrifying dichotomy and doubling is itself redoubled, multiplied, and magnified in torture because instead of the person's pain being subjectively real but unobjectified and invisible to all others, it is now hugely objectified, everywhere visible, as incontestably present in the external as in the internal world, and yet it is simultaneously categorically denied.

This denial, the third major step in the sequence on which torture is built, occurs in the translation of all the objectified elements of pain into the insignia of power, the conversion of the enlarged map of human suffering into an emblem of the regime's strength. This translation is made possible by, and occurs across, the phenomenon common to both power and pain: agency. The electric generator, the whips and canes, the torturer's fists, the walls, the doors, the prisoner's sexuality, the torturer's questions, the institution of medicine, the prisoner's screams, his wife and children, the telephone, the chair, a trial, a submarine, the prisoner's ear drums—all these and many more, everything human and inhuman that is either physically or verbally, actually or allusively present, has become part of the glutted realm of weaponry, weaponry that can refer equally to pain or power. What by the one is experienced as a continual contraction is for the other a continual expansion, for the torturer's growing sense of self is carried outward on the prisoner's swelling pain. As an actual physical fact, a weapon is an object that goes into the body and produces pain. As a perceptual fact, it lifts the pain out of the body and makes it visible or, more precisely, it acts as a bridge or mechanism across which some of pain's attributes—its incontestable reality, its totality, its ability to eclipse all else, its power of dramatic alteration and world dissolution—can be lifted away from their source, can be separated from the sufferer and referred to power, broken off from the body and attached instead to the regime. Now, at least for the duration of this obscene and pathetic drama, it is not the pain but the regime that is incontestably real, not the pain but the regime that is total, not the pain but the regime that is able to eclipse all else, not the pain but the regime that is able to dissolve the world. Fraudulent and merciless, this kind of power claims pain's attributes as its own and disclaims the pain itself. The act of disclaiming is as essential to the power as is the act of claiming. It of course assists the torturer in practical ways. He first inflicts pain, then objectifies pain, then denies the pain—and only this final act of self-blinding permits the shift back to the first step, the inflicting of still more pain, for to allow the reality of the other's suffering to enter his own consciousness would immediately compel him to stop the torture. But the bond between the blindness and the power goes far beyond the practical circles of self-amplification. It is not merely that his power makes him blind, nor that his power is accompanied by blindness, nor even that his power requires blindness; it is, instead, quite simply that his blindness, his willed amorality, *is* his power, or a large part of it. This identification becomes almost self-evident when sadistic forms of power are seen in relation to the benign and legitimate forms of power on which civilization is based. Every act of civilization is an act of transcending the body in a way consonant with the body's needs: in building a wall, to return to an old friend, one overcomes the body, projects oneself out beyond the body's boundaries but in a way that expresses and fulfills the body's need for stable temperatures. Higher moments of civilization, more elaborate forms of self-extension, occur at a greater distance from the body: the telephone or the airplane is a more emphatic instance of overcoming the limitation of the human body than is the cart. Yet even as here when most exhilaratingly defiant of the body, civilization always has embedded within it a profound allegiance to the body, for it is only by paying attention that it can free attention. Torture is a condensation of the act of "overcoming" the body present in benign forms of power. Although the torturer dominates the prisoner both in physical acts and verbal acts, ultimate domination requires that the prisoner's ground become increasingly physical and the torturer's increasingly verbal, that the prisoner become a colossal body with no voice and the torturer a colossal voice (a voice composed of two voices) with no body, that eventually the prisoner experience himself exclusively in terms of sentience and the torturer exclusively in terms of self-extension. All those ways in which the torturer dramatizes his opposition to and distance from the prisoner are ways of dramatizing his distance from the body. The most radical act of distancing resides in his disclaiming of the other's hurt. Within the strategies of power based on denial there is, as in affirmative and civilized forms of power, a hierarchy of achievement, successive intensifications based on increasing distance from, increasingly great transcendence of, the body: a regime's refusal to recognize the rights of the normal and healthy is its cart; its refusal to recognize and care for those in agony is its airplane.

This display of the fiction of power, the final product and outcome of torture, should in the end be seen in relation to its origin, the motive that is claimed to be its starting point, the need for information. Torture is not unusual in giving so prominent a place to so false a motive, for, as noted earlier, other acts of political violence within these same governments such as arrest and gratuitous punishment are frequently accompanied by explanations of motive so arbitrary that they seem intended as demonstrations of contempt. Explorations of other historical moments of brutality such as Camus's of the guillotine in France and Arendt's of Hitler's Germany almost inevitably comment on the obvious erroneousness of the asserted motive: the purpose of an execution cannot be deterrence if the execution is never even publicly announced; the war did not cause but permitted Hitler's mass executions.[68] This false motive syndrome is not adequately explained by the vocabulary of "excuse" and "rationalization," and its continual recurrence suggests that it has a fixed place in the formal logic of brutality. The motive for torture is to a large extent the equivalent, though in a different logical time, of the fictionalized power; that is, one is the falsification of the pain prior to the pain and one the falsification after the pain. The two together form a closed loop of attention that ensures the exclusion of the prisoner's human claim. Just as the display of the weapon (or agent or cause) makes it possible to lift the attributes of pain away from the pain, so the display of motive endows agency with agency, cause with cause, thereby lifting the attributes of pain still further away from their source. If displaying the weaponry begins to convert the prisoner's pain into the torturer's power, displaying the motive (and the ongoing interrogation means that it is fairly continually displayed) enables the torturer's power to be understood in terms of his own vulnerability and need. A motive is of course only one way of deflecting the natural reflex of sympathy away from the actual sufferer. According to Arendt in *Eichmann in Jerusalem*, the speeches of Himmler were full of phrases such as, "The order to solve the Jewish question, this was the most frightening order an organization could ever receive," and she explains:

> Hence the problem was how to overcome not so much their conscience as the animal pity by which all normal men are affected in the presence of physical suffering. The trick used by Himmler—who apparently was rather strongly afflicted with these instinctive reactions himself—was very simple and probably very effective; it consisted in turning these instincts around, as it were, in directing them toward the self. So that instead of saying: What horrible things I did to people!, the murderers would be able to say: What horrible things I had to watch in the pursuance of my duties, how heavily the task weighed upon my shoulders![69]

Concentration camp guards, according to Bruno Bettelheim, repeatedly said to their prisoners, "I'd shoot you with this gun but you're not worth the three pfennig of the bullet," a statement that had so little effect on the prisoners that its constant repetition was unintelligible to Bettelheim until he realized that it had been made part of the SS training because of its impact on the guards themselves.[70]

This last example, because it involves an actual weapon, is paradigmatic of the structure of perception that underlies the false motive even when no overt image of the weapon is present. Every weapon has two ends. In converting the other person's pain into his own power, the torturer experiences the entire occurrence exclusively from the nonvulnerable end of the weapon. If his attention begins to slip down the weapon toward the vulnerable end, if the severed attributes of pain begin to slip back to their origin in the prisoner's sentience, their backward fall can be stopped, they can be lifted out once more by the presence of the motive. If the guard's awareness begins to follow the path of the bullet, that path itself can be bent so that he himself rather than the prisoner is the bullet's destination: his movement toward a recognition of the internal experience of an exploding head and loss of life is interrupted and redirected toward a recognition of his own loss of three pfennig. It does not matter that there is always an extraordinary disjunction between the two levels of need—between being shot and losing three pfennig, between being the victim of the massive concentration camp brutalities and having to watch those brutalities, between extreme and prolonged physical pain of torture and being in need of a piece of information—for the work of the false motive is formal, not substantive; it prevents the mind from ever getting to the place where it would have to make such comparisons. Power is cautious. It covers itself. It bases itself in another's pain and prevents all recognition that there is "another" by looped circles that ensure its own solipsism.

2

THE STRUCTURE OF WAR
The Juxtaposition of Injured Bodies and Unanchored Issues

TORTURE is such an extreme event that it seems inappropriate to generalize from it to anything else or from anything else to it. Its immorality is so absolute and the pain it brings about so real that there is a reluctance to place it in conversation by the side of other subjects. But this reluctance, and the deep sense of tact in which it originates, increase our vulnerability to power by ensuring that our moral intuitions and impulses, which come forward so readily on behalf of human sentience, do not come forward far enough to be of any help: we are most backward on behalf of the things we believe in most in part because, like ancients hesitant to permit analogies to God, our instincts salute the incommensurability of pain by preventing its entry into worldly discourse. The result of this is that the very moral intuitions that might act on behalf of the claims of sentience remain almost as interior and inarticulate as sentience itself.

It is a consequence of the ease with which power can be mixed with almost any other subject that it can be endlessly unfolded, exfoliated, in strategies and theories that—whether compellingly legitimate or transparently absurd—in their very form, in the very fact of occurring in human speech, increase the claim of power, its representation in the world. In contrast, one of two things is true of pain. Either it remains inarticulate or else the moment it first becomes articulate it silences all else: the moment language bodies forth the reality of pain, it makes all further statements and interpretations seem ludicrous and inappropriate, as hollow as the world content that disappears in the head of the person suffering. Beside the initial fact of pain, all further elaborations—that it violates this or that human principle, that it can be objectified in this or that way, that it is amplified here, that it is disguised there—all these seem trivializations, a missing of the point, a missing of the pain. But the result of this is that the moment it is lifted out of the ironclad privacy of the body into speech, it immediately falls back in. Nothing sustains its image in the world; nothing alerts us to the place it has vacated. From the inarticulate it half emerges into speech and then quickly recedes once more. Invisible in part because of its resistance to language, it is also invisible because its own powerfulness ensures its isolation, ensures that it will not be seen in the context of other events, that it will fall back from its new arrival in language and remain devastating. Its absolute claim for acknowledgment contributes to its being ultimately unacknowledged.

Though there may be no human event that is as without defense as torture, others give rise to the same central question—By what perceptual process does it come about that one human being can stand beside another human being in agonizing pain and not know it, not know it to the point where he himself inflicts it?—and once again lead to an answer centering on interactions between the body and voice made possible by a language of agency.

The most obvious analogue to torture is war. The form of torture that leaves the prisoner untouched by the torturer but that requires prisoners to maim one another makes visible the connection between them. Some of the apparent differences between them are partially attributable to the fact that the symbolic and the fictional are much more prominent in torture. War more often arises where the enemy is external, occupies a separate space, where the impulse to obliterate a rival population and its civilization is not (or need not at first be perceived as) a self-destruction. Torture usually occurs where the enemy is internal and where the destruction of a race and its civilization would be a self-destruction, an obliteration of one's own country. Hence there must be more drama in torture: the destruction must be acted out symbolically[1] within a handful of rooms.

War and torture have the same two targets, a people and its civilization (or as they were called earlier, the two realms of sentience and self-extension); the much greater reliance on the symbolic in torture occurs in both spheres. In both war and torture, there is a destruction of "civilization" in its most elemental form. When Berlin is bombed, when Dresden is burned, there is a deconstruction not only of a particular ideology but of the primary evidence of the capacity for self-extension itself: one does not in bombing Berlin destroy only objects, gestures, and thoughts that are culturally stipulated but objects, gestures, and thoughts that are human, not Dresden buildings or German architecture but human shelter. Torture is a parallel act of deconstruction. It imitates the destructive power of war: rather than destroying the concrete physical fact of streets, houses, factories, and schools, it destroys them as they exist in the mind of the prisoner, it destroys them as they exist in the furnishings of a room: to convert a table into a weapon is to set a factory on fire; to hear a confession is to watch from above the explosion of a city block. This same form of substitution occurs in relation to the second target, the sentient source of the first, the human body itself. Whereas the object of war is to kill people, torture usually mimes the killing of people by inflicting pain, the sensory equivalent of death, substituting prolonged mock execution for execution. The numbers involved reinforce this sense of the division between the real and the dramatized. Although the thousands and thousands of political prisoners hurt during the 1970s and 1980s have led Amnesty International to call torture an "epidemic," the numbers of persons hurt are of course vastly larger in war. In torture, the individual stands for "individuals"—huge multiplicity is replaced by close proximity sustained over hours or days or weeks; being in close contact with the victim's hurt provides the sense of "magnitude" achieved in war through large numbers.

But while torture relies much more heavily on overt drama than does war, war too—as is quietly registered in the language of theatres of battle, international dialogues, scenarios, and stages—has within it a large element of the symbolic and is ultimately, like torture, based on a simple and startling blend of the real and the fictional. In each, the incontestable reality of the body—the body in pain, the body maimed, the body dead and hard to dispose of—is separated from its source and conferred on an ideology or issue or instance of political authority impatient of, or deserted by, benign sources of substantiation. There is no advantage to settling an international dispute by means of war rather than by a song contest or a chess game except that in the moment when the contestants step out of the song contest, it is immediately apparent that the outcome was arrived at by a series of rules that were agreed to and that can now be disagreed to, a series of rules whose force of reality cannot survive the end of the contest because that reality was brought about by human acts of participation and is dispelled when the participation ceases. The rules of war are equally arbitrary and again depend on convention, agreement, and participation; but the legitimacy of the outcome outlives the end of the contest because so many of its participants are frozen in a permanent act of participation: that is, the winning issue or ideology achieves for a time the force and status of material "fact" by the sheer material weight of the multitudes of damaged and opened human bodies.

This brief characterization of the structure of war will be unfolded more slowly below and then differentiated from a widely accepted and erroneous account of war with which it might otherwise be confused. Gradually the parallel between what occurs in the interior of torture and what occurs in the interior of war will become visible, as will also a crucial element that differentiates them, endowing war with a moral ambiguity wholly absent from torture. It will become clear why those who wish to outlaw torture have never found it difficult to arrive at an "absolute" formulation of their prohibition, while those who with equal passion work to outlaw the initiation of war have so often stopped short of an unconditional formulation and arrived at the perception that an "absolute" prohibition may be itself morally untenable.[2]

One simple and important formal difference will be visible from the outset. Though the two are structural analogues, the fundamental shape of each comes into being in a different place: the structure of torture resides in, takes shape in, the physical and verbal interactions between two persons, a torturer and a prisoner; the structure of war will again be centered in an extraordinary relation between body and voice but that *relation* will not be itself locatable within the *relation* between any two persons—soldier and soldier, soldier and officer, soldier and civilian—nor even in the *relation* between two large groups of people, such as the hundreds of thousands of persons who face and deface each other across the field of battle.[3] The essential structure of war, its juxtaposition of the extreme facts of body and voice, resides in the *relation* between its own largest parts, the relation between the collective casualties that occur *within* war, and the verbal issues (freedom, national sovereignty, the right to a disputed ground, the extra-territorial authority of a particular ideology) that stand *outside* war, that are there *before* the act of war begins and *after* it ends, that are understood by warring populations as the motive and justification and will again be recognized after the war as the thing substantiated or (if one is on the losing side) not substantiated by war's activity. The central question that is asked here— what is the relation between the obsessive act of injuring and the issue on behalf of which that act is performed—is a question about the relation between the interior content of war and what stands outside it. In order to answer that question, however, it is necessary to back up one step and define the relation between *two* interior facts about war: first, that the immediate activity is injuring; second, that the immediate activity of war is a contest. In participating in war, one participates not simply in an act of injuring, but in the activity of reciprocal injuring where the goal is to out-injure the opponent. The construction, "War is *x*," has, over the centuries, invited an array of predicate nominatives; but there are no two predicate nominatives that have either the accuracy or the definitional totality as the two singled out here, and it is by first understanding precisely how the two qualify one another that it will be possible to arrive at an understanding of the second and more fundamental question about the relation between bodily injury and verbal issues.

Our starting place, then, is the assumption that war belongs to two larger categories of human experience (larger in the sense that each contains war as only one of its terms). First, it is a form of violence; it is a member of a class of occurrences whose activity is "injuring." Second, it is a member of a class of occurrences that are contests. It is in the relation of these two rather than in either individually that the nature of war resides, but for a moment each of the two must be held steadily visible in isolation because each has a way of slipping out of view. Thus it is necessary to back up one more step and make certain that our two "self-evident premises" are indeed self-evident.

I. War Is Injuring

The main purpose and outcome of war is injuring. Though this fact is too self-evident and massive ever to be directly contested, it can be indirectly contested by many means and disappear from view along many separate paths.It may disappear from view simply by being omitted: one can read many pages of a historic or strategic account of a particular military campaign, or listen to many successive installments in a newscast narrative of events in a contemporary war, without encountering the acknowledgment that the purpose of the event described is to alter (to burn, to blast, to shell, to cut) human tissue, as well as to alter the surface, shape, and deep entirety of the objects that human beings recognize as extensions of themselves. In any given instance, omission may occur out of the sense that this activity is too self-evident to require articulation; it may instead originate in a failure of perception on the part of the describer; again it may arise out of an active desire to misrepresent the central content of war's activity (and this conscious attempt to misrepresent can in its turn be broken down into an array of motives, some malevolent, some relatively benign).

The identification of the paths by which injuring disappears from view and not the identification of motive will be attended to here; for any one path is likely to be laden with many motives, and the recitation of all of them in so brief a discussion would be as impossible as the specification of one or two would be misleading. Much more important, regardless of local motives, the structure of war itself will require that injuring be partially eclipsed from view and will invariably bring about that eclipse by one constellation of motives or another. That is, just as torture can be understood to entail three separate steps— the infliction of pain, the objectification of the pain, the disowning of the pain and transfer of its attributes to another location—so, too, it will gradually become evident that war entails a similar structure of physical and perceptual events: it requires both the reciprocal infliction of massive injury and the eventual disowning of the injury so that its attributes can be transferred elsewhere, as they cannot if they are permitted to cling to the original site of the wound, the human body.

It should also be noticed from the outset that while the perpetuation of war would be impossible without the disowning of injuring, this disowning is not necessarily authored (not at any rate exclusively authored) by those who wish to perpetuate war. Although it would not be inaccurate to say that *in general* the physical immediacy of damaged human bodies is more visible in the words of those working to outlaw a particular weapon, to stop a particular war, or to eliminate the universal form of war than it is in the words of those who, because of a political or military or philosophic position, are engaged in its continuation, the generalization would be one so elaborately attended by qualifications and exceptions that it would come to seem unhelpful if not untrue. The qualifications come from three directions. First, active opposition to war does not necessarily require an accurate perception or description of the relation between injuring and political goals. Second and conversely, acceptance of war, or even active sponsorship of war as occurs when a president, prime minister, or statesman must work to ensure the continued participation of his country's population in a given conflict, may in fact be carried out (as can be seen in the political writings of Henry Kissinger and Winston Churchill, or again in the strategic writings of Clausewitz, Liddell Hart, or Sokolovskiy) with varying degrees of attention to and assessment of the centrality of the act of injuring. Third, conventional war entails the participation of a massive number of people, only a small fraction of whom are engaged in the active verbal advocacy of either the elimination or the perpetuation of war; and if injuring disappears, it is its absence in their informal conversation that is perhaps most important.

A deeply tactful, compassionate, and careful account of the alterations that occur in human tissue such as the Stockholm International Peace Research Institute's verbal and visual account of the effects of incendiary weapons in Vietnam, Dresden, Hiroshima, or Nagasaki may place the injured body several inches in front of our eyes, hold the light up to the injured flesh, and keep steady the reader's head so that he cannot turn away.[4] In their attempt to bring about the elimination of such weapons (weapons may be differentiated not by whether or not they injure, nor even by the final extremity of damage since most kill, but by the intensity and duration of suffering before death), such descriptions are crucial; for although in understanding the nature of war the agonized injury of the small Vietnamese girl's burned face and burned off arms—or later her look of terror as she sees in the reflecting surface of window, river, or imported spoon the obliteration of her features—must be multiplied over the thousands and millions of inhabitants of different countries, injury must at some point be understood individually because pain, like all forms of sentience, is experienced within, "happens" within, the body of the individual. Such a study may not, however, specify whether such injury was the intent or accidental effect of the bombing, whether it was within or wholly outside the view of the chemist or corporation who discovered or marketed napalm, and, most important, whether the populations who consented to war consented to this or to something else.

A much more direct account of these questions may occur in writings that endorse, or at least accept, the occurrence of war. Of all writing—political, strategic, historical, medical—there is probably no work that more successfully holds visible the structural centrality of injuring than Clausewitz's *On War*. In his description of invasion, for example, he will say, "The immediate object here is neither to conquer the enemy country nor to destroy its army, but simply *to cause general damage*," as he will often elsewhere specify that the object is to "increase the enemy's suffering."[5] In battle, for example, the soldier's primary goal is not, as is so often wrongly implied, the protection or "defense" of his comrades (if it were this, he would have led those comrades to another geography): his primary purpose is the injuring of enemy soldiers; to preserve his own forces has the important but only secondary and "negative" purpose of frustrating and exhausting the opponent's achievement of his goal.[6] If the visibility of the central fact of damage, and the specification of the particular

form of damage sought in any given tactic, has wrongly contributed to Clausewitz's reputation for "ruthlessness," it has justly contributed to his reputation
for astonishing brilliance: one knows on every page that one is in the presence
of a massive intelligence in part because his powers of description remain available to him at all moments. The written and spoken record of war over many
centuries certifies the ease with which human powers of description break down
in the presence of battle, the speed with which they back away from injuring
and begin to take as their subject the most incidental or remote activities occurring
there, rather than holding onto what is everywhere occurring at its center and
periphery. The enumeration of the paths by which injuring disappears from view
only begins with the one already named here: omission. The character of the
other paths will be illustrated below with passages from formal writings; but
they are most significant insofar as they are recognized as having counterparts
in the informal and unrecorded conversations of the general population, as the
subject of war makes its way into our daily activities and accompanies us as we
walk down the road, sit down to dinner, or return a borrowed book or tool to a
friend.

   A second path by which injuring disappears is the active redescription of the
event: the act of injuring, or the tissue that is to be injured, or the weapon that
is to accomplish the injury is renamed. The gantry for American missiles is
named the "cherrypicker,"[7] just as American missions entailing the massive
dropping of incendiary bombs over North Vietnam were called "Sherwood
Forest" and "Pink Rose,"[8] just as Japanese suicide planes in World War II
were called "night blossoms,"[9] as prisoners subjected to medical experiments
in Japanese camps were called "logs,"[10] and as the day during World War I on
which thirty thousand Russians and thirteen thousand Germans died at Tannenberg came to be called the "Day of Harvesting."[11] The recurrence here of
language from the realm of vegetation occurs because vegetable tissue, though
alive, is perceived to be immune to pain; thus the inflicting of damage can be registered in language without permitting the entry of the reality of suffering into the
description. Live vegetable tissue occupies a peculiar category of sentience that is
close to, perhaps is, nonsentience; more often, the language is drawn from the
unequivocal nonsentience of steel, wood, iron, and aluminum, the metals and materials out of which weapons are made and which can be invoked so that an event
entailing two deeply traumatic occurrences, the inflicting of an injury and the receiving of an injury, is thus neutralized.[12] "Neutralization" or "neutering" (or
their many variants such as "cleaning," "cleaning out," "cleaning up"[13] or other
phrases indicating an alteration in an essential characteristic of the metal, such as
"liquification") is itself a major vocabulary invoked in the redescription of injuring. It begins by being applied only to weapons: it is the other peoples' firepower
(guns, rockets, tanks) that must be "neutralized," but it is then transferred to the
holder of the gun, the firer of the rocket, the driver of the tank, as well as to the
civilian sister of the holder, the uncle of the firer, the child of the driver, the hu-

man beings who must be (not injured or burned or dismembered or killed but) "neutralized," "cleaned out," "liquidated."

Although a weapon is an extension of the human body (as is acknowledged in their collective designation as "arms"), it is instead the human body that becomes in this vocabulary an extension of the weapon. A nineteen-year-old holding a pistol has an arm that is three and a half feet from his shoulder to the tip of his weapon (if the weapon is firing, his reach changes from three and a half feet to five hundred yards). The first three feet are sentient tissue; the last half foot is nonsentient material. An idiom appropriate to an alteration in the surface of the gun's metal is invoked to describe an alteration in the boy's embodied arm as, conversely, an idiom originally invented to describe an unwanted alteration in the tissue of the human arm will be extended to the weapon: so an opponent will find himself in the peculiar position of working to "neutralize" the boy and to "wound" the gun. That the "wounding" language is applied to weapons and arms (that helicopters are injured in the sands of Iran; that the *Sheffield* receives a mortal wound in the waters off the Falkland Islands) would not in isolation be wholly inappropriate since these objects, like libraries and cities, are projections of the human person; but the language is lent to the weapons at precisely the same moment that it is being lifted away from the sentient source of those projections. The language of killing and injuring ceases to be a morally resonant one because the successful shelling of the bodies of thousands of nineteen-year-old German soldiers can be called "producing results"[14] and the death of civilians by starvation and pestilence following an economic boycott is called "collateral effects"[15] at precisely the same time that one turns on a radio and hears the report of an arsenal of tanks that received a "massive injury" or opens a book and reads about the government's hope "to kill a hidden base."[16] Once the populations of two nations consent to devote themselves to damaging each other, the dissolution of their language may not be itself morally disastrous; it may be perceived as inevitable and perhaps even "necessary." These difficult questions are neither raised nor answered here where the object is the relatively modest one of registering two facts: that reciprocal injuring is the obsessive content of war; that its centrality often slips from view.[17]

The habit of mind that is illustrated here as occurring in representative phrases and sentences is a massive one that occurs not in isolated fragments but is formalized into a conventional mode of perceiving the events of war. The exchange of idioms between weapons and bodies has its most serious manifestation in the fact that in many different contexts[18] the central inner activity of war comes to be identified as (or described as though it were) "disarming" rather than "injuring." Although the first term is sometimes intended only as a synonym for the second,[19] it is at other times used explicitly to differentiate the benign activity of eliminating weapons from what is then presented as the only accidental and unfortunate entailment of human injury during those operations. Thus we repeatedly encounter descriptions of war that come astonishingly close in their expressed aversion to weapons to sounding not merely "protective" but nearly "pacifist" in intention. If this confusion has achieved monumental proportions in discussions of nuclear war, it is because our very conception of nuclear war may itself be understood as the culmination of the history of this confusion: it is appropriate that at precisely the moment when weapons are capable of unprecedented injury to the human body (three hundred million persons in the first exchange), and in fact incapable of not inflicting that massive injury (if shot down outside one's own territory they will land in someone's country)—it is appropriate that at precisely this moment those weapons should have names (e.g., anti-missile missile) and be consistently described in such a way that their only target, or only intended target, or only immediate target, appears to be another weapon: that their effect is to "disarm" rather than to injure.[20] It should be understood that the coupling of the two is not an accidental and ironic conjunction but a profound manifestation of a confusion with a long and rich history. In the history of thinking about war, there are probably only one or two other errors at once so persistent, pervasive and deeply wrong as this one.

It is of course precisely because conventional war does in fact include in its interior acts of disarming (a dead soldier may be called a "disarmed" soldier even though his rifle is still functional, and in any war there may be many specific missions that have as their target the destruction of a munitions works, a tractor factory, or an Opel plant) that this particular confusion is more difficult to correct than the parallel confusion that involves importing an occurrence outside war into its interior. That is, a person who believes (perhaps quite rightly) that the *outcome* of a particular war will be greater political freedom for a given population may wrongly think of the interior activity of war as "freeing." But if actually asked to look at several hundred people in a forest slipping behind trees, edging out, lifting a rifle, disappearing, reappearing, bleeding, falling, he would probably agree that the best identification of the immediate activity occurring there would not be "freeing," but reciprocal injuring. Although the exterior occurrence may be imported into his description of the interior (he may begin by describing the activity as "freeing" but then see that those men whose actions he wants to describe as "oppressing" are performing a nearly identical set of gestures), it can still be clearly differentiated from and abstracted back out of what literally occurs there. But any activity that itself actually occurs in the interior of war will be much more difficult for the human mind to assess. Because disarming and injuring accompany one another inside war, it is more difficult for a person watching the men (as they still slip through the light and shade of the forest, all shooting at one another, and all working to get to one of two stockpiles of ammunition that each side knows the other has) to see which activity is central and which is the extension of that center. If the observer were to identify the central act as "injuring," the accuracy of this identification would be confirmed by the similarity between the infliction of injuring here and any instance of it occurring outside war: it might be with a car rather than a tank, a knife rather than a gun, down the steps rather than off a cliff, in solitude rather than in the company of thousands performing the same act, punishable as a crime or exonerated as self-defense rather than accepted as an unremarkable day's work, but in each case a body is being damaged by another body who himself risks danger at the moment he attempts to remain immune. If the observer were instead to identify the central activity as "disarming," he or she would discover little similarity between the immediate activity of the men in the forest and the signing of a disarmament treaty between nations, or the signing of a contract between rival residents of adjoining streets.

It could be argued that just as the original observer could describe the immediate activity as "freeing" if he agreed to call it "freeing by injuring" (or "freeing by injuring those who are oppressing by injuring," or "reciprocal injuring for nonreciprocal outcomes"), so the second could call the activity "disarming" if he stipulated "disarming by injuring" and thus accounted for the profound difference between what is happening here and in a disarmament contract, where it is the very absence of injury that is the motivating force and outcome. But again, unless both terms are understood as synonyms, the phrase "disarming by injuring" still misrepresents war's activity by misidentifying injuring as the subordinate activity. The more accurate formulation of both the substance and purpose of that activity is not "to disarm by injuring" but "to out-injure by injuring and disarming." That is, each side works to out-injure the other and does so in two ways: first, by inflicting injury on the bodies of the opponent; second, by resisting injuries to themselves both by avoiding bullets (running, ducking, diving behind trees, all of which can be called acts of disarming or rendering neutral the enemy's weapons) and by destroying the munitions works or ammunition supplies. To say that each side in a war wishes to disarm the other only expresses the fact that each side wishes to increase its own immunity while inflicting damage on the other.[21] This particular confusion is so fundamental that its clarification will again be required at several later points in this discussion. It is introduced here because renaming injuring "disarming" is one of many ways in which injuring is redescribed and made invisible.

The first two paths by which injuring achieves invisibility—omission and redescription—are, of course, nearly inseparable; they are manifestations of one another. Redescription may, for example, be understood as only a more active form of omission: rather than leaving out the fact of bodily damage, that fact is itself included and actively cancelled out as it is introduced into the spoken sentence or begins to be recorded on a written page. Alternatively, omission may be understood as only the most successful or extreme form of redescription where the fact of injury is now so successfully enfolded within the language that we cannot even sense its presence beneath the surface of that language, or point to the phrase or clause where (as in redescription) it has almost surfaced and then been held in place once more. The difficulty of distinguishing the two can be illustrated by a particular formal convention that occurs within the genre of strategy writing.

With the exception of periodic body counts or "kill ratios,"[22] the intricacies and complications of the massive geographical interactions between two armies of opposing nations tend to be represented without frequent reference to the actual injuries occurring to the hundreds of thousands of soldiers involved: the movements and actions of the armies are emptied of human content and occur as a rarefied choreography of disembodied events. But the quality of abstraction and, above all, the apparent distance of these events from the realm of human pain cannot in any simple way be attributed to the categorical evacuation of the body from the text; for the body, exiled in its ordinary form, is allowed to re-enter in an only slightly unexpected place. Each of the two armies periodically becomes a single embodied combatant, with the real human body's elemental duality of being at once capable of inflicting injury and of receiving it. The ordinary five to six foot vertical expanse of the adult person now becomes a colossus with, for example, one foot in Italy, another in northern Africa, a head in Sweden, an arm pulling back toward the coast of France, then suddenly punching forward toward Germany. The crossing of a river is not now an event enacted by many individuals—some of whom know how to swim and others of whom do not, some vulnerable to wet and cold and some relatively immune, some who have as their worst dream being caught between two banks on a bridge and others who have waited for just this moment of trial—but is rather enacted by a single integrated creature who, if named, takes the name of the division or of the commander, and who steps across in a single step, as when Omar Bradley writes, "Simpson had previously complained of Monty's orders halting him on the west bank of the Rhine when he could have jumped across it against light opposition,"[23] or, similarly, "In stalking us through the Ardennes, the enemy had been forced to expose himself to our fire, especially to the murderous air burst of our proximity fuse. To the 4th Division still nursing its wounds from the Huertgen Forest, this reversal in roles brought a sardonic satisfaction."[24] If such descriptions were sustained over pages or even whole paragraphs, the text would become a mythology of giants lumbering across rivers and stalking through forests, but of course the text only periodically and momentarily breaks out of abstraction into this form of description.

It is precisely because this form of description is a widely shared convention that it need not often be sustained over an entire passage but can instead be invoked with a single word or phrase, the fragments of a story whose outlines are familiar to all. Thus, one's own army may become a single gigantic weapon, a "spearhead" or a "hammer"; a certain territory or part of the army may become an "appendix" or an "underbelly"; each army has an "Achilles heel," or a vulnerable "hinge" or "joint" or a "rear" that may be "penetrated"; two divisions may be attacked where they stand "shoulder to shoulder," and so forth. Although a shared convention rather than a construction introduced by a single writer, some are more masters of it than others. It is used, for example, with great frequency and agility by B. H. Liddel Hart. Ludendorff's "long-cherished idea of a decisive *blow* against the British in Belgium" in World War I becomes a blow that will be delivered or not delivered with his own magnified hands: "He had failed to *pinch* out the Campiègne buttress on the West.... The tactical success of his own *blows* had been Ludendorff's undoing .... He had driven in three great wedges, but none had penetrated far enough to sever a vital artery."[25] Whether he is describing the strategy of Alexander, Napoleon, or the Battle of the Marne, one of the large combatants may begin to enact, with all the grace of slow-motion photography, a dance of shifting weight dispersed across the lift and fall of giant limbs: "He first drew the enemy's attention and resources to their left flank; then pressed hard on their right and center . . . he turned this frustration to ultimate advantage by a pretense of swinging his weight farther to their left, while actually swinging it to their right and center."[26]

It should be stressed that this convention, whether occurring in strategic, military, or political writings, arises not out of any attempt to obscure human hurt but out of purposes appropriate to those writings. The convention expresses the fact that the fate of the overall army or overall population, and not the fate of single individuals, will determine the outcome; it also has the virtue of bestowing visibility on events which, because of their scale, are wholly outside visual experience. It is, however, a convention which assists the disappearance of the human body from accounts of the very event that is the most radically embodying event in which human beings ever collectively participate. It is not that "injury" is wholly omitted, or even that it is, strictly speaking, redescribed, but rather that it is relocated to a place (the imaginary body of a colossus) where it is no longer recognizable or interpretable. We will respond to the injury (a severed artery in one giant, a massive series of leechbites in another[27]) as an imaginary wound in an imaginary body, despite the fact that that imaginary body is itself made up of thousands of real human bodies, and thus composed of actual (hence woundable) human tissue.

The wound thus becomes a way of articulating and "vivifying" (literally, investing with life) the idea of the strategic vulnerability of an armed forces, and will in most instances, if noticed, be accepted as it is intended, as only a "colorful" form of description: a colossal severed artery, if anything, works to deflect attention away from rather than call attention to what almost certainly lies only a very short distance behind the surface of that image, a terrifying number of bodies with actually severed arteries. In fact, when this descriptive convention occurs in close proximity to a sentence referring to actual body damage, it tends to appropriate attention away from that sentence since it is itself, by its very scale, so visually compelling and, at the same time, so easy to contemplate. Unlike a real wound, it will not, however visually startling, stupefy us into silence or shame us with the shame of our powerlessness to approach the opened human body and make it not opened as before. Describing the first German use of chlorine gas at Ypres on 22 April 1915, Liddell Hart writes that it left a gap four miles wide "filled only by the dead and by those who lay suffocating in agony from chlorine gas-poisoning." One sentence later that four-mile-wide gap becomes a gap in the jaw of a giant, and the embodied soldiers are now teeth in that jaw: "With the aid of gas the Germans had removed the defenders on the north flank of the salient as deftly as if extracting the back teeth from one side of a jaw. The remaining teeth in front and on the south flank of the salient were formed by the Canadian Division (Alderson), nearest the gap, the 28th Division (Bulfin), and the 27th Division (Snow), which together comprised Plumer's 5th Corps. The Germans had only to push south for four miles to reach Ypres, and loosen all these teeth by pressure from the rear."[28] Similarily, when in Churchill's 6 June 1944 address to the House of Commons he announces and assesses the liberation of Rome, he specifies the body count for the two sides as 20,000 and 25,000. But if our attention (or the attention of those originally addressed) does not move to the 45,000 dead it may be because it is still lingering with the striking image that immediately preceded the body count, the image of eight or nine German giants, the eight or nine divisions who were diverted to Italy and "were repulsed, and their teeth broken, by the successful resistance of the Anzio bridgehead forces in the important battle which took place in the middle of February."[29] The maimed colossus will typically require neither our sympathy nor our anger nor our shame, and it is often, as here, made to look slightly ludicrous in the midst of its mighty catastrophe.[30]

In these first two forms of description, the action of injuring and the injury it accomplishes are invisible, absent from view in the case of omission, and actively escorted out of view in the case of redescription. But they cannot always achieve and maintain invisibility—there were, for example, at the end of World War I thirty-nine million corpses and at the end of World War II between forty-seven and fifty-five million corpses—and more remarkable, perhaps, than those forms of description already looked at are the particular vocabularies that arise once the injuries are seen, and that assign them to an accidental, incidental, or subordinate position: human wounds are not, as earlier, escorted out of view but are instead escorted from the center of view to the margins. There are four main paths by which injuring can be relegated to a still visible but marginal position.

In the first of these, the injuries and deaths and damage are referred to as a "by-product" of war: the term is usually preceded by an adjective ("terrible by-product," "necessary by-product," "acceptable by-product," "inevitable by-product," "unacceptable by-product"); which one appears in a given instance is determined by the particular argument that is being made. But if injury is designated "the by-product," what is the product? Injury is the thing every exhausting piece of strategy and every single weapon is designed to bring into being: it is not something inadvertently produced on the way to producing something else but is the relentless object of all military activity. Although we may become lost in the intricacies of this helicopter's ability to hover, and that one's inclusion of a particular type of radar, or the M113's ability to move over six hundred yards in a given number of seconds less than the M150, what is being indicated at all times is the relative degree of the object's power to injure (to sight the opponent, to approach and reach the opponent, and to damage the opponent) or the relative degree of its power to remain immune while injuring so that it can go on injuring (hence to sight the opponent, to approach the opponent, to damage the opponent, and to go on damaging the opponent the maximum length of time before it is itself knocked out of the war). So, too, the complexities of strategic decisions are complexities about where best (not necessarily maximally) and how best (not necessarily maximally) to inflict injury, or where and how to inflict injury while keeping one's side immune in order to sustain its injury-inflicting powers. It is not the language of "production" or "creation" that is false here, for that vocabulary acknowledges the fact that something not in existence, not naturally occurring, was brought into being by conscious human agency and inventiveness. Nor, if one wanted to include within the production metaphor the goals that are exterior to war (e.g., political freedom,[31] territorial legitimacy, assertion of authority in a given hemisphere) would it be false to designate injury an "intermediate product" that will somehow (that "somehow" will be discussed at a later point) become one day a different product, political freedom. But while it is accurate to designate injury the "product" of the immediate activity of war or in this second case, the "intermediate product" in the long-range outcomes, it is in no case accurate to identify it as the by-product. One may say, for example, that one's object is to make paper and not to kill trees, but the acres of felled trees are then an intermediate product (actively brought about by concentrated acts of labor), not a by-product.

The language of "by-product" denotes "accidental," "unwanted," "unsought," "unanticipated," and "useless."[32] The last meaning is the most crushing, for while the others are only an abdication of responsibility, the last asserts that the deaths on neither side were centrally useful to whatever it was that was being sought through the war's activity. The only thing more overwhelming than that a human community should have a use for death, the extreme "use for" that is signalled by the shift from "it is needed" to "it is required" (and soldiers understand that it is *this use* to which they have been summoned, *this* to which they have consented: that they are going either "to die for one's country" or "to kill for one's country")—the only thing more overwhelming than the fact that it will have this use for death is that the community will then disown that use and designate those deaths "useless." The millions who stood on the streets of Nagasaki and Dresden, the twenty million Russians who died in World War II, the generation of French and British maimed in World War I, the fifty-seven thousand Americans who died in Vietnam—whatever a war's use, whatever its aspirations, whatever its accomplishments, its deeds and outcomes, they have nothing central to do with these dead and injured who are war's by-product.

A second major metaphor again emphasizes the notion of the accidental and unanticipated while moving wholly out of the language of production: here the injuries are seen as having occurred on the road to another goal. While the first vocabulary makes injury the unintentional outcome in the process of making, the second identifies injury as an unforeseeable interruption on the path of arrival. Driving down the road to X (freedom, authority), we suddenly found many people had stepped onto the road (or the road widened and so ran through the field where they worked) and they were run over. To describe those who die as an "accidental entailment" is as dismissive as to say they are a not particularly useful by-product in the first vocabulary. It is interesting that in the road metaphor, as in the production metaphor, the human mind comes very close to articulating and understanding the nature of injury in war, but then shifts the metaphor so that the very thing that must have been intuited when the mind reached for that metaphor now slips from view. If one is talking of the interior activity of war, then injury is not something on the road to a goal but is the goal itself. If one is including within the road metaphor goals that are exterior to the activity of war (freedom, territorial sovereignty), then it is appropriate to designate these things as the goal or destination beyond injury; but it is crucial to see that, in this use of the metaphor, the injured bodies would not be something *on* the road to the goal but would themselves *be the road* to the goal. If a country's leaders decide that there is no way to reach a desired outcome except by war, they are saying what is said when someone wants to reach a city in a remote geography that is only connected to the individual's present geography by a single path down which he or she must move. Injured bodies are the material out of which the road is built (and again, we are postponing the question of how the road of injury can end up in the town of freedom, just as it was necessary to postpone the analysis of how injury can be an intermediate product that in a later transformation becomes the final product, freedom). As injured bodies were the product or the intermediate product but not the by-product, so they are the destination at the end of the road or the road across which one can move to a destination, but they are not something that inadvertently stepped onto the road.

The production and road metaphors are described here as though war were a diffuse, global phenomenon; but each of the metaphors works within, and thus can be recognized within, specific descriptions of specific moments in particular wars. The idea of "accidental" injury is omnipresent throughout war descriptions. The spatial metaphor of the road, for example, with its emphasis on accidental entailment or spatial collision, becomes in aerial bombing the unanticipated expansion of the ground included in the bomb's impact. As one moves from the linear image to the circular one, the goal at the end of the road is the precise point at the center of the circle, and the unintentional collision with people and objects stepping onto the road becomes the unintentional swelling of the circle's circumference to include massive amounts of sentient and nonsentient ground. There are endless concrete instances illustrating this habit of mind. In his study of John Kennedy, for example, Theodore Sorensen describes how during the Cuban Missile Crisis military advisors repeatedly named a "surgical strike" as one option, but it eventually became clear that the precision asserted in the word "surgical" was impossible; Kennedy realized that, acting on the military fiction of "surgical strike," he might have brought about a massive and harrowing catastrophe.[33] Similarly, Michael Walzer reveals that Allied planes during World War II were incapable of targeting their bombs with any more precision than a five-mile radius, yet the misleading term "strategic bombing" was habitually used and the massive, wide-of-the-mark damage was then designated "unintentional," even though it was in all instances "foreseeable."[34] In both these instances, the formal terms used for the bombing, "surgical strike" and "strategic bombing," already contain within them an anticipatory account of the resulting injuries as beyond the purposes of those doing the bombing. A particularly striking instance of this is the designation of cities, civilians, and economic targets as "indirect targets,"[35] a term whose legitimacy resides in its differentiation of those targets from exclusively military targets (the relentlessly noncivilian entailing battlefields of World War I), yet a term that nevertheless creates the peculiar situation of requiring the agents of the damage to aim directly for the indirect target (bringing about injuries that, however horrifying, will not be or are not or were not after all precisely the military's object). These parallel moments have endless equivalents in any military engagement and need not be multiplied here. What is crucial to see is that just as any one form of injury may be understood as having accidentally occurred "on the road" to inflicting injury elsewhere (on the road to injuring soldiers, some civilians were massacred; on the road to injuring civilians, some children were accidentally killed), so all the injuries collectively, those to soldiers, civilians, adults and children alike, are through this metaphor ultimately understood to be injuries that occurred on the noninjuring road to a noninjuring destination.

A third vocabulary is that of cost: injury is the cost of, to complete the metaphor, the thing that was purchased. This idiom has a great deal in common with the "production" vocabulary, since both belong to the inclusive vocabulary of production and exchange. If one perceives injury as "the intermediate product" that will (once one crosses the boundary of war into the territory of goals external to war) eventually be transformed into the final product, freedom, then "cost" is an almost exact structural equivalent. Just as felled trees are an intermediate product that will be transformed into paper, so one may "make money" (or, for example, make cloth in one's daily work which is then transformed into money) with which one can then purchase paper. In that purchase, the money has undergone a transformation into the final product: paper. The money, like the trees, is an intermediate product. Thus, as one goes into the cloth factory and makes money that will be transformed into paper, so one goes into war and makes the injuries that will somehow be transformed into freedom (again this problematic but ultimately valid assertion of transformation will be described later). The idiom is so close to the earlier one that there would be no reason to introduce it as a separate vocabulary here were it not for the fact that its invocation has a characteristic not found in the production metaphor. What typically happens is that injury (and other negative outcomes of war) keeps receding to become the "cost" of a smaller and smaller unit; what one purchases with injury increasingly diminishes. One may begin with the large claim that "war (injury) is the cost of freedom (or better boundaries, or whatever issue the participants believe themselves to be fighting for)." But now the scope of this claim may begin to contract so that "war" itself, first conceived of as the cost, now becomes the thing purchased by, for example, battle. One may turn with dismay from the spectacle of massive injuries and, finding that one has not completely assimilated what one has seen by saying, "War is the cost of freedom," one now tries again, sighs, and consoles oneself, "Ah well, battle is the cost of war." The first word in each of the two sentences means "injury," for this particular construction, "X is the cost of Y," comes up precisely at the moment when some anguishing spectacle is being absorbed and explained to oneself or to others. (That is, at times when one is recalling moments of camaraderie, or medical repair, or heroism, one never stops and invokes the construction "X [camaraderie] is the cost of Y".) But in the second sentence injury has been assigned a smaller place: it is no longer war; it is the cost of war. Now in turn the first term in our second sentence becomes the last term in a third sentence, as in the statement, "Well, slaughter is the cost of battle." And in turn, that first term in this third sentence will become the last term in a fourth sentence, as in the statement, "Blood is the cost of slaughter."[36] It is not that any one person moves through the four but that a population as a whole, in their separate murmurings, keep articulating back and forth across their entirety the full series as though to keep the words in the air, to keep them from landing where they can be seen and assessed.

So the cost vocabulary permits and encourages the receding series

War (injury) is the cost of freedom.
Battle (injury) is the cost of war (formerly, injury).
Slaughter (injury) is the cost of battle (formerly, injury).
Blood (injury) is the cost of slaughter (formerly, injury).

which not only generates tautology (reading back up through the series, injury is the cost of injury is the cost of injury is the cost of injury is the cost of freedom; or, with injury one can purchase injury with which one can purchase injury with which one can purchase injury with which one can purchase freedom), but does so in such a way that precisely that tautologically self-evident centrality of the act of injuring will itself be steadily minimized. The injury which in the first sentence is recognized as massive (by analogy, the destruction of a city) is folded within itself until by the second construction it seems only the destroyed house within the otherwise standing city, and by the third only a closet within that house, and finally by the fourth a shelf in the closet. It is not that we will cease to perceive and feel the power of the injury. The wound on the shelf, a damaged head, a torn off arm, an open belly will stare out at the observer by the closet door and flood him with the nausea of awe and terror, overwhelm him, bring him even to his knees as though it were a gun rather than an open gash poised in his direction. But at least he knows that if he could just unfix his gaze, raise or drop his eyes in a small arc of vision, there would be other objects on the shelves and other closets and other rooms filled with sunlight and newspapers and a sleeping cat, rather than having to know that the injury is here and there and there and there and everywhere he can turn his eyes, that all the shelves and all the rooms and all the streets up and down the city are covered with blood, slaughter, battle, and war.

As in this third vocabulary injury may recede further and further from view by being tucked into successively smaller units, so there is a fourth vocabulary that distances the injury by a continual act of extension, as though it were the umbrella on an ever-extending shaft.[37] Injury becomes the extension or continuation of something else that is itself benign. Clausewitz's famous dictum, "War is the continuation of politics by other means," achieves its authority and authenticity in the brilliant ease of assertion with which a complicated and elusive phenomenon is suddenly made to stand before one as though it had always been self-evident.[38] Nevertheless, it is a statement which, when cited in isolation, as it so often is, sometimes seems to assert that "war is the continuation of peace (or peacetime activities) by other means" and thus to ally and elide it with a benign activity. Its continuity with peace, the predicate nominative, grammatically dominates the "by other means" of injuring. It would seem that such a dictum would sponsor as its counterpart only delicate parody: Dying is living only different; bleeding is breathing only not exactly. But its equivalent is not in such wistful nonsequiturs but in very serious, often very intelligent if less famous claims. Precisely the same structure, for example, is involved in Liddell Hart's remarkable definition of military strategy: the aim of strategy, he writes, is to bring about a situation so advantageous "that if it does not itself produce the decision, its continuation by battle will."[39] This is a breathtaking definition: the sentence, in its own elegant turn of thought as well as in its familiar Clausewitzian cadence, may work to suspend critical thought by seeming itself to have carried out the act of thinking so successfully. But it is a statement that though not in itself wrong—it after all summarizes a tradition of strategy[40]— can mislead in a number of directions. Taken literally, it may seem to describe injuring as a failed or lesser form of strategy and thus make the record of twentieth-century slaughter not only a record of what in Clausewitz would be lapsed politics, but what now in Liddell Hart becomes lapsed or inadequate military strategy. Tannenberg, the Marne, Gallipoli, Warsaw cease to be the work of the military and become the failed work or the breakdown of the military. While one can reasonably describe the occurrence of civilian crime—murder, rape, theft—as the deterioration of the legal or the police enforcement systems, it is not equally reasonable to understand battle as the dissolution of the military system, as the thing that happened in spite of their presence rather than the fact of the presence itself.

Although in the immediacy of its enunciation the definition appears to hold ideal military strategy separate from the realm of injury, it may of course be fairly objected that it only works to separate strategy from battle, to separate strategy from reciprocal injuring, and that certainly it assumes the one-directional injuring of capitulation, imprisonment, even physical wounding. The mere fact that an opponent retreats, pulls back, or breaks off the engagement has itself no military advantage, as Clausewitz points out: "Getting the better of an enemy— that is, placing him in a position where he has to break off the engagement— cannot in itself be considered as an objective [since no one is keeping an abstract score card], and for this reason cannot be included in the definition of the objective. Nothing remains, therefore, but the direct profit gained in the process of destruction. This gain includes not merely the casualties inflicted during the action, but also those which occur as a direct result of his retreat."[41] To say that successful strategy is one in which the decision is brought about without a battle is, then, to say that successful strategy is one in which the injuring occurs in only one direction: the lesser or back-up military form (the battle) is one in which the injuring is reciprocal, two directional, and only by one side's eventually *out-injuring* the other will battle approximate the perfect case, which is *one-directional injuring*. Thus, the original definition, which seems to posit noninjuring against injuring, instead posits one-directional injuring against two-directional injuring. If this is what Liddell Hart's definition meant all along, it is deeply accurate; but it should be noticed that this reformulation has neither the benign sound nor the elegance of the original, and it was certainly for the sake of its benignity and elegance that that formulation was arrived at.

A third way of assessing both the strength and the weakness of the original formulation (which is being used here as a model for the much wider habit of describing wounding as an extension of some more innocent activity) is to attend to the temporal variations that strategy discovers in the act of injuring. A commonly accepted way of understanding what happens in the end of war is to say, as Paul J. Kecskemeti so effectively phrases it, that the participants agree to forgo the next round, that each can imagine the outcome well enough that they need not enact it: one side succumbs.[42] This description applies equally to a unit within war such as a battle: the two sides may forgo the last round, let us say the last five hours of the battle. But this moment of assessment can be pushed back further and further into the battle so that the participants agree to forgo the whole second half of the battle (two days, for example), or the whole last three quarters (three days), until finally we arrive at Liddell Hart's position: before the battle even begins, the participants assess their relative situations and forgo the actual physical locking of arms because the outcome is so clear. Overwhelmed by the display of the opponents' superior injuring ability, one side surrenders. What it is crucial to see is that this situation is one of two fundamental temporal relations that can exist between the infliction of injury on an opponent and the opponent's perception of that infliction of injury. In the situation described here, the perception precedes and anticipates the injury (and modulates the form of but by no means eliminates the fact of the injury since those who surrender do not turn around and go home but are shot or imprisoned). Anticipated injury has surprise injury as its strategic opposite. In the first case, the injury-inflicting capacity of one side is displayed to the other before the actual infliction so that the opponent will capitulate; in the second case the injury-inflicting capacity is kept invisible from the opponent who must *not* be allowed to see it *before* it is enacted and ideally will not even see it *while* it is being enacted (or at least not in the first few minutes in which it is being enacted: there should be at least a few minute lag between the inflicting of injury and the injured side's awareness that it is being injured), but will only be allowed to see it *after* it has begun to be enacted. It is important to recognize "anticipated injuring" (in which injury is judged to have the greatest effect if *foreseen*) and "surprise injuring" (in which injury is judged to have the greatest effect if wholly *unforeseen*) as counterparts of one another, for it again underscores the fact that a strategy so superb as to eliminate the battle is not a strategy that eliminates injuring but enacts one temporal form of injuring, anticipatory injuring.

The description of the end of war that was transferred to the end of the battle and was then pushed back up through the battle's middle to its beginning may now be retransferred back to the war. There, too, the forgoing of the next round may be pushed back through successive stages until one eliminates battle after battle and at last arrives at a situation where the war itself never starts: each side merely displays their weapons, their injury-inflicting capacities (the arms race), and one side backs down, succumbs to the other before the occurrence of physical damage.[43] It is again crucial to stress that the anticipatory injury is an actual injury, that the hypothetical situation invoked here is one in which one side has surrendered a piece of territory, or succumbed to the other's ideology as superior, and so forth. If, for example, one country were the only country with a nuclear weapon and through that possession required its opponents to capitulate to all its wishes, it would be reciprocal injuring (battle, war) that would have been avoided and replaced not by noninjuring but by one-directional anticipatory injury. This situation would be very different from that in which both sides had equal weapons and both backed away equally, for here the anticipated injury is *only* anticipatory, imagined and never enacted, since the conflict was averted not by unilateral capitulation but by shared retreat. It should also be differentiated from the situation in which neither side has arms, where not only is there no actual physical war, nor the actual injury of capitulation through anticipation, but instead shared rejection of even the display of the capacity to injure: both sides exempt themselves from the contemplation of either wounding or being wounded. The two situations are radically different from that in which a massive inequality of arms is achieved and perpetuated by a solitary country, and the rival country's ambition for equality of arms is scorned by the first country as war-mongering (a single weapon ensures peace; two bring war) or dismissed as motivated by senseless sibling envy (we have a weapon so now they have to have one too). The dream of an absolute, one-directional capacity to injure those outside one's territorial boundaries, whether dreamed by a nation-state that is in its interior a democracy or a tyranny, may begin to approach the torturer's dream of absolute nonreciprocity, the dream that one will be oneself exempt from the condition of being embodied while one's opponent will be kept in a state of radical embodiment by its awareness that it is at any moment deeply woundable.

The centrality of the act of injuring in war may disappear—the centrality of the human body may be disowned—by any one of six paths. First, it may be *omitted* from both formal and casual accounts of war. Second, it may instead be *redescribed* and hence be as invisible as if omitted: live tissue may become minimally animate (vegetable) or inanimate (metal) material, exempt from the suffering that live sentient tissue must bear; or the conflation of animate and inanimate vocabularies may allow alterations in the metal to appropriate all attention, as in the designation of "disarming" as central; or the concept of injury may be altered by relocating the injury to the imaginary body of a colossus. Third, it may be neither omitted nor redescribed and instead acknowledged to be actual injury occurring in the sentient tissue of the human body, but now *held in a visible but marginal position* by four metaphors that designate it the by-product, or something on the road to a goal, or something continually folded into itself as in the cost vocabulary, or something extended as a prolongation of some other more benign occurrence.

Crucial to the analysis that follows is not the intricacy of the paths by which it disappears but the much simpler and more fundamental fact that injuring is, in fact, the central activity of war. Visible or invisible, omitted, included, altered in its inclusion, described or redescribed, injury is war's product and its cost, it is the goal toward which all activity is directed and the road to the goal, it is there in the smallest enfolded corner of war's interior recesses and still there where acts are extended out into the largest units of encounter. As a major premise of the analysis, the centrality of injuring is an extremely modest premise because it is self-evident and so elementary that it is itself anterior to an array of intricate and morally sophisticated questions often asked about war that will not even be touched on here. Bertrand Russell, for example, calls attention to the morally problematic human habit of saying, "I am going off to die for my country" rather than acknowledging that "I am going off to kill for my country,"[44] just as Mouloud Feraoun summarizes the universal self-description of all participants in war: "defending a just cause, killing for a just cause, and risking an unjust death."[45] But in the present discussion, it is not the difficult distinctions among these phrases that will be attended to but only their common denominator that will from this point forward be assumed. Whether a boy announces that he is going off "to die" for his country or going off "to kill" for his country, he is saying that he is going off "to alter body tissue" (either his own or another's) for his country, and the eventual destination here is to understand the structural logic of an event in which alterations in human tissue can come to be the freedom or ideological autonomy or moral legitimacy of a country. For now, it is only the centrality of injuring that is designated a given. "War kills; that is all it does," writes Michael Walzer in the midst of a complex analysis of just and unjust wars.[46] "Being shelled is the main work of the infantry soldier," writes Louis Simpson about World War II; "Everyone has his own way of going about it. In general, it means lying down and contracting your body in as small a space as possible."[47] Though this premise may be disowned in endless ways, it may also be reowned, both by looking directly at a war and by looking at the echo of words of those who have looked, moral philosopher, foot soldier, poet, strategist, general, painter. "But see," begins Leonardo da Vinci in the last sentence of his long instruction to painters on how to represent a battle, a verbal treatise that creates a large visual canvas of dust and sun and horse bodies and human bodies, faces in physical pain and faces drawn in exhaustion—"But see," he at last concludes, bringing the long rush of instructions to a sudden halt, "see that you make no level spot of ground that is not trampled over with blood."[48]

War is relentless in taking for its own interior content the interior content of the wounded and open human body.

II. War Is a Contest

Our second premise, that war is a "contest," invokes a predicate nominative that is far less concussive than "injuring" and thus a much less difficult fact to hold steadily visible before one's eyes. Insofar as there is a reluctance to identify war as a contest, the reluctance originates in an impulse almost opposite to that which helps bring about the eclipse of injuring: when the identification is avoided or explicitly rejected, it is so because "contest" is always attended by its near synonyms of "game" and "play," thus allowing war's conflation not only with peacetime activity but with that particular form of peacetime activity that is least consequential in content and outcome. In fact, in insisting on the accuracy and importance of identifying war as a contest, the present analysis may appear to be subverting itself, since it was only a moment ago emphasizing the error of descriptive conventions that permit and encourage such conflations. The dangers of the identification will be briefly summarized here before going on to show why, despite these dangers, it is an identification that must be made.

The extreme inappropriateness of importing connotations of playfulness into war is not adequately accounted for simply by pointing to war's "seriousness" and designating "play" and "war" opposites, since an absolute opposite for play—work—exists even within the boundaries of peace. But through this familiar peacetime opposition, the extremity of the even greater tonal distance separating games and war can be articulated. Although play is often sensuous (for in play the senses become self-experiencing), work entails a far deeper embodiment: the human creature is immersed in his interaction with the world, far too immersed to extricate himself from it (he may die if he stops) and thus almost without cessation he enacts a constant set of movements across the passing days and years. In contrast, the very nature of play requires that the person be only half submerged in the world of his activity, that he be able to enter and exit from it freely: the activity, even if never engaged in before, can be started in seconds, or ended just as quickly. The person at play, protected by the separability of himself from his own activity, does not put himself at risk: he acts on the world outside his body with less intensity than the person at work, and if he sometimes puts that world at risk, it is because his own immunity from risk makes him inattentive to the forms of alteration he is bringing about. It is in the very nature of work—as is dramatically visible in forms of physical labor and craft such as coal mining, farming, building, or inventing—that the worker "works" to bring about severe alterations in the world (relocating a rock; creating a piano or a hayfield or a house where there was none) and only brings about those alterations by consenting to be himself deeply altered (that his muscles, posture, gait will be altered is certain; that he will undergo the more severe alteration of injury is at least risked).

The activity of war is, viewed within the framework of this opposition, the most unceasingly radical and rigorous form of work.[49] The soldier's survival is at stake not in the real but diffuse way it is for the worker who out of his labor creates his own sustenance and will, if he stops, eventually starve; it is more immediately and acutely at stake; it is another soldier's direct object to kill him and his own work to be for the other a target yet to keep himself alive. The form of world alteration to which he devotes himself does not simply entail the possibility of injuring but is itself injuring, and it is this form of self-alteration to which his own body is at every moment subject. He cannot will his entry into and exit from the activity on a daily basis. There is not, as there is for most workers, a brief interval of exemption at the end of the day when he is permitted to enact a wholly different set of gestures; the timing of his eventual exit will by determined not by his own will but by the end of the war, whether that comes in days, months, or years, and there is of course a very high probability that even when the war ends he will never exit from it. Although in all forms of work the worker mixes himself with and eventually becomes inseparable from the materials of his labor (an inseparability that has as only its most immediate sign the residues which coat his body, the coal beneath the skin of his arm, the spray of grain in his hair, the ink on his fingers), the boy in war is, to an extent found in almost no other form of work, inextricably bound up with the men and materials of his labor: he will learn to perceive himself as he will be perceived by others, as indistinguishable from the men of his unit, regiment, division, and above all national group (all of whom will share the same name: he is German) as he is also inextricably bound up with the qualities and conditions—berry laden or snow laden—of the ground over which he walks or runs or crawls and with which he craves and courts identification, as in the camouflage clothing he wears and the camouflage postures he adopts, now running bent over parallel with the ground it is his work to mime, now arching forward conforming the curve of his back to the curve of a companion boulder, now standing as upright and still and narrow as the slender tree behind which he hides; he is the elms and the mud, he is the one hundred and sixth, he is a small piece of German terrain broken off and floating dangerously through the woods of France. He is a fragment of American earth wedged into an open hillside in Korea and reworked by its unbearable sun and rain. He is dark blue like the sea. He is light grey like the air through which he flies. He is sodden in the green shadows of earth. He is a light brown vessel of red Australian blood that will soon be opened and emptied across the rocks and ridges of Gallipoli from which he can never again become distinguishable.

The extreme difference in the degree of a person's separability from his own activity in play, in work, and in war is one manifestation of the distance that separates them. The severe discrepancy in the scale of consequence makes the comparison of war and gaming nearly obscene, the analogy either trivializing the one or, conversely, attributing to the other a weight of motive and consequence it cannot bear. The conflation may occur as a flat assertion of equality—war is a game, games are war—or, more often, as the importing of the attribute of one into the other's sphere. The transfer may occur in either direction. The hatred that in war grants nothing to an opponent is sometimes imported by analogy into descriptions of peacetime games and contests that now become in their competitive urge and obsession disguised forms of the passion to destroy others. Conversely, the optimistic interpretation (as occurs, for example, in Social Darwinism) of competition in peacetime games as contributing to human "progress" may in turn be imported into descriptions of war, now allowing war itself to become, quite remarkably, a contributor to human progress and evolution.[50] As the inappropriateness of these importations indicates, there are profoundly legitimate reasons for avoiding "contest" language altogether, or for using it with extreme reluctance and care.

Nevertheless the identification must be made here because war *is* in its overall structure of action a contest. The benign reluctance to use the language entails the possibility that the most important facts about the activity will be unseen: that the men described above as inextricably engulfed in the materials and labor of their task are moving across the land toward others who, equally engulfed, move toward them and like them, work to *out-perform* the other in their appointed labor (or deconstructed labor since it is world unmaking rather than worldmaking to which they devote themselves). The "contest" language is crucial because it registers the central fact of reciprocity, ensures that reciprocal and nonreciprocal, one-directional and two-directional are major categories of description: without it, as was suggested earlier, alternatives between two-directional and one-directional acts of injuring can be misrepresented or misunderstood as an opposition between injuring and noninjuring. The perceptual problems accompanying the sign of agency (as introduced in Chapter I) are here vastly magnified because the sign is doubled and reversed as two weapons face each other, and the complications can only be sorted out by registering from the outset the overall framework of mutual and reversed action. Thus the analysis that follows will insist on the structural fact of contest, while at the same time insisting on the exclusion of the tonal fact of "play."

It should be noted that although play and game and contest share enough ground that they may be perceived as synonyms, the first two terms differ in the degree to which their activity is formalized, and the second two differ in the area of their formalized activity that is emphasized. The term "play" indicates only the person's separability from his own activity but does not stipulate the presence or absence of any formal structure—there may be a solitary player or a dozen, there may be rules or there may be none, there may be a defined beginning and end of the play or it may come out of nowhere and dissolve just as imperceptibly. The term "game" indicates play that is formalized, play in which there is a specified degree of organization (a start, a finish, and a center that proceeds according to specific rules) and in the most common use of the word entails two sides who share an activity but will not share the outcome (both, for example, will run but only one will be called "best runner"). The term "contest" again entails reciprocal activity for nonreciprocal outcomes, but in the shift from "game" to "contest" there is a shift in emphasis from the "reciprocal activity" stressed in the first to the "nonreciprocal outcome" stressed in the second. War is in the structure of its activity a contest because it entails reciprocal activity for nonreciprocal outcomes, with the weight of intention and motive located in the final facts, the nonreciprocal outcome, the unique form of "ending" that, more than any other part of war, makes it what it is and compels people to seek it as a form of arbitration when all else has failed.

When formal and informal descriptions of war do include (and they often do) allusions either to specific peacetime games or to the generic attributes of the universal form of games, it is the "contest structure" of war that is being indirectly acknowledged and that may be understood as having occasioned the comparison. Such allusions habitually occur—as becomes hauntingly visible in Paul Fussell's account of forms of verbal consciousness and memory in World War I—in the language and actions of the young combatants themselves. Near the football once kicked into the No Man's Land at Somme and now preserved in a British museum, there may still seem to hover in ghostly outline the solitary frozen image of a jubilant leap into the air, the high-spirited kick of a lone British boy moving into the day's battle; or there may instead cling to its leather surface not just this solitary image but the shouts and frozen gestures, the verbal fragments and moments of mime, of hundreds of such boys, invoking in unrecorded moments of battle on the Western Front as at the Turkish lines near Beersheba as at Gallipoli the familiar formulas from football, cricket, or track;[51] or there may instead seem to cling to that same British football the images of hundreds of thousands of the near-children of every country and century who in the language of voice and bodily gesture, an outflung arm, an urge "to win," intuitively reach for and discover the sign and signal (however tonally inappropriate) of war's contest structure. If the war is looked on favorably and at a distance, the "innocence" of the mimed analogue will be recognized, as in Fussell's compassionate account of World War I; if the war is out of favor, the same repertoire of gestures may, fairly or unfairly, appear to the "spectator" as cruel, crude, obscene, as in the merged actions of hurling frisbees and hurling bombs in Vietnam;[52] or the response may be divided as some persons must have cheered while others winced with shame when, during the early weeks of the 1979–80 confrontation between Iran and the United States, a newsservice photograph picked up and passed from place to place the image of a theatre marquee announcing that day's occurrences between the two countries as a football score.

As allusions occur in the language and gestures of the immediate participants, so they may be invoked by military strategists, as when Liddell Hart uses the difference in chess between a knight's move and a queen's move to describe the difference between air mobility and that form of mobile attack introduced with the tank,[53] an example which shows that the particular game cited is not always, as earlier, a team sport but may instead be a board game, as it may also be dueling or, especially if the risks are incalculable, gambling (the phrase "nuclear gambling'' or ''the throw of the nuclear dice'' has become in recent years a familiar one[54]). One way of registering war's kinship with other forms of contest without allowing it to collapse into or be contaminated by the diminutive connotations of those other contests is to use it in a ''not'' construction, which allows the allusion to be both invoked and rejected, as when on 8 June 1944, Churchill asks the members of the House to caution the British people ''against over-optimism, against the idea that these things are going to be settled with a run.''[55] Another version of this ''not'' construction is the use of the allusion only when the form of the military encounter is being discredited, as when according to World War I historian Barbara Tuchman a colleague of the Russian General Jilinsky described Jilinsky's strategy at the Battle of Tannenberg as a strategy ''designed for *Poddavki,* a Russian form of checkers in which the object is to lose all one's men.''[56] Although statesmen, sensing the inappropriate tone of the vocabulary, tend to avoid references to specific games, their descriptions of war of necessity include fragments of the generic attributes of games, ''opponents,'' ''contestants,'' ''beating,'' ''winning'': ''The *prize* of *victory,''* begins a speech by Alexander Haig;[57] ''Control was assumed to be in the hands of one of the *contestants,''* writes Kissinger of the conflict in Vietnam;[58] ''We have still *to beat* the Japanese,'' Churchill reminds the British electorate in the spring of 1945.[59] Such terms inevitably occur because they designate characteristics that belong as intimately to war as to any other contest. If the language common to soldier, strategist, and statesman registers the fact that war is a contest, it need not be relied on for certification of the fact: identifying war as a contest only because the language of participants or commentators invokes the idiom familiar from peacetime contests would be the equivalent of proving the Alps are mountains only because one discovered verbal allusions to the Alleghenies in the names of certain alpine valleys and passes rather than because of the monumental fact and form of the Alps themselves. War is in the monumental fact of its structure a contest. Thus, even in definitions in which there has been an attempt (in all probability motivated by deep tact) to sidestep the identification by replacing the predicate nominative, ''War is a contest or conflict in which . . .'' with a circumlocution (''War is a condition in which . . . ,'' ''War is a state in which . . . ,'' even ''War is an institution in which . . . ''), the clause following the predicate nominative almost immediately carries us back to the idea of contest, as when Hugo Grotius, the father of international law, writes that war is not a contest but ''a state of contending [contesting] parties.''[60]

Just as the identification of the central activity of war as injuring is a modest premise because so simple, massively evident, and prior to the complications of war's other characteristics, so the identification of war's formal structure as a contest is equally elementary and modest, for it does not require that one enter into and perform difficult assessments about the intricacies of war's formal structure. It is here not necessary, for example, to determine whether the contest (in either its ''ideal'' or its ''usual'' form) takes place through a form of engagement other than battle, through one decisive battle, or through many battles (six hundred and fifteen battles in World War I; one hundred in the 1936–39 Spanish Revolution; eight in the 1935–36 Ethiopian War; zero in the 1921–25 Riffian War, and so forth[61]). Nor is there any entry here into difficult and important questions about the particular status of various forms of international rules operating throughout the duration of the conflict.[62] Nor is there an attempt to identify those elements of its formal structure that can be altered with each new revolution—machine gun, tank, helicopter—in the technology of weapons (such as the way that nuclear arms not only dramatically contract the temporal duration but also, according to Soviet military strategist Sokolovskiy, fundamentally alter the spatial configuration of war by wholly eliminating the conventional distinction between front and rear; or again fundamentally alter what had previously been regarded as a universal attribute of war by eliminating one of the military's major tasks, the concentration of a massive force, since now ready-made within the weapon is that very concentration[63]). Nor is there any attempt here to judge the complicated relations between any two of its parts, the relation between the form of the initial declaration and the form of the surrender, or in turn the form of the surrender and the nature of the treaty.[64] The single structural fact about war registered here, that it is a contest, is elementary and constant regardless of other contingencies of form: what is described here will be true whether war takes place through battle or some other arena of injuring, whether its duration is hours or years, whether the weapons are conventional or nuclear, whether it happens on the territory of one of the contestants, on the neutral ground of land or sea, or on the "high ground" of outer space, whether it proceeds according to international rules and conventions or further compounds its already active contribution to the dissolution of civilization by disregarding them.

The recognition of the contest structure of war obliges one to recognize only a small constellation of attributes summarized in the sentence: War is a contest where the participants arrange themselves into two sides and engage in an activity that will eventually make it possible to designate one side the winner and one side the loser (or more precisely, makes it possible for the loser to identify itself so that the other side will recognize itself as the winner by default). In other words, in consenting to enter into war, the participants enter into a structure that is a self-cancelling duality. They enter into a formal duality, but one understood by all to be temporary and intolerable, a formal duality that, by the very force of its relentless insistence on doubleness, provides the means for eliminating and replacing itself by the condition of singularity (since in the end it will have legitimatized one side's right to determine the nature of certain issues). A first major attribute here is the transition, at the moment of the entry into war, from the condition of multiplicity to the condition of the binary; a second attribute is the transition, at the moment of ending the war, from the condition of the binary to the condition of the unitary. There are, for example, in the opening moments of war, no longer the diffuse five hundred million persons, projects, and concerns that existed immediately prior to war's opening, because those five hundred million separate identities have suddenly crystallized into *two* discrete identities, Russian and American; and even if the number of national identities is more than two or (as in a civil war[65]) less than two (the twelve of the 1900–01 Boxer expedition; the four of the 1902–03 Venezuelan War; the two of the 1904–05 Russo-Japanese War; the four of the 1906–07 Central American War; the two of the 1910–20 Mexican Revolution; the two of the 1911–12 Italo-Turkish War; the five of the 1912–13 First Balkan War; the thirty-eight of World War I; the one of the 1916–36 Chinese Civil Wars; the two of the 1916 Irish Rebellion; the six of the 1917–20 Russian Revolution and so forth year by year up to the fifty-seven national participants in World War II[66]), they will become two during the war. The issues, too, may begin by being either multiple or unitary: there may be six or sixty claims on one side and two on the other, neither or both of which overlap with two claimed by the first side; or there may be a single issue shared. But in either event the multiplicity or unitary nature of the issues will during the course of the war also subside into the double, for what will be "at issue" is each side's right to its own issues. Until the end of the war, the state of doubleness will reign. The distinction between "friend" and "enemy"— identified by Carl Schmitt as the fundamental distinction in politics equivalent to good and evil in moral philosophy and beautiful and ugly in aesthetics[67]—is in war converted to an absolute polarity, whether that polarity is registered in some version of the us-them idiom (what Henry Kissinger calls the "our side-your side formula" visible in the familiar military pairs of offense-defense, aggressor-defender since no participant in war ever identifies itself as the aggressor) or instead in the more neutral naming of pairs (the colors of the red and the white or the blue and the grey; the East and West of cold war; the North and South of the United States, Korea, Vietnam; the Union and the Confederacy; the twofold coalitions of the Allied and Central powers, or the Allied and Axis powers), and the doubleness will also become an extensive world view applicable not only to all persons in the universe of friends and enemies, but to all objects and all places, as in Paul Fussell's description of the ominipresent binary categories of World War I, the visible friend and the invisible enemy, the normal (us) and the grotesque (them), the division of the landscape into known and unknown, safe and hostile.[68] This insistent duality will reign until the end of the war when it will become clear that the concussive state of doubleness was all the while in the process of eliminating itself, the condition of two was moving forward to the condition of one, the belligerent equality transforming itself into the peaceful inequality that entails the designation of one as "winner."

With these two premises in place—that the central acitivity of war is injuring; that war is in its formal structure a contest—it is possible to begin to assess the nature of war by approaching it through the question, what is it that differentiates war from other kinds of contest? In any contest, the participants[69] perform some activity X and must out-X each other—out-swim each other, out-debate each other, out-bake each other, out-spell each other, out-think each other. (One may say that they do not always perform the same activity, as in a talent contest where one sings, another dances, another performs a monologue, but all these actions may be understood as the same X, the displaying of talent, and the participants must out-display or out-talent one another.) In war, the shared activity, the "X," is injuring, and the participants must work to out-injure each other. Although both sides inflict injuries, the side that inflicts greater injury faster will be the winner; or, to phrase it the other way around, the side that is more massively injured or believes itself to be so will be the loser. The qualification "believes itself" is an important one, for countries will differ in the level of injury that represents the borderline between tolerable and intolerable damage. It sometimes happens in war that the side that is in absolute terms injured less has reached its own cut-off point of unacceptable injury before the other side (in absolute terms more greatly injured) has arrived at its own level of unacceptability: here the second side is victorious. For any contestant in war, there is, as Clausewitz notes, an intensity and scale of damage more oppressive than the territorial or ideological sacrifices the other side is demanding.[70] Each side begins the war by perceiving physical damage as acceptable and ideological and territorial sacrifices as unacceptable; through the war each side tries to bring about in the other the fundamental perceptual reversal—damage as unacceptable and sacrifices as acceptable. Thus, to make the description here applicable to the greatest number of actual wars, it should be understood that what is meant by the term "out-injuring" or by the phrase "each side works to out-injure the other" is more precisely the sentence, "each side works to bring the other side to the latter's perceived level of intolerable injury faster than it is itself brought to its own level of intolerable injury."

War involves, of course, thousands of other skills (making weapons, mining fuel, raising food, tending the sick); but however much each contributes to the outcome, no one of these is the basis of the contest; that is, no one of them is what injuring is, the means of identifying a winner and a loser. If it were the ability to grow food, for example, a spring and a summer, or perhaps seven springs and seven summers, could have been designated the allotted period in which the allied and axis powers were required to devote themselves to growing food: the side that brought forward the better cumulative harvest would be the winner and would have the right to prescribe to the watching world the outcome to certain issues. So, too, if it were medical care, a way of comparing two rival systems could be devised, one that only entailed the cure of naturally occurring illness or the repair of peacetime injuries and did not require the manufacture of a massacre in order to demonstrate medical prowess. If it were strategy, two sides could simply submit war plans, and the more elegant maneuvering, the more brilliant path of choices, could be determined and a winner designated without ever having had to enact those plans. An abundant harvest, a generous and effective medical system, strategies as sturdy as they are ingenious, will contribute to war precisely by contributing to their side's ability to out-injure the other side, out-injuring accomplished in part by remaining as healthy and whole, well-fed and medically repaired as possible; but it is the activity of injuring that is the substantive and determining act, and of course the out-injuring side may well have had the lesser harvest as well as the less theoretically elegant strategy. It is, as was noticed in an earlier section, seductive but wholly inaccurate to describe the determining activity of the contest in terms of the external issues (the men are not out-freeing each other, or out-performing each other in the act of liberation, or out-proving to each other the legitimacy of their side's historical claim to a certain strip of land) just as it is seductive and inaccurate to describe it in terms of an activity interior to but not central to war (the men are not primarily out-performing each other in mining coal; nor are they out-disarming each other except insofar as that is a synonym for out-injuring; nor are they out-believing each other in God; nor are they out-loving each other in their thoughts of the families they have left behind, though all these acts and attributes may be present, and may well give them a superior capacity to inflict wounds and to withstand their own wounds). It is, then, once again, the depth, massiveness, intensity, or speed of injuring that is central and the feat of out-injuring that determines the winner.

What is most crucial to see is that so far nothing differentiates war from any other form of contest. Injuring has made it possible to arrive at a winner and a loser; but the work of arriving at a winner and a loser is an achievement common to *every* act and attribute on which a contest has ever been based as even, for example, the activity of roping calves makes it possible to arrive at a winner and loser in a calf-roping contest, or as the designing of a spectacular building makes it possible to differentiate a winner and a loser in an architectural competition, or as the number of baskets makes it possible to designate a winner and a loser in a basketball game. If, then, the only function of going to war is to provide the means for determining a winner and a loser, that work could be as easily accomplished by roping calves, imagining beautiful buildings, or lifting balls and letting them fall through rings in the air.

We must identify a *second* function accomplished by out-injuring, a function other than determining a winner and a loser, and so answer the question, Is there something that differentiates war from all other contests, is there something that differentiates injuring from every other act or attribute on which a contest can be based? One of two possibilities is true: either there is nothing or there is something. If there is "nothing," then another form of contest could perform the function of war just as well and far less painfully: though this would of course necessitate the heartsickening recognition that all previous wars might have had a substitute, so too would it entail the recognition that future wars might have a substitute. If, on the other hand, there is "something," then in turn one of two things will be true. We may be required to conclude that wars, of the past and of the future, are necessary and must be accepted as performing a work that has and can have no equivalent in any other form of activity. Or, it may instead be the case that being able to identify and articulate that "something" could enable us to locate an equivalent that perhaps only at first appeared not to exist, or that it could enable us (once the "something" is precisely defined and understood) to invent its equivalent if none already exists. If, then, the question, "What is it that differentiates injuring from any other act on which a contest can be based?" is not a question that can be easily answered, neither is it a question that can be easily unasked.

III. What Differentiates Injuring from Other Acts or Attributes on Which a Contest Can Be Based

A simple element that, perhaps more than any other, complicates the answering of this question is the discrepancy between the small number of participants in most peacetime contests and the massive numbers in war.[71] As will become evident later, the most pervasive error in answering the question may in fact come about by taking as a conceptual model for war the image of single combatants: that is, imagining the determining activity as injuring, but contracting the scale down from an extremely large number of persons to two persons. The inappropriateness of this model will become visible at a later moment, but rather than beginning with the complications that arise from conforming war to the diminutive sale of other contests, it will be helpful to begin with the opposite movement and imagine an ordinary form of contest based on some activity other than injuring now magnified until it conforms to the scale of participation normally occurring in war.

Although the numerical discrepancy is itself an important difference between war and other means of determining a winner and loser, it is not itself the critical differentiating element (and thus does not in and of itself provide a path of substitution for war, though it might in combination with other elements provide a form of substitution). That is, one can imagine a contest based on some relatively benign activity, distributed out over two disputing populations: all members of each civilization—or a large proportion, say all young adults between the ages of eighteen and thirty-five—could be paired off in a massively extended sequence of chess matches or tennis matches, a sequence of pairings that would take over a thousand days, perhaps two or three thousand, during which the record of wins and losses would be steadily added up with the cumulative successes and failures announced at frequent but somewhat irregular intervals (three days, ten days, thirty days) so that the two entire populations would be at all times imaginatively engaged in the progress. Or, if a single talent—chess, tennis—assumes in a population an inconceivable homogeny of skill (even more extreme than the narrow and uniform set of skills required of a population in war) there could instead be scheduled a large array of contests: all conscripted citizens could be allowed to choose the arena of competition—mechanized invention, embroidery, boxing, singing, running, swimming, skiing, chess, dance, and so forth[72]—and each person's chosen activity could be coordinated with that of a partner from the enemy nation in a feat of organization that would probably not be that much greater than that necessitated by war. This extraordinary organizational requirement would, in fact, be one of its benefits, again increasing the depth of engagement of the overall population. The energies of the unconscripted citizens would in large part be absorbed in support activities, such as transporting and escorting the contestants, or covering their jobs while they were in an intensive period of training for, participating in, or recovering from their specific contests. The massive scale of participation (not only in terms of the number of persons, both conscripted and unconscripted, but in terms of the degree and depth of mental attention that each day and every day over endless days would have to be directed toward this exhausting process) would be critical. It is national rather than individual consciousness that is at stake and that (in one case certainly and probably to a large degree in both cases) must over the course of events become altered, and cannot be altered by proxy, cannot be altered by altering the fate of a small number of participants such as those who take part in the Olympics.

In the course of war at least one side must undergo a perceptual reversal (what Paul Kecskemeti calls a "political reorientation"[73]) in which claims or issues or elements of self-understanding that had previously seemed integral and essential to national identity will gradually come to seem dispensable or alterable, without seeming (as it once would have) to cancel out, dissolve, or irreparably compromise the national identity. The surrogate form of contest imagined here differs from war in that there is no injury either to the bodies of persons or to cities, buildings, bridges, factories, houses, the material signs and extension of personhood. One may thus say that the world deconstruction so essential to war has been replaced with neutral activity (one that will certainly prevent, or at least retard, additional projects of world-building because of the absorption of the nation's energy in the contest, but one that does not actively destroy what has been conceived of and brought into reality by the civilization thus far). But for the work of war to be accomplished, the world deconstruction must still be occurring in one place, in the interior of human consciousness itself, because aspects of the country's self-description that it had seemed impossible to be without must now be disowned without causing the dissolution of the country. War destroys persons, material culture, and elements of consciousness (or interior culture): the third in this triad must in the benign form of contest also be destroyed, even if the first two are left intact. The losing country must erase part of the slate and begin to re-imagine itself, re-believe in, re-understand, re-experience itself as an intact entity, but one not having some of the territorial or ideological attributes it had formerly (including sometimes its very name or its form of government). Thus, without a deep and massive commitment on the part of the two populations to the contest—to, that is, the very process whose whole raison d'être is to guarantee as an outcome this form of loss to one side and to determine to which of the two sides it will happen—the imaginary contest cannot even begin to duplicate the contest of war.

But even if this surrogate contest should *begin* to duplicate the contest of war, it will not *end* by duplicating war, for it is in the nature of its ending that war is remarkable among contests, remarkable in its ability to produce an outcome in one kind of activity (injuring) that is able to translate into a wholly different vocabulary, the right to determine certain territorial and extraterritorial issues. Even if both countries consented to the international chess match, or tennis match, or match based in a whole array of activities, it is almost inconceivable that at the moment when the final locus of loss was apparent, the loser would allow this outcome to determine who had the greater right to a certain set of islands, or who had the right to determine the form of government that would now reign in the defeated country. It might happen that the countries would participate and then suddenly and summarily refuse to abide by the outcome; or, as would be more psychologically realistic, they might in the midst of the competition begin to call into question the legitimacy of, for example, using "singing" as a reasonable basis for the contest, or accuse the opponent of giving its athletes outlawed hormones or of warming the metal runners of the sleds, or come to believe that the other side's forests (or the forests of a third country supplying wood to the competitor) made better violins and thus made impossible any comparison between the two populations' levels of skill in playing the violin, or they might begin to suspect that chess matches were to the opposition's advantage because of its superior computer technology, and so forth. Whatever the particular form of the disavowal, the attributes of the contest would now themselves become elements and issues in the very dispute it was the work of the contest to resolve: they might themselves be one day listed as among the very issues over which the two countries eventually went to war.[74] In contrast, the designation of a winner and a loser in war is accepted as determining the nature of external issues.

The utter lack of connection between the interior nature of the contest activity and the interior nature of the external issue is not important here. That is, while it is certainly true that a population's earning of the title "best national swimmers" has nothing intrinsically relevant to its now having "a greater right to have its own way on the question of international oil leases," neither is there any interior connection between the designation "best injurer" and the right to determine such issues. The act of injuring itself is no more appropriate to the issues (and is therefore less desirable) than any other act, accident, or talent on which one can base a contest in order to arrive at a winner and a loser. Only when the issue is strength does injuring become inherently appropriate, an appropriateness that is misleading and tautological since at the point where the issue is physical power the arena of injury has already been entered. Conceivably, two countries could in fact find a competitive activity or attribute that happened also to be integral to both countries' conceptions of nationhood: such an attribute might at first appear to provide an interior connection between the substance of the contest and the substance of the external issues. In a conflict between the United States and the Soviet Union, for example, it would be unacceptable to base a competition on a comparative analysis of the degree of "individuation" in the two populations, since that attribute would be consonant with the fundamental political philosophy of one country more than the other. But both democratic and socialist forms of government conceive of the provision and distribution of food as essential, and of their own form of distribution as morally superior: thus the contest based on a harvest, imagined earlier, might seem a legitimate way of both determining a winner and a loser, and determining which government had a greater authority to specify certain issues. Again, both countries have in their founding conceptions a desire to create a political structure that provides an answer to the question in different ways asked by both the Federalist Papers and the writings of Marx, "What kind of political structure will create a noble and generous people?" Consequently, they might agree to a form of competition judging these qualities, such as the matching of medical systems, the matching of provisions for the injured, the matching of the level of violent crimes (presumably a noble and generous population will not be given to crime), the matching of their respective means of distributing and sharing wealth with other populations, and so forth.

Three alternative readings are possible here. One might argue that the title of winner in such a contest would have an intrinsic connection to the authority now accorded to that country to determine certain issues, such as whether a third emerging nation should, as a result of the contest, take on a democratic or socialist form of government. That is, it might seem appropriate that the form of government that had just been shown to be most effective in providing food or most generous in spirit should become the form of government in the country whose form of government had been an open question and source of dispute. Alternatively, one could instead argue that there is no more compatibility here between the contest and the external issues than there was in the wholly irrelevant swimming race (why would having the most food make legitimate the annexing of a certain territory that further increased the winner's economic well-being and wealth); there would in fact be an internal contradiction in taking some generous or benign attribute (food growing, medical care) and using it as a basis for denying the legitimacy of another population's position in a dispute. Or again, one could take a third position and argue that such an attribute, because it is integral to an internationally shared conception of nationhood, does have an interior connection to a disputed issue but no more so than does "injuring," since the capacity to injure, or as it is more typically phrased, the presence of an armed forces, is also integral to an internationally shared conception of "sovereignty" and "nationhood." But which of these three positions one believes to be most accurate does not matter because in any event such a form of contest would no more produce an outcome by which the contestants would abide than would the earlier contests based on some wholly arbitrary activity such as racing: like them, it would only compound the dispute by amplifying the number of issues that would eventually be settled by war. Would the designation of "best provider" be earned by the quantity of one staple or the number and variety of foods; is it sheer bulk or nutritional excellence that matters and how is that measured; is the most just, effective, or bountiful form of distribution the one which has the greater excess in a part of its population, the rate at which more and more segments of the population enter the "excess sector," or is it registered instead in a small, steadily growing margin of increase spread evenly and almost invisibly over the entire population; is medical prowess registered in the number of extraordinary illnesses that can be cured or in the percent of the population that is guaranteed some minimal degree of care? Like the earlier contests, the new contests would be more likely to produce more material for the dispute than to provide the means of "choosing" between the disputants. The internal relevance of the contest activity is not, then, necessarily a virtue: it might, in fact, be argued that it is the very virtue of a contest like injuring or swimming that its central activity is utterly irrelevant to the external issue, and thus provides a way of stepping away from the endlessly self-amplifying intricacies of dispute and makes available a wholly arbitrary but (in the case of injuring) agreed upon process of choosing between disputants.

Only at the end of the war does the benefit of injuring occur. The answer to the central question here—what differentiates war from any other contest?—is that the designation of a winner and a loser is accepted as an abiding designation by the two contestants and thus carries over to the enactment of the winner's issues. Whatever the temporal duration of the moment of "victory" (the moment when the condition of equality and duality suddenly became the condition of singularity and inequality), whether that should be understood as having occurred on a day, an hour, or a minute, it is an outcome that endures long beyond that brief moment of transition that is war's "end," crosses over the final temporal boundary and becomes permanently objectified and memorialized in the disposition of postwar issues. But to say this is only to displace the first question by a second question, or to allow the first question to reappear in a slightly altered form: what is it in the nature of injuring that ensures the duration of its own outcome? It does not help to identify the "issues" as the structural equivalent of what in an ordinary peacetime contest is called the "prize,"[75] for that only introduces a slightly different idiom that again necessitates the reiteration of the same question: why in an international contest based on injuring does the loser allow the victor his prize as that loser almost certainly would not had the contest entailed food growing, tennis, medical care, or artistic design; having lost by any other means it would have disowned the contest rather than agreeing to disown elements in its own system of self-belief. Each new idiom, each new metaphorical construction, only reintroduces the same problem: in the sentence, "Whoever wins, gets to determine the issues," what is it that explains the transition between the second and third words, that explains the phrase "wins, gets"? What is it that allows the force of injuring to survive beyond the termination of its own activity? What is it (to return to constructs encountered at an earlier point in the analysis) that allows the translation of open bodies into verbal issues such as freedom? How is it that the road of injury arrives in the town of freedom, or that the intermediate product of injury is transformed into the final product of freedom?

There is essentially only one answer (either explicitly articulated as in the writings of Clausewitz, or simply assumed to be the case as in many other historical and political descriptions of war[76]): a military contest differs from other contests in that its outcome carries the power of its own enforcement; the winner may enact its issues because the loser does not have the power to reinitiate the battle, does not have the option further *to contest* the issues or *to contest* the nature of the contest or its outcome or the political consequences of that outcome. Thus injuring as the activity on which a contest is based not only designates a winner and a loser and in so doing brings about the cessation of its own activity (a description that would so far apply to most contests), but also (unlike other contests) ensures that one of the two participants will no longer have the ability to again perform the activity. If this were true, it would indeed make war necessary, for it would have no substitute or equivalent. It would only have an equivalent if there were another contest that destroyed the capacity for the activity on which the contest was based as well as all other powers to contest by any means: it would be as though the loser in a song contest were unable to request one more round of song because he were, as a result of the competition, no longer able to sing;[77] or as though the losing chess player's spatial imagination were now permanently deranged; or as though the contestant in embroidering lost forever the intricate play of the small muscles in her fingertips or her capacity for visual play of subtle colors; or as though a dancer were to move across the floor through the competition to the final moment and, stepping across the final threshold, arrive not only in the space of loss but a space in which she was never again able to walk. Of course, this description is not true of the song contest, chess match, cloth-making or dance competition—but it may be that it is also untrue of war.

Our understanding of war is deeply conditioned by this view that permeates not only the formal language of strategy but also the language of casual description. There are certainly recorded in history particular wars in which the defeated were wholly deprived of the power of retaliation either because, as in ancient Greece, whole populations were sometimes annihilated or enslaved or, as in medieval Northern Europe, because the population was sometimes too dispersed to raise a new army after the (therefore decisive) battle.[78] But such descriptions are exceptional. Clausewitz himself in his brilliant phenomenology of war designated this condition as war's essential characteristic, but then went on to find it missing from a troublingly large number of instances. He writes, for example, in the second chapter of Book I:

> But the aim of *disarming the enemy* (the object of *war in the abstract*, the ultimate means of accomplishing the war's political purpose, which should incorporate all the rest) is in fact not always encountered in reality, and need not be fully achieved as a condition of peace. On no account should theory raise it to the level of a law. Many treaties have been concluded before one of the antagonists could be called powerless—even before the balance of power had been seriously altered. What is more, a review of actual cases shows a whole category of wars in which the very idea of *defeating the enemy* is unreal: those in which the enemy is substantially the stronger power. . . . If war were what pure theory postulates, a war between states of markedly unequal strength would be absurd, and so impossible. At most, material disparity could not go beyond the amount that moral factors could replace; and social conditions being what they are in Europe today, moral forces would not go far. But wars have in fact been fought between states of *very unequal strength, for actual war is often far removed from the pure concept postulated by theory.*[79]

To these categories of problematic war—those that end before the defeated is powerless, those that end before there is even a serious alteration in the balance of power, those initiated against an obviously stronger enemy, those fought between states of dramatically unequal strength—he adds other descriptions in the course of his long analysis that further discredit the claim of the "ideal." Until the retreat, for example, the winning and losing side in an engagement may differ very little in their number of casualties; sometimes the winner's casualties are greater.[80] Finally, even when there is in fact an absolute defeat of the loser's military, that absolute defeat is only an absolute if lifted out of time, and may instead be accurately perceived by the defeated as a "transitory evil" until the reassumption or recreation of power is possible.[81] In such passages, Clausewitz is openly troubled: although he repeatedly describes the problem as the failure of the real to achieve the ideal,[82] it may be that he recognized in "ideal" and "real" the presence of a false description and a true description, and recognized also that what was at stake was not a falling away of practice from theory but a falling away of the whole intelligible basis for war, a falling away of the single element that differentiates it from other forms of contest and legitimates its use over its alternatives. The uneasy tone of such passages may express his recognition that if war does not carry the power of its own enforcement, there ceases to be any attribute that makes sense of, let alone justifies, its use.

Battles and wars of the twentieth century would have provided Clausewitz with as many examples of his problematic categories as his own century did. There were, for example, many only barely decisive battles of World War I, and there was in the Russo-Japanese war the famous siege of Port Arthur in which the losing Russians suffered 31,306 killed, wounded, and missing while the victorious Japanese had 57,780 killed, wounded, or missing with 33,769 sick of beriberi; or again the battle of Mukden in which Russia lost 60,000 killed and wounded (and then 25,000 prisoners after surrender) whereas the Japanese lost 71,000, each side having entered the war with approximately 300,000 men.[83] More important, the end of major wars provides contemporary illustrations of why Clausewitz would begin to doubt that the basis of war is that it carries the power of its own enforcement.

The positions of the losers in the Vietnam War and in World War II (what are for many the two most familiar wars) illustrate the absence of this base in dramatically different ways. The defeat of the United States by North Vietnam did not entail the loser's inability to continue or to renew military hostilities: its military prowess was in the period beginning with its defeat many times that of North Vietnam's. Although the position of the United States seems anomalous, it conforms to some of Clausewitz's categories (a war between greatly unequal powers, a war concluded before a drastic shift in power, a war concluded before the defeated is powerless), and has parallels not only in the wars on which Clausewitz drew but in contemporary wars such as the 1954 defeat of France in Vietnam, the 1962 defeat of France in Algiers, the 1956 withdrawal of Britain from the Suez. Although all these may in turn be grouped together as an anomalous category under a rubric of their own (such as "colonial wars"), it may be that what occurred there can more accurately be understood as a magnification of an ordinary outcome than as an exception. The position of Germany at the end of World War II is very distant from that of America at the end of Vietnam; and perhaps of all recent wars, World War II has an ending that most closely approaches the complete neutralization of the opponent's capacity to injure required of war in its ideal form. Although the unconditional surrender demanded by the Allies is not perceived as unusual by most Americans (in part because it was an objective in the Civil War), it is in fact a rare form of surrender.[84] But even here, in the extremity of defeat, occupation, and disarmament, the notion of the absolute is, as Clausewitz anticipates, eliminated if the temporal frame is altered from one of days and weeks to one of weeks and months. What is striking about the end of this war is not only the speed with which the defeated began their return to health but also the attitude of the victors to this return. Among the many things for which the Marshall Plan both as an historical fact and as an imaginative construct is remarkable is the way in which it illuminates the structure of war, illuminates the absence from that structure of any requirement that the defeated be without the power of renewal. There is throughout Marshall's speech before the Senate Committee on Foreign Relations (8 January 1948) and Truman's Message to Congress (19 December 1947) a refusal to indulge or even to acknowledge the notion that it is dangerous, absurd, or even odd to include West Germany in the program for European economic recovery.[85] The issue is not overtly addressed except for one brief moment late in Truman's speech.[86] Perhaps even more revealing is the fact that these two speeches—which (as acts of persuasion rather than as studies) take account of but do not include any of the detailed substantive material found in the technical reports of the Committee of European Economic Cooperation—together contain, in the midst of their haunting generalities, only one specific piece of data, a piece of data offered as an illustration of the feasibility of collective recovery. That single fragile detail, which in its isolation carries a tremendous resonance, is taken not from one of what had been the European Allied powers (United Kingdom, France, Belgium, Netherlands, Luxembourg, Norway, Greece), nor from one of the neutral countries (Switzerland, Sweden, Turkey, Denmark, Iceland), nor even from one of the lesser Axis powers (Italy), but from Germany: "In the last few months coal production in the Ruhr district of Western Germany has increased from 230,000 tons a day to 290,000 tons a day."[87] Rather than locating the image of its own victory in the impotence of its former opponent, or even basing its postwar confidence in its economic superiority to that opponent, a plan is being brought about to encourage the full recovery of the opponent's strength and, even more startlingly, its already existing strength is with open admiration pointed to as a source of assurance that the plan will be for all participants a success.[88] The end of this war, like the radically different end of the Vietnam War, and in different ways like the end of other twentieth-century wars, demonstrates that even if there have been historical moments in which war carries the power of its own enforcement, it is not essential to its structure that it do so.

Such endings, like those on which Clausewitz drew, show that the character of injuring is altered when the context is altered from two people to two multitudes of people, and the first is not an accurate model for the second. That is, it may be that the widely shared assumption that war carries the power of its own enforcement arises from the mental reflex of thinking about war by holding steady the contest activity as injuring but conceiving of that activity as occurring between two people each working to kill the other. In mortal combat between two persons, the outcome *does* carry the power of its own enforcement; the designation of a winner and a loser also endures and transfers to the enactment of the no longer contested issue because the contest has eliminated the capacity of the loser to protest or even wonder about his own version of those issues which, importantly, have died with him. If one visualizes a silver and black moonlit terrain innocent of all human inhabitants but two, each with a claim to a certain (thus disputed) rock or tree, or each with a conception of god that he (once dreamed and now) insists the other should share, there will be, once they have physically contested the issue, only one man, one claim to the rock, one idea of god. The claim and the conception survive along with the embodied survivor: the second claim, second conception, and second embodied combatant have ceased to exist.[89] It will not occur to anyone to be puzzled here about why there is only one surviving issue, for the other issue only existed as aspects of world-consciousness, fragments of self-extension of the embodied person, attached to him, appearing when he appeared, disappearing when he disappeared. The situation is radically different and thus the model becomes useless when in the shadows of the same midnight terrain one now envisions two populations, each with a claim to a disputed peninsula it does not want to share, or each with a conception of god or a political utopia it insists the other should share. After the end of their combat, there will be, as in the first landscape, a single reigning set of issues, one no longer disputed claim to the peninsula, one unrivaled political philosphy; but now, unlike the first vision, the singularity of issues cannot be attributed to the unitary nature of the surviving population, for there will be, in varying numbers, embodied survivors in both populations.

If the two situations were to be summarized in a formula that represented embodied persons by a letter (X, Y) and represented the disputed ideas, culture, issues, forms of disembodied self-extension, ideas about property, land or heaven, by the letter primed (X', Y'), then in the first picture the elements present before combat are $X + X' + Y + Y'$ and after the war $X + X'$, whereas in the second picture the elements present before combat are $X + X' + Y + Y'$ and after the war $X + X' + Y$. The disappearance of Y' cannot be explained (as it is in the power of enforcement argument, and as it was in the first model) by the disappearance of Y. Only if war regularly included genocide (or enslavement) would the second situation conform to the first, and would the first be an appropriate model for the second. So far from being true is this that not only do wars not include genocide (with few historical exceptions), but the approximation of this act (when it has occurred) has been perceived to be outside war and in the realm of atrocity. So, too, the possibility of genocide that arises with nuclear weapons has led humanity to the cry for their elimination. That is, so little is genocide (or the permanent elimination of the opponent's capacity to injure) a structural requirement of war that it in fact seems a deconstruction of war, a deconstruction of a deconstruction.

That the widely shared and erroneous understanding of war, most clearly articulated by Clausewitz, may originate in the two-person model is suggested by three factors: first, by the fact that this description is accurate when conceived of as occurring between two solitary figures; second, by the fact that even in a relatively confined war the events are happening on a scale far beyond visual or sensory experience and thus routinely necessitate the invocation of models, maps, and analogues; third (a specific and immediately relevant instance of the second), by the existence of the descriptive convention, richly elaborated by strategists, historians, political philosophers, and perhaps all who have occasion to speak about war, of conceiving of two national armed forces as two colossal single combatants. Thus the convention, which exists primarily to assist the visualization of troop movements and acts *during* war, is (inappropriately) available at the moment when the mind turns to thinking about the nature of war's *ending*. Sometimes, of course, the link to the model of single combatants may be overt, as when Freud in "Why War?" writes both that "That purpose [of compelling the other side to abandon its claim] was most completely achieved if the victor's violence eliminated his opponent permanently, that is to say, killed him," and later, "Wars of this kind [between different units: cities, provinces, races, nations, empires] end either in the spoliation or in the complete overthrow and conquest of one of the parties."[90] Of course, what is of crucial importance is the error of the power of enforcement idea, and not the question of whether or not it originates in the model of single combatants.

Clausewitz may be alone in the overt lucidity with which he articulates his conception of the formal structure of war and then articulates his concern about the absence of this formal property from many different types of war. But he is only alone in his lucidity; he is not alone in either the assumption or the suspicion that the assumption is wrong, for there seems to be a widespread—almost a collective human—tendency simultaneously to believe and disbelieve that war carries the power of its own enforcement, to know that it certainly must be the case (otherwise some benign contest could and should be substituted), yet at the same time to know that it is almost certainly not the case, to believe it just enough to permit actual wars to be occasioned and to permit the notion of war to appear legitimate, but not so much that after-the-fact descriptions of war can include this crucial element. The complicated act of believing and disbelieving is not usually, as it is with Clausewitz, articulated by any one person but is instead fractured into five or six different positions, each adopted by a separate person, thus allowing the overall problem to appear and disappear, show now an edge, now a full surface, stand now exposed and now eclipsed in its full significance. The idea that war carries the power of its own enforcement may, first of all, simply be assumed in what a writer or speaker says but not itself be announced; or second, it may itself be explicitly articulated; third, it may, having been clearly articulated by one person then be recognized by another person as utterly false (with or without this person making the further recognition that the intelligible basis of war may have just disintegrated in his hands); fourth, a person may notice that it is absent from one particular war, or even one particular kind of war, but rather than beginning to suspect its absence from the generic fact of war, the person may attribute its absence to the geographical, political, or military peculiarities specific to that war (the particular war, rather than the idea about war is perceived to be defective); fifth, seeing its general absence, one may propose an alternative explanation of the end of war without wondering whether this substitute explanation does not now call into question the relevance of the activity of injuring—that is, if there is to be a substitute, why make the substitution only in the final moments rather than displacing injuring altogether?

Each of these positions has so many individual exponents, and the endlessly repeated complications and nuances of interaction between the positions create over time and the expanse of a population such a thick texture of conversation, that it may be idle to represent each with any one example, yet the specific instances may make the familiarity of each position more immediately recognizable. The first position, where the principle is assumed but itself unannounced, is often present in strategic analysis where there may be a discussion of the relative advantages of bringing about the effect by a first means or a second means—such as Hans Delbrück's alternatives of rout or exhaustion—without ever examining or even precisely naming the effect that is instead simply assumed to exist as a reachable goal. That a decisive and absolute outcome is possible is a premise, and what is actually analyzed, questioned, debated are the means of bringing it about. Its occurrence as an unarticulated assumption is not, of course, confined to strategic discourse and may appear in any form of cultural or philosophic inquiry: in, for example, Freud's essay, referred to a moment ago, neither the problematic nature of war's end nor the power of enforcement idea is Freud's subject; but, as is evident in the passages cited, it is an idea that is assumed in passing to be true.

The second position, the overt identification of the principle, in turn makes the falsity of the principle easier to identify, as when Secretary of State Alexander Haig in a speech at the Center for Strategic and International Studies announced that historically, societies have always risked "total destruction if the prize of victory was sufficiently great or the consequence of submission sufficiently grave,"[91] an observation which one commentator, Theodore Draper (carrying us forward to our third position here) calls a "fatuity," pointing out, "There has not been a war of total destruction since the Third Punic War of 146 B.C."[92] The exponent of the second position, who believes that war entails the total destruction of one opponent, may well have arrived at that position out of historical ignorance, or he may have arrived there (as exponents of the third position often imply) out of ruthlessness, but he may also have been assisted in his arrival there by his conscious or unconscious recognition that without this total destruction war itself is a brutal fatuity because it has not only brought about the heartsickening damage that would not have occurred in any other substitute contest, but at the same time has not in fact accomplished any outcome that could not have been accomplished by the other contest. Conversely, the exponent of the third position may rightly see the error of his opponent with or without seeing the structural collapse of war as a privileged event. The complexity of the second and third positions, and the rhythm of alternation between them, are visible, for example, in debates over unconditional surrender and more moderate forms of surrender, and are most pervasively present in debates over unlimited war and limited war, whether occurring on an academic or theoretical plane, or instead occasioned by a particular form of war (nuclear, for instance), or instead occasioned by an actual war (such as the extensive disagreement on this issue between military and civilian authorities during the Korean War [93]).

Familiar instances of the fourth position, in which the absence of the ''power of enforcement'' outcome is attributed to the peculiarities of a particular war, include analyses of the Vietnam War that explain its aberrant form in terms of the idiosyncrasies of Southeast Asian terrain, the structure of village life, and the nature of guerilla warfare,[94] or again in analyses of the Korean War, out of which came the famous pronouncement of General MacArthur's 5 April 1951 letter to the United States House of Representatives, ''There is no substitute for victory,'' a pronouncement that in part became famous in the literature of war precisely because there was no ''victory'' in MacArthur's sense, and was in fact a ''substitute.''[95] Again, Liddell Hart, personally anguished both by the suffering and by the indecisiveness of trench warfare in World War I, saw that Clausewitz's idea of ''decisive victory on the battlefield'' was wrong, but instead of suspecting that the error lay in the idea contained in the first two words, ''decisive victory,'' he instead concluded that it was the last three words, ''on the battlefield,'' that were wrong and thus went on to advocate decisive victory through the bombing of economic and industrial targets.[96] Of course, precisely the same problematic indeterminacy may be characteristic of the new arena of injuring. It has been argued that dropping the atom bomb on Hiroshima and Nagasaki is not only not the event that brought about victory over Japan, but that it probably did not even hasten the arrival of that victory[97] (and though this position is itself challengeable, that the point can be argued in either direction demonstrates the absence of clarity in the event itself); so, too, it has been argued that the worst firestorm bombing of German cities occurred after victory was secured.[98] One can imagine Liddell Hart's solution to the indeterminacy of the trenches repeated over a chronological sequence of alternative arenas, each as indecisive as the last but each time leading observers to the conclusion that it is the arena that is preventing (or, at least, failing to contribute to) the decisive outcome, rather than an error in the very notion of decisiveness, and thus each time generating suggestions for a new target that will in turn be one day recognized as indeterminate, rather than each ever being recognized as one in a series and thus itself an argument for the cessation of war.

As the many variations in the fourth position suggest, the particular, historically problematic event may be that the given war as a whole lacks a decisive ending (e.g., Korea); or instead, the war has a decisive ending, but it is impossible to locate the kind of event (e.g., battle) that produced that quality of decisiveness; or, since the attribute is not visible in an observed event (e.g., trench warfare), it is assigned to a less closely observed event (e.g., it did not happen in the trenches, so it must have been brought about by the suffering that was inflicted on the civilians). It is crucial to underscore the fact that while these cumulative instances repeatedly call into question the existence of the power of enforcement phenomenon, the reverse is *not* true: that is, the occurrence of battles with a wide margin of victory or of a war with a wide margin of victory would *not* certify the existence of the phenomenon. Unless the "margin of victory" or "decisive victory" approaches the situation in which the losing side is permanently deprived of the capacity to injure (by being either annihilated or enslaved or permanently occupied and policed), these terms do not overlap the phenomenon in question: a war's having a decisive ending is very different from its ending working through the power of enforcement phenomenon, though the absence is even more visible in an only marginally decisive outcome. Moreover, the existence of a war with a victory so absolute that it did conform to, or at least approach, the "power of its own enforcement" condition would certainly not even then demonstrate that this condition is essential to war—it would have merely shown that it may sometimes be a characteristic of war, that it is an occasional or accidental attribute but not that it is a structurally necessary attribute. In identifying *the structure* of something—in this case, war—those elements that are common to all, or almost all, of the instances of the thing constitute its structure, those minimal elements that must be present for the thing to be what it is, a war.

The fifth position, the introduction of an alternative explanation for the end of war, more than any one of the other positions, makes visible the two-directional habit of believing and disbelieving in the power of enforcement phenomenon simultaneously. By far the most frequently singled-out alternative is the morale or moral (the two are usually conflated) element. Up to this point, we have been saying that injuring is the activity that designates a winner and a loser but not one that carries the power of its own enforcement (and thus one not having a differentiating attribute making it preferable to another contest); but now, with the introduction of the argument that victory is determined by morale, the activity of injuring is no longer credited with even having performed the work of designating a winner and a loser (thus all the more reason to have displaced the activity of injuring from the very start with some benign activity, though remarkably enough this is never a conclusion drawn by exponents of the morale position). In victory, the ratio of moral to physical factors, goes Napoleon's famous dictum, is three to one, a claim which, especially among military writings,[99] is widely accepted in its general outlines, even if the actual ratio is often moderated downward or at other moments upward, as in Field Marshall Montgomery's assessment, "I consider morale the greatest and only factor in war."[100] Though no one can doubt that there will be required of soldiers—both those who will eventually be named among the losers and those who will be named among the winners—unthinkable strength, courage, camaraderie, alertness, self-sacrifice, pride, exhilaration, a whole array of genuinely spirited and spiritual attributes, and though no one can doubt that these attributes and the general level of self-belief among soldiers can vary considerably, these two acknowledgements are not the same as the much more problem-laden assertion that victory belongs to, reveals, the side with the superior morale. There are four bases on which this assertion is objectionable, the last of which is the most important to the present analysis.

Its invocation is, first of all, uneven, inconsistent. It tends to be invoked by military historians or political rhetoricians on the winning side, not on the losing side, and is almost never flanked by a series of parallel instances that would quickly expose the tawdriness of the translation of physical prowess into morale, and, worse of all, morality. Although, for example, it is theoretically conceivable that in the first half of the 1940s the Allied spirit may have been morally superior to the spirit of the German people, one would only want to say that the Allied military victory itself was an objectification of a discrepancy in morale or moral character if one were willing to cite Germany's rapid victory over Poland, its quick forcing of the capitulation of France, and its almost uncontested imprisonment of Jews as a manifestation of its superiority over these three populations in the element of morale, an interpretation that Hitler, at that moment the victor, would have cheerfully accepted. Similarly, to identify the end of the American Civil War as an objectification of the stronger morale or moral character of the Union troops would then oblige one to identify the original subjugation of blacks in the South as a manifestation of the superiority of white morale, again an interpretation that Southern pro-slavery leaders would have accepted. This is not to say that the moral character of the North was not superior, for on at least one overwhelmingly important issue it was; it is to say that military victory is neither the proof nor the sign of that supremacy. In war as in peace, the seductiveness of physical and political power is in nothing so apparent as in its ability to oblige observers to redescribe it as a moral superiority.

As problematic as, and implied by, its inconsistent invocation is the second basis, its own internal characteristics. Even if one were successfully able to hold separate the connotations of "morale" and "moral" and claim only the former (since the two may be irrelevant to one another and may even work at cross-purposes; for example, the single-minded sense of self-rightness entailed in high morale is likely to be diminished by elements of moral character, the introspective contemplation of the relative merits of the two sides or of the virtue of the acts one is, during war, participating in), the notion of "morale" still tends to have an aura of the spiritual, to signal some capacity for self-transcendence or form of consciousness different from the physical events. The most familiar formulations, such as Napoleon's, tend to single it out as something separable from and even poised against physical acts. But if one were to watch the final hours of a battle, both groups of men exhausted from their unbroken thirty-six hour confrontation, themselves hungry, some wounded, almost all in the acute state of loss that will come from having seen in recent hours the maiming of comrades whose dead bodies may still lie nearby, and *if* the one group of men were *visibly* able to carry on more than the other, the superiority would be most factually or literally described as this: the ability to go on injuring even when you are yourself badly injured. Although it might be that one would want to be able to do this, the "this" is not separable from the activity of injuring but is a name for the sustained capacity to injure. Nor does it seem to have the aura of self-transcendence, the capacity to live beyond, to produce work that is itself free of, the pressures and terrors of the body that one might attribute to a person able to nurse another despite the fact that he was himself in great pain, or to a person able to go on composing music even when he or she lost the ability to hear, or more simply to people carrying out the events of everyday life despite losses to their own bodies or those of members of their families.[101]

A third basis on which it is an objectionable form of explanation is the apparent freedom of the claim from any requirement for evidence. Although military reports from the front, especially those requesting reinforcements or a change in plan, will include descriptions of morale, the assertion that the final outcome was determined by morale is not ordinarily accompanied by a comparison of eyewitness accounts of morale on the two sides.[102] In fact, what tends to be cited as evidence is the lack of any physical event (such as a wide discrepancy in the margin of deaths on the two sides) that could explain how it is that one side won. The invocation of morale as the winning determinant is an enduring instance of circular reasoning: its active presence is inferred from the presence of a victory coupled with the absence of any other visible locus of victory.

The fourth and most important basis of objection is structural.[103] If "morale" is taken as a shorthand summation of an army's injuring capacity, its invocation as an explanation does not create any special problem in the structural logic of war; but if it is asserted as something separate, and as something appropriately conflated with "morality," then it creates a self-refuting structure in which the act of belief-disbelief is now built into war's very form. In order to visualize what occurs in this explanation, the diffuse *attribute* of "morale" can be translated into an *activity*: it is spoken about as separate from injuring, and picturing it as a discrete activity makes it possible to hold steadily visible the asserted fact of that separation. Because morale has connotations of the human spirit, the capacity to live beyond the body, the capacity to dwell in the realm of symbols and substitutes rather than the raw physical events of survival, it is at least as often associated with world-building as with world-destroying, with creating as with killing, and thus can be taken to reside in the other benign activities invoked earlier, the composition of music, singing, nursing, house building, chess playing, or a hundred others, and can be represented by any one of them, say singing. When war is described as turning in its final stage on the element of morale (separable from injuring), what is drawn is a model of war constructed along the lines in the following narrative. A dispute arises between two populations. In order to determine a winner, they agree to have a contest. They could have either an extravagant three-year-long song contest or instead a three-year-long war. They choose the second because, though each would allow the designation of a winner and a loser, injuring—unlike singing—will carry the power of its own enforcement. But after moving through three autumns, three winters, three springs, and two summers during which they butcher one another (if the word is ugly, the acts it represents are far uglier) they begin to approach the third summer, and they realize that not only will injuring not carry the power of its own enforcement but it will not even make possible the distinction between the winner and the loser: despite fluctuations, the body count on each side tends to approximate that on the other side, and thus to continually re-establish the equality of the two sides rather than to expose their inequality. Thus here, at the end of war, at the very place where the exceptional virtue or the exceptional contribution of injuring was to have occurred (and for the sake of which injuring was chosen over any alternative), it is suddenly necessary to make arrangements for the insertion of the song contest into the overarching frame of war. Like an architectural detail from one period appropriated into the building of another period, it becomes the portal through which the final exit out of war will occur. This is the equivalent of the morale argument: the acceptance of the brutalities of war with the eleventh-hour insertion of the chess match or tennis match or talent contest contracted down into the period immediately preceding its ending; the abbreviated contest does not displace or provide a substitute for the injuries, for thousands of injuries have by this time already occurred and will continue to occur in the final weeks; it instead substitutes for the single element that was thought to necessitate and hence justify the injuring. The fragile song contest (which no one precisely saw, though everywhere here and there it is said voices were heard) is like a small jewel placed down in the midst of a three-year massacre and relied on to perform the very work for the sake of which its own activity had been originally rejected.

Thus, the fifth position, the alternative description of war's ending, is—especially when placed in the company of the other four—important as a manifestation of our collective capacity at once to believe in and to doubt the power of enforcement phenomenon, which appears not to be the answer to the central question raised here and leaves us back at that original question.

The conclusions that emerge when war is compared with other contests may appear to go in contradictory directions. On the one hand, war continues to have no analogue and hence to be undisplaceable: neither vastly magnifying some other benign contest to the scale of war nor contracting the arena of injuring down to the familiar scale of peacetime contests (the two most available paths of substitution) replicates the outcome of war. On the other hand, what differentiates it from potential substitutes remains elusive, since it seems only to perform the work of designating a winner and a loser, and not a loser incapable of reperforming the activity by which the issue might be recontested. But there is less of a contradiction than there might at first appear, for the differentiating characteristic, the fact that the losers and winners abide by the outcome of those designations to an extent that they would not in any other contest, has not been called into question. All that has been questioned is the account of that outcome as happening through the power of enforcement principle, the explanation according to which the losers abide because they have no choice. It is not that they abide, but that they are compelled to abide, that is untrue. Furthermore, although the power of enforcement principle is not at work in the way that it is widely believed to be, the very fact that it is widely believed to be at work may be in the end the occurrence that lets it work. If populations, whether out of shared opinion, self-conscious judgment, or unselfreflecting impulse and intuition, assume this to be the case, it will be the case not because it had to be, but because it was believed it had to be. The outcome of war endures long beyond the temporal moment and is translated into the disposition of issues because it is believed to and hence *allowed to* carry the power of its own enforcement.

Though the principle does not in itself literally exist, its effect is just as literally brought into existence by its being assumed, unquestioned, if questioned reaffirmed, and most importantly, acted upon. Thus, the question that confronts us is not how does injuring (once extended from two figures to two multitudes of figures) create an incontestable outcome, but how does it—or why does it—give rise to the fiction that its outcome cannot be (or should not, or must not be) contested? Whether there is a wide or a narrow margin of victory, once the war ends it will be *as though* war carried the power of its own enforcement, and it is the *"as though"* mechanism, the *"as if"* reflex, that may at last expose the terrifying resources of war as two populations assume their respective designations as "winner" and "loser," pass over the final boundary, and stand on the narrow piece of terrain that separates the activity of world-destroying that has just ended from the activity of world-rebuilding that is about to begin.

This "as-if" function is the subject of the next section.

IV. The End of War: The Laying Edge to Edge of Injured Bodies and Unanchored Issues

The extent to which in ordinary peacetime activity the nation-state resides unnoticed in the intricate recesses of personhood, penetrates the deepest layers of consciousness, and manifests itself in the body itself is hard to assess; for it seems at any given moment "hardly" there, yet seems at many moments, however hardly, *there* in the metabolic mysteries of the body's hunger for culturally stipulated forms of food and drink, the external objects one is willing habitually to put into oneself; *hardly* there but *there* in the learned postures, gestures, gait, the ease or reluctance with which it breaks into a smile; *there* in the regional accent, the disposition of the tongue, mouth, and throat, the elaborate and intricate play of small muscles that may also be echoed and magnified throughout the whole body, as when a man moves across the room, there radiates across his shoulder, head, hips, legs, and arms the history of his early boyhood years of life in Georgia and his young adolescence in Manhattan.

The presence of learned culture in the body is sometimes described as an imposition originating from without: the words "polis" and "polite" are, as Pierre Bourdieu reminds us, etymologically related, and "the concessions of politeness always contain political concessions."[104] But it must at least in part be seen as originating in the body, attributed to the refusal of the body to disown its own early circumstances, its mute and often beautiful insistence on absorbing into its rhythms and postures the signs that it inhabits a particular space at a particular time. The human animal is in its early years "civilized," learns to stand upright, to walk, to wave and signal, to listen, to speak, and the general "civilizing" process takes place within particular "civil" realms, a particular hemisphere, a particular nation, a particular state, a particular region. Whether the body's loyalty to these political realms is more accurately identified as residing in one fragile gesture or in a thousand, it is likely to be deeply and permanently there, more permanently there, less easily shed, than those disembodied forms of patriotism that exist in verbal habits or in thoughts about one's national identity. The political identity of the body is usually learned unconsciously, effortlessly, and very early—it is said that within a few months of life British infants have learned to hold their eyebrows in a raised position. So, too, it may be the last form of patriotism to be lost; studies of third and fourth generation immigrants in the United States show that long after all other cultural habits (language, narratives, celebrations of festival days) have been lost or disowned, culturally stipulated expressions of physical pain remain and differentiate Irish-American, Jewish-American, or Italian-American.[105]

What is "remembered" in the body is well remembered. When a fifteen-year-old girl climbs off her bike and climbs back on at twenty-five, it may seem only the ten year interval that her body has forgotten, so effortless is the return to mastery—her body, however slender, hovering wide over the thin silver spin of the narrow wheels. So, too, her fingers placed down on piano keys may recover a lost song that was not available to her auditory memory and seemed to come into being in her fingertips themselves, coming out of them after the first two or three faltering notes with ease, as though it were only another form of breathing. Even these nearly "apolitical" examples are not wholly apolitical, for at the very least there is registered in her body the fact that she lives in a culturally stipulated time (after the invention of bicycles and pianos) and place (a land where these objects are available to the general population rather than to the elite alone, for she is not a princess); someone from an earlier century or from a country without material objects might think—hearing the description of a girl gliding over the ground on round wings, her fingers fanning into ivory shafts that make music as they move—that it was an angel or a goddess that was being described. There exist, of course, forms of bodily memory that are anterior to, deeper than, and in ordinary peacetime contexts beyond the reach of culture. The body's self-immunizing antibody system is sometimes described as a memory system: the body, having once encountered certain foreign bodies, will the next time recognize, remember, and release its own defenses. So, too, within genetic research, the DNA and RNA mechanisms for self-replication are together understood as a form of bodily memory.[106]

What is remembered in the body is well remembered. It is not possible to compel a person to unlearn the riding of a bike, or to take out the knowledge of a song residing in the fingertips, or to undo the memory of antibodies or self-replication without directly entering, altering, injuring the body itself.[107] So, too, the political identity of the body is not easily changed: if another flag is placed in front of British eyes, it will be looked at or looked away from with eyes looking out from under eyebrows held high. To the extent that the body is political, it tends to be unalterably political and thus acquires an apparent *apolitical* character precisely by being unsusceptible to, beyond the reach of, any *new* political imposition.[108] It is not surprising, for example, that China's national birth control goals have not been easily accepted, "embodied," by the residents of Guangdon Province, where the seven-thousand-year-old feudal philosophy of child-bearing often makes ineffective the verbal advocacy of one-child-to-a-couple, even after ten visits, twenty visits, or a hundred visits to the couple from family planners, as well as pledge programs, the promise of bonuses to couples who comply, and the threat of forms of deprivation to erring couples, such as taking away a sewing machine or other important tool from the family household.[109] If a new political philosophy is to be absorbed by a country's population, it is best introduced to those who have not yet absorbed the old philosophy: that is, it is most easily learned by the country's children, whether the shift is in the direction of radical justice (the teaching of racial equality to United States children through school integration[110]) or instead in the direction of radical injustice (the teaching of racial hatred to German children in the Hitler Youth Corps). As Bourdieu writes of even the passing on of cultural "manners" from one generation to the next, "The principles em-bodied in this way are placed beyond the grasp of consciousness, and hence cannot be touched by voluntary, deliberate transformation, cannot even be made explicit; nothing seems more ineffable, more incommunicable, more inimitable, and therefore, more precious, than the values given body, *made* body by the transubstantiation achieved by the hidden pedagogy, capable of instilling a whole cosmology, an ethic, a metaphysic, a political philosophy, through injunctions as insignificant as 'stand up straight' or 'don't hold your knife in your left hand.' "[111]

Of the many things that might be said about the nature of injury in war, a small number may begin to lead to an explanation for the overarching question that confronts us here: the question of how injuring creates an abiding outcome, an outcome that is "as though" the losers were deprived of the capacity to renew the activity of injuring, even though in almost no case is the losing side actually placed in that position. First, it is not the case that the body is normally apolitical and only becomes political at the moment of war. Not only is a specific culture absorbed at an early age by those dwelling within its boundaries, but (particularly if there is no change in political philosophy) the nation-state will without notice continue to interact on a day-to-day basis with its always embodied citizens. It might even be argued that the attributes of a particular political philosophy, its generosities and its failures, are most apparent in those places where it intersects with, touches or agrees not to touch, the human body—in the medical system it formally or informally sponsors that determines whose body will and whose body will not be repaired; in the guarantees it provides or refuses to provide about the quality and consistency of foods and drugs that will enter the body; in the system of laws that identify the personal acts toward another's body that the state will designate "unpolitical" (unsocial, uncivil, illegal, criminal) and that will thus occasion the direct imposition of the state on the offender's body and the separation of that unpolitical or uncivil presence from contact with the citizens.

It may be that the degree to which body and state are interwoven with one another can be most quickly appreciated by noticing the most obvious and ongoing manifestation of that relation such as the fact that one's citizenship ordinarily entails physical *presence* within the boundaries of that country, a relation between body and state that can be overlooked by being too obvious. Or it may instead be that it can best be appreciated by noticing almost random instances of the intricate and specific locations of contact between them. In the United States law of torts, for example, rulings about product liability first began with objects that entered the human body (food, drink) or were directly applied to the body's surface (cosmetics, soap) before being extended to objects in less immediate relation to the body (the container for the food; the lights in a shopping market parking lot there to assure vision and visibility to the shoppers).[112] In United States criminal law, people accused of committing crimes cannot be compelled to incriminate themselves verbally, but can be compelled to incriminate themselves physically (to be physically present for identification in the courtroom, for example, or to provide a sample of blood or hair that may match fragments of these substances found near the person hurt[113]). In United States constitutional law, again to take only one specific example, cases in which the Supreme Court has invoked the "right of privacy" have tended to be on subjects directly connected to the human body (issues of conception, contraception, definition of generational relations) rather than issues of psychology, religion, or profession.[114] The specific and intricate interaction of body and state would have equivalents in any nation. It is precisely because political learning is, even in peacetime, deeply embodied that the alteration of the political configuration of a country, continent, or hemisphere so often appears to require the alteration of human bodies through war.

But the fact that the human body is political in peace as well as in war (and here we arrive at a second, more important point) does not mean that the body-state relation is in the two conditions continuous. The nature of that relation in ordinary life, far from normalizing what occurs in war, makes compellingly visible by contrast the exceptional nature of going to war. What is first of all visible is the extremity with which or the extreme literalness with which the nation inscribes itself in the body; or (to phrase it in a way that acknowledges the extraordinary fact of *the consent* of the participating populations in conventional war) the literalness with which the human body opens itself and allows "the nation" to be registered there in the wound. While in peacetime a person may literally absorb the political reality of the state into his body by lifting his eyebrows—by *altering* for the sake of and in unselfconscious recognition of his membership in the larger political community the reflex of a small set of muscles in his forehead—now in war he is agreeing by entering a certain terrain and participating in certain acts to the tearing out of his forehead, eyebrows, and eyes. That is, though in peace and in war his existence as a political being entails (not simply disembodied beliefs, thoughts, ideas but also) actual physical self-alteration, the form and degree of alteration are incomparable, as is the temporal duration of the alteration. The nation may ordinarily be registered in his limbs in a particular kind of handshake or salutation performed for a few seconds each day, or absorbed into his legs and back in a regional dance performed several days each year; but the same arms and legs lent out to the state for seconds or minutes and then reclaimed may in war be permanently loaned in injured and amputated limbs. That the adult human being cannot ordinarily without his consent be physically "altered" by the verbal imposition of any new political philosophy makes all the more remarkable, genuinely awesome, the fact that he sometimes agrees to go to war, agrees to permit this radical self-alteration to his body. Even in the midst of the collective savagery and stupidity of war, the idiom of "heroism," "sacrifice," "dedication," "devotion," and "bravery" conventionally invoked to describe the soldier's individual act of consent over his own body is neither inappropriate nor false.

The answer to the question about the duration of the outcome of war *in part* resides in the duration of the contest activity. What is remembered in the body is well remembered; the bodies of massive numbers of participants are deeply altered; those new alterations are carried forward into peace. So, for example, the history of the United States participation in numerous twentieth-century wars may be quietly displayed across the surviving generations of any American family—a grandfather whose distorted feet permanently memorialize the location and landing site of a piece of shrapnel in France, the feet to which there will always cling the narration of a difficult walk over fields of corn stubble; a father whose heart became an unreliable pennywhistle because of the rheumatic fever that swept through an army training camp in 1942, at once exempting him from combat and making him lethally vulnerable to the Asian flu that would kill him several decades later; a cousin whose damaged hip and permanent limp announce in each step the inflection of the word ''Vietnam,'' and along with the injuries of thousands of his peers assures that whether or not it is verbally memorialized, the record of war survives in the bodies, both alive and buried, of the people who were hurt there. If the war involves a country's total population or its terrain, the history may be more widely self-announcing. Berlin, orange and tawny city, bright, modern, architecturally ''new,'' confesses its earlier devastation in the very ''newness'' necessitated by that devastation, and in the temporal discrepancy between its front avenues where it is strikingly ''today,'' nineteen sixty, seventy, now eighty-five, and the courtyards immediately behind these buildings where time, as though held in place by the still unrepaired bullet holes, seems to have stopped at nineteen forty-five. Berlin, bombed from above and taken block by block. Or again Paris, architecturally ancient, silver-white and violet-blue, announces in the very integrity of its old streets and buildings (their stately exteriors undisturbed by war except by the occasional insertion here and there of a plaque to a fallen member of the Resistance) its survival, its capitulation; just as the very different history of World War I, in which two out of every three young Frenchmen either died or lost a limb, is still visible in all the windows of the subway cars—''LES PLACES NUMÉROTÉES SONT RÉSERVÉES PAR PRIORITÉ 1° AU MUTILÉS DE GUERRE . . . ''—an inscription that each day runs beneath the standing city as though in counterpoint to and partial explanation for that later story recorded above.

The physical signs cited here are those that survive the physical activity that produced them by ten years, forty years, sixty years. They only suggest in ghostly outline how deep, how daily, how massive is a population's experience of the residues of war in its immediate aftermath, the first two years, the first four months, when the survivors are immersed, engulfed, in those signs. This immediate period—the war has just been declared over, the designations of ''winner'' and ''loser'' have just been received, the terms of peace and the disposition of postwar issues are in the process of being negotiated and accepted and acted on—is what we are here trying to understand.

There are, to return to a point arrived at earlier in the discussion, three arenas of damage in war, three arenas of alteration: first, embodied persons; second, the material culture or self-extension of persons; third, immaterial culture, aspects of national consciousness, political belief, and self-definition. The object in war (as in any imaginable surrogate contest through which an international dispute was to be settled) is the third; for it is the national self-definitions of the disputing countries that have collided, and the dispute disappears if at least one of them agrees to retract, relinquish, or alter its own form of self-belief, its own form of self-extension. In war, the first and second forms of damage are the means for determining which of the two sides will undergo the third form of damage. Both sides will suffer the first and second kinds of damage, but only one will undergo the third, and it is the designation of "winner" and "loser" that determines which side will undergo that change in the third arena. But, in addition, once the war is declared over, the first and second arenas of damage function as an abiding record of the third, surviving long after the day on which the injuring contest ended, objectifying the fact that such a contest occurred, that there was a war, that there was a winner and a loser. It is crucial to notice here that the damage does not objectify or specify *who* was the winner and *who* was the loser but only that there *was* a war, that there was a winner and a loser. The very endurance of the record partly explains why the outcome is abided by. That is, it is not that "injuring" provides an outcome that cannot itself then be contested: not only are there on the losing side many persons who though wounded are still capable of holding a gun, but in addition, their siblings, thirteen, fourteen, or fifteen years old when the war first began, are now eighteen, nineteen, or twenty—the population of soldiers, even in the midst of devastation, renews itself. But once the end is agreed to, that the war occurred and that the cessation of its occurrence was agreed to are etched into their bodies and their material culture.

In the hypothetical, surrogate contests invoked earlier, nothing performed the parallel act of memorialization. If one were to imagine substituting for war the massive and extended form of a tennis contest, it would be necessary as well to provide a vehicle of memorialization. It would not be enough to build a miniature of a tennis court (or, if a weaving contest, a loom) in the town green of every city or village in each of the two countries, for that would only be the equivalent of war statues and placards conventionally placed in town greens and not the equivalent of the sentient memorialization in the damaged bodies and the continuously experienceable memorialization in the land. There would need to be a record in the more immediate and intimate vicinity of the body, such as a blue thread worn around the upper arm of all participants for a period of eight years. It would be a fragile visible presence, barely noticed but in its constant light pressure against the skin, unalterably there. The thread would be blue for winners and losers alike, for again physical damage in war memorializes without specifying—it is not (after a war) as though the winners are alive and the losers are dead; nor as though the winners are alive and uninjured and the losers alive but injured; nor as though the injured winners all lost a leg and the injured losers, a hand or an eye. Thus any surrogate form of objectification should also be constant across the two populations, only certifying that a process entailing the performance of a reciprocal acitivity for a nonreciprocal outcome took place. Or, rather than a thread, the memory of the resolution process could be absorbed into and objectified by a small bodily gesture incorporated into daily life: all persons in the two populations could learn to turn their instep out immediately before beginning to walk and again immediately before stopping the walk; the gesture would become habitual and would be unselfconsciously carried forward into peacetime activities, noticed only now and then but steadily accompanying them in all their renewed work of world-building.

These imagined signs, the fragile blue thread or the small arc of an outward kick, are obviously inadequate equivalents for the "signs" of war, but they may by their very inadequacy work to illuminate the nature of the wounds for which they are poor substitutes and thus one day suggest what an "adequate" equivalent would be, a substitute at once benign and compelling. They are therefore not lightly invoked here; for if it is awkward, even embarrassing, to contemplate such substitutions, how much more humanly awkward it is not to notice that the single advantage war has over other contests is that its outcome is abided by, that injuring works in part through the abiding signs it produces; or how much more awkward, humanly embarrassing, would it be, having recognized that war's outcome holds through the process of memorialization, to fail to search for or invent a set of signs that would perform the same work in its place. It is not only war's horrifying inner activity of injuring and being injured that is consented to, but its end, its being over, that is consented to; and it is the human capacity for consent, for making contracts, pledges, promises, processes that have outcomes that are then (as though physically necessary) abided by, and in particular the way in which that capacity for making promises or outcomes that will be abided by is assisted by the nature of injury that has been our implicit subject from the start, and will be the wholly explicit subject of the discussion with which this chapter now ends.[115]

The characteristic of injuring noted in passing above—that the injuries memorialize without specifying winner or loser—must be attended to more closely; for the nature of injuring in war is remarkable not only for the extremity and endurance of the relation between body and belief, but simultaneously for the complete fluidity of referential direction. This third characteristic of injuring is perhaps the most crucial in understanding the working of the resolution of war. Injuring, the contest activity, has no relation to the contested issues: if the wounded bodies of a Union and a Confederate soldier were placed side by side during the American Civil War, nothing in those wounds themselves would indicate the different political beliefs of the two sides, as in World War II there would not be anything in the three bodies of a wounded Russian soldier, a Jewish prisoner from a concentration camp, a civilian who had been on a street in

Hiroshima, to differentiate the character of the issues on the Allied and Axis sides. But neither would those injuries make visible who had won and who had lost. In the instances invoked earlier, nothing in the distorted bones of the American grandfather's feet and the American cousin's hip shows that the United States was on the winning side in World War I while on the losing side in Vietnam; again the wounds in the bodies of a generation of Frenchman or the visible damage in the streets of Berlin do not indicate whether France won or lost in 1918, whether Germany won or lost in 1945; again there is nothing in the craters of the terrain of Vietnam that signals the locus of victory. In each case, all that is indicated is that there was a "war," that there was "the reciprocal activity of injuring for a nonreciprocal outcome."

The activity of injuring in war has, then, two separable functions. It is the activity by means of which a winner and a loser are arrived at. It also, after the war, provides a record of its own activity. In the first of these two functions the injuries are referential: it very much matters to which of the two sides injuries and damage occurred, and thus the disputing nations and watching world will attend to the "kill ratios."[116] However, in the second function, the location of the injuries will cease to be dual and will collectively substantiate that the war occurred and is now over. The difference in the direction of referentiality in the two functions can be seen by looking at the different place that casualties have *during* a conflict and *after* that conflict. If in the midst of the American Civil War, one learns that a battle is occurring in which there are so far 20,000 casualties, it will very much matter to know whether the North has lost two thousand and the South eighteen thousand, or instead the South has lost two thousand and the North eighteen thousand, or instead that North and South have each lost ten thousand. Such differences will determine who is victorious in the battle, and over the various battles will eventually help to determine which of the two "out-injures"[117] the other, and thus, which of the two is "winner" and which is "loser." But once those final labels are designated and the war is over, it will cease to matter how the casualties in that battle (or in the war as a whole) were distributed, and all 20,000, whether originally suffered very unequally or almost equally, will now substantiate the winner and winning issues or (to phrase it in an almost equally accurate way) will collectively substantiate the nonexistence or disappearance of the losing issue. Late in the Civil War, a Southern family might say, "The autonomy of the South, the survival of its economic and racial ideas, is crucial to us: 94,000 have died in substantiation of that importance"; or a family in the North might say, "The unity of the country, the supremacy of the industrial North and its beliefs about human justice are crucial: the cost of those beliefs is the lives of 110,000 of our lads." But once the war is over, these verbal constructions will tend to be replaced by one in which the casualties—not now 94,000 and 110,000 but 204,000 (or 534,000 if including death from disease)—collectively substantiate, or are perceived as the cost of, a single outcome: "Racial justice and national unity were to be crucial to the United States as a nation: 534,000 died in the Civil War," or "The young America was maimed by the slavery of which it was necessary to rid itself violently: 534,000 died in the Civil War." Thus a Southern boy who may have believed himself to be risking and inflicting wounds for a feudal system of agriculture, and until the end of the war will have suffered much hardship and finally death for those beliefs, will once the war is over have died in substantiation of the disappearance of that feudal system and the racial inequality on which it depended.

The fluidity of the injured body's referential direction is here manifest in the verbal habit of invoking all casualties as a single phenomenon once the war is over. A second and kindred verbal habit is to keep the injuries of the two sides separate, but to cite those that occurred to one side *either* to make manifest the fact that that side won *or* to make manifest the fact that they lost. That is, juxtaposed to the assertion that a country won or lost is the assertion of the number of injuries that occurred, showing that injuries are perceived to be demonstrable "proof" of *either* victory or defeat: "They won: ten thousand died there!" and "They lost: ten thousand died there!" are constructions appearing with almost equal frequency, as in the two sentences: "The Russians lost the Battle of Tannenberg: thirty thousand Russians died there," and "It was the Russians who won the second great war with Germany: twenty million Russians died in World War II." These sentences do not cancel each other out; both are meaningful statements, but the special character of injury is indicated by the fact that it can be perceived as (and thus invoked as) substantiating disparate phenomena, even phenomena so disparate that they are opposites, as with "winning" and "losing." Both the citation of the collective deaths as a single reality and the citation of one side's death as certification of either its victory or its defeat arise because of the referential instability of the body that allows it to confer its reality on whatever outcome occurred. It may be in part precisely because *once the war has ended* the physical alterations no longer belong to two sides but seem to belong to neither or both simultaneously, that the erroneous idea can arise that injury was not, *during the war*, the central activity, or that it had nothing central to do with determining a winner and a loser, and thus comes in descriptions of war to be unrepresented (as in the categories of omission and redescription looked at earlier) or misrepresented (as in the metaphors of by-product, accidental entailment, receding cost, or extension of an innocent activity).

Thus the body's relation to political belief in war is not only much more extreme and enduring (and thus more "substantial," more intensely and experiencably "real") than it is in peace, but it also differs—and this attribute of injuring is as responsible as the other for the "as if" phenomenon that occurs at the end of war—by being much less culturally specific. Although, for example, the self-alteration that occurs when a person gives his eyes and forehead "for his country" is much more extreme than what occurs when he unselfconsciously raises his eyebrows "for his country," it is in its interior content much emptier of any reference to that country: it is not more British than Greek or American or Russian in a way that holding those eyebrows in a certain habitual position *is* more British than Greek, American, or Russian. Although the referential instability of the body reaches an extreme in injuring, it characterizes even the uninjured body in war. In his study of the technological changes in warfare over the course of civilization, for example, William McNeill finds the creation of the modern army in the seventeenth century "as remarkable in its way as the birth of science or any other breakthrough of that age,"[118] and lists as a major effect of drill—the rhythmic movement of marching in step with many men or of firing a gun by following a precise series of forty-two successive acts performed identically by all participants—the disappearance from the soldier's body of the signs of a particular region or country: "the psychic force of drill and new routines was such as to make a recruit's origins and previous experience largely irrelevant to his behavior as a soldier."[119] Such drills and routines, like fundamental strategic concepts,[120] tend to become international and thus to become shared by the two nations in any war.

This same emptying of the body of cultural content is even more true of injury, for there is nothing in the interior of what had been a boy's face, nothing in the open interior of what had been a torso, that makes the wound North Korean, German, Argentinian, Israeli. Though a moment before he was blown apart he himself had a national identity that was Chinese, British, American, or Russian, the exposed bones and lungs and blood do not now fall into the shape of five yellow stars on a red field, nor into the configuration of the union jack, nor the stars and stripes, nor the hammer and sickle; nor is there written there the first line of some national hymn, though he might have, up to a moment ago, been steadily singing it. Only alive did he sing: that is, only alive did he determine and control the referential direction of his body, did he determine the ideas and beliefs that would be substantiated by his own embodied person and presence.

The wound is empty of reference, though its intended referent can be inferred by the uniform over which the blood now falls, or by some other cultural insignia, a symbol and fragment of disembodied national identification. At this moment, the extremes of body (the always embodied body) and culture (the usually partially embodied and partially disembodied but now wholly disembodied forms of self-extension), the extremes of body and polis, the extremes of sentience and self-extension, are made to have something to do with one another by sheer proximity, the proximity not of body and ground (for he may be an Iranian on the terrain of Iraq, an American in North Korea) but of body and uniform, body and flag, or body and verbal assertion as when a comrade finds him in the brush, kneels beside him, and locating some fragment of identification, announces to those who stand nearby, "He is American."

But when the referential direction is determined by proximity or juxtaposition, what is proximate, what is juxtaposed, can be changed: a different symbolic counterpart or cultural fragment can be placed beside the wound whose compelling reality may now work on behalf of the different constellation of beliefs clinging to that new fragment. This shift explicitly happens when the enemy hears of his death and reads it not as a death but as a successful kill, reads it as belonging to their side, substantiating their cause. One may wish to think that the wound is specifically "French" because it resides in the body of a French boy, but it is the nature of injury that its attributes can be lifted away from the site, as though the wound in the chest were the severing of that tissue's relation with the rest of the body. If from that wound he dies, his whole body is deeply affected, radically altered; his whole body is now the wound. Does this dead boy's body "belong" to his side, the side "for which" he died, or does it "belong" to the side "for which" someone killed him, the side that "took" him? That it belongs to both or neither makes manifest the nonreferential character of the dead body that will become operative in war's aftermath, a nonreferentiality that rather than eliminating all referential activity instead gives it a frightening freedom of referential activity, one whose direction is no longer limited and controlled by the original contexts of personhood and motive, thus increasing the directions in which at the end of the war it can now move. If the wounded boy lives, the intensely real pain in his chest may for a time preoccupy all his attention and thus eclipse his beliefs; but it may, as he begins to recover, recertify those beliefs, bestow its intense reality on those convictions. So, too, if he dies, the compelling spectacle of the open body may confer its compelling character on the political beliefs or national allegiance of the comrades who find him: they may, as they look with horror at what the opponents have done to his flesh, feel with renewed vividness their commitment to their task. So, too, his death— perhaps coupled with many others in a single figure—may when contemplated by the enemy, revivify their own self-belief: Ten Brits died in the explosion— long live Ireland!; All of Dresden is burning—long live the Allied cause! The extremes of body and culture are still being placed side by side in these sentences, as originally they were placed side by side in the blood that fell across the uniform.

The fluidity of reference in the solitary wounded boy's experience or in others' experience of the solitary boy becomes most apparent in its overall occurrence, as in the earlier example of the collective casualties from the Civil War substantiating a single outcome. This would be equally true of any other war. The fifty million deaths of World War II after the war substantiate the outcomes determined by the winners rather than half of them substantiating the only once ''wished-for'' issues or ''once believed-in beliefs'' of the loser.[121] Together they confirm the abiding reality of one set of issues and the disappearance of another. So, too, in the midst of ongoing contemporary wars—such as that between Iraq and Iran or between Israel and the Arab nations—the casualties belong to two sides, and their distribution and separate locations work to determine a winner and a loser; but when in their turn these wars, too, one day end, people from both sides will be coupled together in collective substantiation of the outcome.

The attributes of injuring now partially visible should be placed within the frame of the overall analysis that will be briefly summarized here; those attributes will then be themselves elaborated in more detail. The hurt body's instability of reference may be so unpleasant to contemplate—particularly since it denies the population's ability to ensure what finally will be confirmed by its own violent and physically costly display of will—that one may be tempted to insist that the function of injuring ceases once the contest ends. But this temptation must be avoided because its claim is untrue. If it were true, if injuring only worked to provide a means for designating a winner and a loser, any other contest activity could be substituted in its place, since any one of them is equally able to provide a means for deciding a winner and a loser. To argue that it only has the first function makes the collective human failure to provide a surrogate for war inexplicable, since its work could be so easily and benignly duplicated. But unlike injuring, the substitute activity would only provide ''designations'' and would not provide compelling, ''abiding'' designations—designations that would be ''abided by.'' It would not provide an end to its own activity that was received as though it were itself incontrovertible, ''uncontestable.'' Thus the question is how precisely does injuring do this? It does not, as has been stressed, do this by eliminating from the losing side the power to contest, but instead by providing an outcome that is as real *as if*, as absolute *as if*, as compelling *as though*, the losers no longer had the power to contest, and it is the nature of the ''as if'' function that is being explored in this present section.

The overarching frame of the discussion has, then, four central parts. The first and second are that injuring has two functions: the first function is to determine which population and set of disputed beliefs will be the winner, which population and set of disputed beliefs will be the loser; the second function is to substantiate whatever outcome was produced as a result of the first function. As it becomes clear that these are two very different, quite separable functions, it is possible to back up for a brief moment to the erroneous ''power of enforcement'' argument in which the defeated no longer have injury-inflicting capacity, and see that the error here is in assigning all the work of injury to the first of the two functions and wrongly believing that what differentiates injuring from any other contest activity is *the completeness* with which it establishes the categories of winner and loser, in effect pushing ''winners'' to mean ''alive'' and losers to mean ''dead.'' Injuring instead has a wholly distinct second function which is to make the outcome produced by the first function seem absolute, *seem real* until there is time for the issues to be universally acted on and in this way *made real*. Injuries-as-signs point both backward and forward in time. On the one hand they make perpetually visible an activity that is past, and thus have a memorialization function. On the other hand they refer forward to the future to what has not yet occurred, and thus have an as-if function. This might be called their "fiction-generating" or "reality-conferring" function, for they act as a source of apparent reality for what would otherwise be a tenuous outcome, holding it firmly in place until the postwar world rebuilds that world according to the blueprint sketchily specified by the war's locus of victory. That this function entails fictitiousness does not mean that it entails fraudulence: what it substantiates is *not untrue*: it is just *not yet* true. The work of this second, reality-conferring function depends in turn on two attributes of injury, and it is these two attributes that are the third and fourth of the central conclusions here, the third and fourth elements of what is understood here to be war's essential structure. First, the visible and experienceable alteration of injury has a *compelling and vivid reality* because it resides in the human body, the original site of reality, and more specifically because of the "extremity" and "endurance" of the alteration. Second, this reality can be conferred on either set of disputed issues (or as sometimes happens, a mixture of issues from the two sides) because of the nonspecificity of reference, or the *referential instability* of the hurt body. Each of these two will be briefly elaborated below, so that the shape of the overall activity of war, as well as the shape of any potential substitute, will gradually become more clear.

Referential Instability

The nonreferentiality of the hurt body, or the referential instability of the hurt body (for the two are a single phenomenon) can be understood by returning to the soldier's own universal declaration that he is going "to die for his country" or "to kill for his country." By looking at the three phrases—"to die," "to kill," and "for my country"—it will become clear that there is a deterioration of national reference in the first two that is then reclaimed by yoking each to the third phrase. The referential direction that is destroyed in "dying" and "killing" is then rescued or re-established by appending a direction in the phrase "for the country." If at the end of war the person is on the losing side, his acts will not have substantiated the country specified in his original declaration of motive, and how this happens can best be understood by looking at that declaration of motive itself.

To Kill. It has often been observed that war is exceptional in human experience for sanctioning the act of killing, the act that all nations regard in peacetime as "criminal." This accurate observation acknowledges that the act of killing, motivated by care "for the nation," is a deconstruction of the state as it ordinarily manifests itself in the body. That is, in consenting to kill, he consents to perform (for the country) the act that would in peacetime expose his unpoliticalness and place him outside the moral space of the nation. What in killing he does is to wrench around his most fundamental sanctions about how within civilization (and this particular civilization, his country) another embodied person can be touched; he divests himself of civilization, decivilizes himself, reverses not just an "idea" or "belief" but a learned and deeply embodied set of physical impulses and gestures regarding his relation to any other person's body. He undoes the learning in his body as radically as he would if he were suddenly required to abandon the upright posture and move on four limbs as in his pre-civilized infancy. He consents to "unmake" himself, deconstruct himself, empty himself of civil content "for his country." That this is actually done "for his country" is not being questioned here, and certainly it is not being made light of. All that is being said is that *that it is for his country* (for civilization, for his country's version of civilization) is not visibly *interior* to the act itself, is not interior to the embodied gestures he performs. When during peacetime he touches gently his neighbor, or keeps a five-inch space between himself and an acquaintance encountered on the street, one can "see" civilization inside the gestures and postures themselves, see it literally residing within him, as will be especially apparent if one then observes his restraint when he comes upon someone he deeply dislikes and avoids him rather than shattering him. He need not look up from any of these three acts and specify that this is for civilization, for to do so would be superfluous. Because his act of killing does not itself contain civilization in its interior, the fact that it is being done for a particular civilization, the referent for his act, is re-established and carried by the appended assertion (either verbalized or materialized as in the uniform), "for my country."

To Die. The "unmaking" of the human being, the emptying of the nation from his body, is equally characteristic of dying or being wounded, for the in part naturally "given" and in part "made" body is deconstructed. When the Irishman's chest is shattered, when the Armenian boy is shot through the legs and groin, when a Russian woman dies in a burning village, when an American medic is blown apart on the field, their wounds are not Irish, Armenian, Russian, or American precisely because it is the unmaking of an Irishman, the unmaking of an Armenian boy, the unmaking of a Russian woman, the unmaking of an American soldier that has just occurred, as well as in each case the unmaking of the civilization as it resides in each of those bodies. The arms that had learned to gesture in a particular way are unmade; the hands that held within them not just blood and bone but the movements that made possible the playing of the piano are unmade; the fingers and palms that knew in intricate detail the weight and feel of a particular tool are unmade; the feet that had within them "by heart" (that is, as a matter of deep bodily habit) the knowledge of how to pedal a bicycle are unmade; the head and arms and back and legs that contained within them an elaborate sequence of steps in a certain dance are unmade; all are deconstructed along with the tissue itself, the sentient source and site of all learning.

Thus the roll call of death should always be taken as it was first taken by Homer in the record of war that stands at the beginning of western civilization. Here each death, whether Trojan or Greek, comes before one's eyes in four aspects: the name of the person; the weapon ("freighted with dark pains") as it approaches the body; the site of entry and the slow motion progress of the widening wound (for we are to understand that it is the deconstruction of sentient tissue that is taking place, and that this deconstruction always occurs along a specific path); and fourth and finally, one attribute of civilization as it is embodied in that person, or in that person's parent or comrade, for the capacity for parenting and camaraderie are themselves essential attributes of civilization.[122] Each attribute is invoked into the center of the wound, for each is implicated there and itself unmade: so the spear that cuts through the sinew of Pedaeus's head, passing through his teeth and severing his tongue, passes also through the work of goodly Theano who "reared him carefully even as her own children"; the bronze point that enters Phereclus through the right buttock, pierces bladder and bone, and pierces as well the shipbuilding and craftsmanship bodied forth in this son of Tecton, Harmon's son; in the lethal fall of Axylus from his car is the fall of the well-built Arisbe, a home by the high road where entertainment was given to all; the huge jagged rock that cuts and crushes through the great-souled head of Epicles cuts its way too through his gradually shattered camaraderie with Sarpedon.[123] So, too, the twentieth-century litany of war deaths occurs in the same way: for the United States, the Vietnam War is not 57,000 names but names, bodies, and embodied culture—not Robert Gilray but Robert Gilray, from the left the artillery shell approached, entered his body and began its dark explosion, exploding there, too, the image of the standing crowd that each week watched his swift run across the playing fields of Chatham; not Manuel Font but Manuel Font, around his fragile frame the fire closed in, burning into his skin, and skull and brain, burning even into the deep, shy corners where he studied at school. So the list would continue through tens of thousands of others. That the war deaths occurred on behalf of a terrain in which pianos could be played and bicycles could be pedalled, where schools would each day be entered by restrained and extravagantly gesturing children alike, must be indicated by appending the direction of motive, "for my country," since the deaths themselves are the unmaking of the embodied terrain of pianos and bicycles, classmates, comrades, and schools.

For My Country. Thus "to kill and to die"—or in the idiom that embraces both simultaneously, "to hurt" (to hurt within one's own body; to hurt an opponent's body) or "to alter body tissue"—are alike in having no interior referent and, if they are to have one, requiring a separate specification. But precisely because there is nothing "interior" that itself stipulates and in doing so *limits* its referent, the act of "dying" or "killing" can be lifted away and coupled with a different referent. What was "for his country" may at the end be "for the opponent's country" or "for the postwar world coinhabited by two countries and having certain characteristics (for example, boundaries) determined by one of the two more than by the other." That civilization is itself unmade is perhaps in nothing so evident as in the fact that the participants' national motives can be themselves deconstructed and replaced by outcomes deeply antagonistic to those motives. At the same time, however, it is clear that what the bodies of those slain in war ultimately substantiates will not (even if they were on the losing side) be contradictory to their original motives if by the motivational description, "to kill and to die for my country," what was meant was "to kill and to die in substantiation of my country's commitment to a process of resolving a dispute that requires of both sides the performance of a reciprocal activity and requires of both sides the acceptance of a nonreciprocal outcome." Thus if in the end the open bodies are juxtaposed to a country's right to a certain territory (whether that "right" is claimed to arise from a historical precedent, from an international rule, from the culture's idea of god), they will lend force and conviction to that; if they are juxtaposed to the idea of the international privilege of a colonizing power, they will substantiate that; if instead they are juxtaposed to the right of self-determination, they will substantiate that; and so forth. What is in each case substantiated is the structure of war itself that from the day on which it began required of both sides the performance of a single reciprocal activity and from the day on which it began was understood to require of both sides the acceptance of a nonreciprocal outcome.

Reality Conferring

Thus injuring in its second function is relied on as a form of legitimization because, though it lacks interior connections to the issues, wounding is able to open up a source of reality that can give the issue force and holding power. That is, the outcome of war has its substantiation not in an absolute inability of the defeated to contest the outcome but in a process of perception that allows extreme attributes of the body to be translated into another language, to be broken away from the body and relocated elsewhere at the very moment that the body itself is disowned, made to disappear along one of the six paths described earlier. The force of the material world is separated from the fifty-seven thousand or fifty million hurt bodies and conferred not only on issues and ideologies that have as a result of the first function been designated the winner, but also on the idea of winning itself.[124]

The intricacies of the process of transfer that make it possible for the *incon-*

testable reality of the physical body to now become an attribute of an issue that at that moment has no independent reality of its own have already been looked at in their most radical and deconstructed form in torture, and will in later chapters be looked at in this very different and benign occurrence within civilization. What gradually becomes visible (by inversion in its deconstructed form in torture and war and straightforwardly in its undeconstructed civilized form) is the process by which a made world of culture acquires the characteristics of "reality," the process of perception that allows invented ideas, beliefs, and made objects to be accepted and entered into as though they had the same ontological status as the naturally given world. Once the made world is in place, it will have acquired the legitimate forms of "substantiation" that are familiar to us.[125] That an invented thing is "real" will be ascertainable by the immediately apprehensible material fact of itself: the city (not the invisible city asserted to exist on the other side of the next sand dune, but one within the sensory horizon) has a materialized existence that is confirmable by vision, touch, hearing, smell; its reality is accessible to all the senses; its existence is thus confirmed within the bodies of the observers themselves. A made thing may instead not take a materialized and hence physically experienceable form (the idea of justice, a theory of gravity, a description of electricity); still that it is "real" or "true" will be certified by materialized or experienceable instances of it, supporting facts (the theory of gravity cannot itself be seen but the falling apple can). But in the very earliest moments of invention (or when as in war there is a crisis of belief that suddenly carries a population back to the early stages because the world must be quickly reinvented) a disembodied idea that has no basis in the material world (*either* because its material form has not yet been discovered and thus it is not-yet true *or* because it can never have a material form since it is fundamentally untrue) can borrow the appearance of reality from the realm that from the very start has compelling reality to the human mind, the physical body itself.

That is, instead of the familiar process of substantiation in which the observer certifies the existence of the thing by experiencing the thing in his own body (seeing it, touching it), the observer instead sees and touches the hurt body of another person (or animal) juxtaposed to the disembodied idea, and having sensorially experienced the reality of the first, believes he or she has experienced the reality of the second. So, for example, a city, though invented, is real; a blueprint of a city is still experienceable through the senses but much less so than the built city, and thus is judged less real, is more immediately recognized as invented; but, finally, the city might exist only as a verbal assertion that on this very ground a city will next year exist. The prophet may, as he speaks these words, cut open a body and read in its entrails the exact date on which the city will appear: the coming of the city may be believed, received as compelling truth, because the open body has lent it its truth. So, too, an idea of god, or an explanation of lightning, or the asserted power of a ruler over the winds, could be juxtaposed to a body part that "demonstrates" or "substantiates" the truth of the assertion by having itself indisputable "substance" that somehow is read as belonging to its counterpart. It is as though the human mind, confronted by the open body itself (whether human or animal) does not have the option of failing to perceive its reality that rushes unstoppably across his eyes and into his mind, yet the mind so flees from what it sees that it will with almost equal speed perform the countermovement of assigning that attribute to something else, especially if there is something else at hand made ready to receive the rejected attribute, ready to act as its referent. Why the body should, at a certain moment in the process of creation, have this frightening power of substantiation may for a time be unfathomable; but that it has this power can be demonstrated in many different contexts, some of which are ancient and some contemporary.

Although for the time being it is only this phenomenon as it occurs in war that is being pointed to, it will be helpful to recall here the difference between this form of bodily translation and that form of bodily translation that is its antithesis and occurs benignly in civilization.[126] The two sometimes occur in the same cultural moment, and the difference can be seen there. For example, after sacrifice was no longer central to the Brahmin religion in India, it occasionally re-occurred because it was needed in order to make impregnable important points within the city: the body of the one slain in sacrifice was buried in the foundation of the crucial gate, bastion, or dam.[127] Though this act of sacrifice involves a translation of the material fact of the body into a disembodied cultural fiction (the assertion of impregnability), so does the building of the city itself: basic attributes of the human body are projected in and protected by the finished arrangement of wall and gate; further, the construction is literally brought into being by the body's outward translation of itself into artifact through labor. But the translation that occurs within the sacrifice differs from the other in three ways. First, it requires that someone be hurt. Second, there is a twisting of terms. The attribute of the body before the translation is the opposite of what the attribute is called after the translation: thus pain is relied on to project power, mortality to project immortality, vulnerability to project impregnability. Only insofar as the dead body and the impregnable gate both enjoy a form of non-sentience that places them beyond the reach of further hurt do they possess parallel attributes of invincibility.[128] Third, the translation is a shorthand version of what occurs in civilization. The extreme fact of the body—not some of its attributes (its rigidity and defensiveness in the case of the wall) or some of its power (such as the focusing of energy that occurs in work) but the body itself, the corpse—is laid edge to edge with an extreme of sublimation, not a partially materialized and thus self-substantiating construction but a wholly verbal and disembodied assertion of impregnability.

This third element, the laying edge to edge of the extremes of the material and the immaterial is, when separated from the first two, not necessarily negative.

It recurs in many forms in primitive thought, as though the unmediated juxtaposition of body and voice is a rehearsal for or (as in the Indian city above) a
commemoration of the capacity for self-transformation out of the privacy of the
body into richly mediated forms of culture. It frequently appears, for example,
in early forms of oaths, a form of implicit contract that is relevant to an understanding of war because war, like a pledge, entails the capacity to abide by a
certain not-yet substantiated outcome. As he gives his oath about the selection
of a wife for Isaac, the servant of Abraham places his hand under Abraham's
thigh.[129] An unsubstantiated statement (unsubstantiated because its realization
belongs to the future) is given substantiation by being placed immediately beside
the material reality of the body. The place touched by the servant is so intimate
that it is almost interior to the body, and it is in oaths often the interior of the
body that is exposed, usually through some form of wounding, in attempts to
bestow the force of the material world on the immaterial.[130] The Arab accompanied his statement with the act of dipping his hand in the blood of a camel;
the Homeric oath was taken standing on a slaughtered steed; as the Roman
uttered his words, a hog was slain; two Nagas taking a mutual oath held between
them a dog that was chopped in two.[131] Sometimes the swearer's own body is
used in confirmation, as when the Zuni Indian put an arrow down his throat to
show that the words emanating from his mouth had their source in the realm of
material substance;[132] or the swearer's own body might be wounded, opened,
as when the Sema Naga taking an oath bit off his own forefinger.[133]

What is striking about such unmediated juxtapositions, and relevant to the
way in which at the end of war opened bodies and verbal issues are placed side
by side, is that in most instances the verbal assertion has no source of substantiation other than the body. The truth claimed in the verbal statement is remote
either because, as in the oaths above, it belongs to the future or to the unwitnessed
past; or because, as in the instance of the augur who locates his prophecy in the
disposition of entrails, it originates in a religious realm inaccessible to the senses;
or because, as in the rites of blood brothers who by opening the body and mixing
their blood acquire the relationship normally acquired through the interior mechanism of biology alone, the rite itself is openly acknowledged to create rather
than to confirm the truth of the verbal assertion it accompanies (''Now we are
brothers.''). The body tends to be brought forward in its most extreme and
absolute form only on behalf of a cultural artifact or symbolic fragment or made
thing (a sentence) that is without any other basis in material reality: that is, it
is only brought forward when there is a crisis of substantiation. As a result of
this unanchored quality, the disembodied cultural fragment has a fluidity not
shared by its physical counterpart—in war the damage inflicted on bodies is
unalterable, whereas the symbolic claims or issues change with great ease. There
may, in any given war, be specific reasons why the issue is unanchored. Like
an oath, its confirmation may belong to the future: the regime, once in power,

will acquire the support of the people; the war, once over, will have made the world safe for democracy. Or, like augury, its claimed substantiation may originate in a metaphysical or historically inaccessible sphere: divine sanction for a claim to the disputed land may reside in an unconsultable source, or may depend on the mystical inaccessibility of historical dialectic. Or, like the pacts of blood brothers (in which the ritual of interior blood-mixing performed on one day is believed to have created a shared sameness of blood for all time thereafter), it may describe itself as a universalization of its own activity: this war is all wars and the end of this particular war is the end of all wars. Or, finally, it may be unanchored because—as has been stressed in many disparate accounts of nineteenth- and twentieth-century wars—it is simply untrue: the constructs, claims, and issues may be lies. It is, in fact, the unanchoredness of even the most legitimate issues in war that makes them only with difficulty distinguishable from and therefore so easily substituted by lies.

Whatever the specific causes of this unanchoredness, there is always one general cause. *The dispute that leads to the war involves a process by which each side calls into question the legitimacy and thereby erodes the reality of the other country's issues, beliefs, ideas, self-conception. Dispute leads relentlessly to war not only because war is an extension and intensification of dispute but because it is a correction and reversal of it. That is, the injuring* not only provides a means of choosing between disputants but also *provides, by its massive opening of human bodies, a way of reconnecting the derealized and disembodied beliefs with the force and power of the material world.*

In the dispute that leads to war, a belief on each side that has "cultural reality" for that side's population is exposed as a "cultural fiction": that is, by being continually called into question, it begins to become recognizable to its own population as an "invented structure" rather than existing as it did in peacetime as one that (though on reflection invented) could be unselfconsciously entered into as though it were a naturally occurring "given" of the world. As the dispute intensifies and endures, the exposed "cultural fiction" may seem in danger of eroding further into a "cultural fraud," in danger of eroding from something that is uncomfortably recognizable as "made" into something potentially identifiable as "unreal," "untrue," "illegitimate," "arbitrary." The more the process of derealization continues, the more desperately will each side work to recertify and verbally reaffirm the legitimacy and reality of its own cultural constructs. Although at a distance human beings take pride in being the single species that relentlessly recreates the world, generates fictions, and builds culture, to arrive at the recognition that one has been unselfconsciously dwelling in the midst of one's own creation by witnessing the derealization of the made thing is a terrifying and self-repudiating process.

The outlines of this process are visible in any historic account of the conflict preceding a given war. It is not that a country's population actively wishes to discredit the other population's forms of belief and self-description but, rather, that their own beliefs and descriptions contradict the other population's; and thus by merely continuing to believe in and reaffirm their own constructs, they inevitably contribute to the deconstruction of the competing construct. Whatever the particular descriptions that have collided, what has always collided is each population's right to generate its own forms of self-description. Prior to World War I, for example, Germany may believe that the series of treaties between Britain, Belgium, France, and Russia are the "encirclement" of Germany, while these other countries may believe that it is the "encirclement" of France by Germany that is prevented by the treaties; France may perceive Alsace-Lorraine as a deep and abiding part of her national integrity temporarily separated from her at Versailles in 1871, while Germany may see France's yearning toward Alsace-Lorraine as territorial lust for land that has long and rightfully been part of Germany, and as a dangerous extension of French presence toward the German heartland.[134] Thus as each country reasserts its description, it denies the authenticity of its neighbor's description; and for either to revise its conception of alliance or territory is to relinquish its own autonomy of self-description, self-belief. Although all men create, what they create differs from country to country, and when the competing creations collide, the revelation that these were "only" creations is intolerable. No matter how deeply interior to one country an event seems to be, it will also be a very differently understood event in its neighbor's self-construction: when in some future generation, East and West Germany reunite, no one can doubt that the impulse to do so will have arisen out of the depths of Germany's own interior national consciousness; but neither can one doubt that that event, though occurring safely within her own reconstituted boundaries, will also cut very deeply into the interior national self-description of France and into that of Russia.

Thus in a dispute, each side reasserts that its own constructs are "real" and that only the other side's constructs are "creations" (and by extension, "fictions," "lies"). In certifying the "reality" of its own descriptions, each will bring forward and place before its opponent's eyes and, more important, the eyes of its own population, all available sources of substantiation. For example, in the 1979–80 dispute between Iran and the United States, each deeply believed not simply that its own descriptions were right and the opponent's were wrong, but that its own descriptions were "real" and the opponent's, "unreal." Did not—the United States protested in astonishment—Iran see that the taking of hostages was self-evidently wrong: that the guarantee of an embassy's immunity has international *reality*? Did not—Iran protested in astonishment—the United States see that providing sanctuary to a man who had inflicted hurt on the Iranian people and who would do so again was self-evidently wrong, that the damage done to the Iranian people was *real* and would again be *real*? The answer to the United States' question was no. The answer to Iran's question was no. Thus

there took place for a time a reality duel, each side attempting to bring forward
more and more signs of the density, weight, scale—simply, the substantiveness—
of its own beliefs.

The counterpoint of authentification becomes self-amplifying: the United States
cites international law, focusing attention on the fact that these rules have been
consented to (brought into existence and invested with reality) not only by its
own population, but by the population of many countries; Iran, in turn, places
before international cameras the image of thousands of Iranians certifying the
guilt of the Shah; now the United States asks other countries to cite the international rules, making visible the fact that these laws were not simply once
agreed to but are (despite Iran's action) at this very moment still in effect; now
in turn Iran places before the same cameras not simply Iranians but American-
residents-in-Iran and other foreign (i.e., international) residents who also bear
witness to the wrongs of the ousted Shah. History is invoked to substantiate the
descriptions of both sides: the United States recalls to its own population the
meaning of sanctuary during revolutions; Iran recalls to its population the United
States' backing of the Shah in 1951 to demonstrate the legitimacy of its fear
about his reimposition in the present. Neither side recites to its own people
historical precedents that would provide a parallel for the opponent's self-description—the Boxer Rebellion does not often enter into the conversations of
Americans; historical (and prophetic) evidence that religious mullahs are as
capable of inflicting pain and terror in people as western shahs does not enter
into the conversation of Iranians.

Above all, the incontestable reality of the body is continuously reinvoked by
both sides.[135] This man is dying of cancer; here are the doctors' reports; do the
Iranians not believe in the reality of the body? Thousands of people were tortured
by the Shah; here are first-hand accounts and the internationally respected accounts of Amnesty International; do not the Americans believe in the reality of
the body? The numbers support Iran's side: you are talking about one man's
hurt; we are talking about the hurt of thousands. No, the numbers support the
side of the United States: you are talking about rules that hold sway within your
own boundaries; we are talking about rules that hold sway in all countries in the
international community. A temporal perspective favors the beliefs of the Americans: you are talking about hurts inflicted in the past; we are only attending to
a sickness that is happening right now. No, a temporal perspective favors the
Iranians: here are the enduringly damaged bodies of those hurt under his regime
who will be hurt again if he, or a western surrogate, is returned. But as each
side works to authenticate its own description, it also works to devalue the
opponent's description. If the history of the Shah's reign in Iran and the United
States' responsibility in that history has no place in the United States' self-
description, then the international laws that protect the United States' power of
self-description must be derealized: the hostages are taken in Iran. In turn, if

the international laws have no reality (the quality of compelling authority) in Iran, then Iran must be derealized: the freezing of Iranian assets and the placing of visiting Iranian students in limbo is a way of saying Iranian money and citizens henceforth have a provisional unreality within U.S. borders. Central events are appropriated into the self-understanding of each side: above all, the hostages "belong" to each side. For the United States, they are American citizens wronged in Iran. For the Iranians, they are surrogate Iranians: that is, they are Americans-in-Iran whose suffering the United States will attend to and who will, by an act of proxy, draw attention to the otherwise unattended to sufferings of Iranians hurt under the Shah. Thus each day in the United States the church bells ring at noon to commemorate the absent American hostages; and Ayatollah Khomeini presents a full page Christmas message in the *New York Times* noting that the sound of those midday church bells commemorates all those Iranians who were hurt under the previous regime. The contest to out-describe continues.

This dispute did not lead to war because each side successfully certified its own constructs while gradually also crediting those of its opponent.[136] But in the early days of conflict, many Americans confessed their readiness, even eagerness, to go to war. What is the structure of the feeling of being eager to go to war? To invoke psychological and motivational words like "aggression" or "pride" does not answer the question, but only introduces an alternative vocabulary that reoccasions the same question. The timing and context of such feelings here and in other international disputes suggest that when the system of national self-belief is without any compelling source of substantiation other than the material fact of, and intensity of feeling in, the bodies of the believers (patriots) themselves[137] then war feelings are occasioned. That is, *it is when a country has become to its population a fiction that wars begin*, however intensely beloved by its people that fiction is.

In understanding this, it is crucial to notice the difference in the ontological status of the word "country" in the phrases "longing for one's country" and "killing for one's country," in order to hold visible the fact that the object in the first phrase has an objective reality independent of the intentional state that it does not have in the second phrase. If one is away from one's homeland and misses Ireland or longs for Israel, Ireland and Israel have an independent existence apart from the longing (however much they may also be altered—enhanced or diminished—by the citizen's own image of them). But they do not independently exist when one dies for Ireland or kills for Israel: if the Ireland "for which one dies" existed, an Ireland independent of British rule, one would not be dying; if the Israel "for which one kills" existed, an Israel secure within boundaries acceptable to her population and the population of her neighbors, one would not be killing. If the Britain for which one dies existed, a Britain whose territorial self-definition included the Falkland Islands, one would not be dying. If the Germany for which one kills, a Germany with an homogenous Aryan population, a Germany internationally recognized as the dominant state in Europe, a Germany whose sense of humiliation acquired in an earlier war had been healed, one would not be killing. If the democracy for which one dies existed in a world safe for democracy, one would not be dying to make the world safe for democracy. If the country for which one kills existed in a world in which there was an end of wars, one would not be killing to make a war that ends all wars. It is to the degree that the object does not exist, or is perceived to be in danger of ceasing to exist, that the bodily alteration (asserted in dying and killing to be on behalf of that object) is occasioned and necessitated; the object is understood to be being brought into existence by the act. The Russia that one ''yearns for'' exists; the Zimbabwe that one ''pays taxes for'' exists and must be kept in place; the United States that in 1861 one ''dies for,'' a geographically and morally unified nation, does not exist; the Northern boy kills because it does not exist; he brings it into being by his acts which are productive of the thing; just as the country for which the Southern boy dies, one securely separate and autonomous from the North, does not exist and he brings it into being by his death. In this and every war, the injuring contest is a contest to see which one of the two not-yet existing countries will be produced as an outcome.

Thus the process of de-realization in the dispute that leads to war is both *continued* in war and *reversed* by it. The relentless work of de-realizing the competing cultural constructs is on the one hand intensified as each population actively struggles to eliminate the other's beliefs that have come into conflict with its own: it is not by eliminating, de-realizing, or annihilating the opponent that one will have salvaged one's own national self-definition, but by oneself winning the contest and earning by that designation of winner the right to enact one's own constructs. Thus the first function of injury (the establishing of the categories of winner and loser, the one that can be replaced with any contest activity) *continues* the de-realization process of dispute. But injuring (in its second function) is a *reversal* of dispute because injuries provide the radical material base for the winning issues, investing them with the bodily attribute of reality until there is time for both of the populations to consent to them, enact them, make them real.

In the first moments and days and weeks when the war has just been declared over, the competing fictions (*both* the one on the losing side that is now officially unreal and the one on the winning side that is to be accepted as henceforth real) have even less reality than they did before the war started; and the sudden abrupt assertion, ''Now one of these constellations of beliefs will be accepted as true'' would be a vacant absurdity (as was earlier seen when this assertion was imagined as occurring at the end of one of the benign contests) if there were not something giving it a compelling quality, not only memorializing that a process of resolution indisputably took place, but also lending those attributes of ''compelling reality'' and ''indisputable facticity'' to the outcome as well.[138] Thus the relentless and extraordinary physical alterations occurring in war itself are framed on either side, at its opening and at its close, by unanchored issues, issues deprived of substantiation: this framing unreality of the exterior constructs appropriates the reality of the interior content of war; the relation of inside to outside is the relation that, for example, exists in a very abbreviated form in oaths between the juxtaposed extremes of the opened body and the verbal assertion.

Verbal Unanchoredness Inside War

Although it is the unanchoredness of the *exterior* framing issues that is of crucial importance here, it should also be recalled that all forms of language *within* the interior of war tend to have this same unanchored quality. The utter derealization of verbal meaning occurring there, the presence of fictions or, more drastically, "lies" has often been commented on. Strategy, to begin with just one major form of language interior to war, does not simply entail lies but is essentially and centrally a verbal act of lying—the goal of every strategic design is to actively withhold meaning from the opponent, as is persuasively summarized in Stonewall Jackson's strategic motto, "Mystify. Mislead. Surprise." The enemy must believe you are telling the truth when you are lying and, equally important, must believe you are lying when you are telling the truth.[139] Strategy, or military language, is a large phenomenon itself made up of many smaller parts, many of which have rubrics that actively announce their purpose of withholding meaning. *Codes*, for example, are attempts to make meaning irrecoverable, or, at the very least, to embed that meaning in multiple tiers of arbitrarily sequenced signs in order to divert the opponent's energies into hours of incomprehension. In *camouflage*, the principle of lying is carried forward into the materialized self-expression of clothing, shelter, and other structures: you wear what you wear because the enemy must think you are not there when you are there; or must believe to have seen you toward the east when you are coming from the west. Camouflage also works in language: tanks, as Liddell Hart points out, were originally christened "tanks" to verbally camouflage them as "cisterns" until their first use in World War I (What are all those things? Tanks.).[140] The rubric, *bluffing*, also makes visible the centrality of verbal fictions. In surprise or unanticipated injuring, the opponent must perceive your immediate power over them as much less than or much more distant than it actually is; conversely, in anticipatory injuring (when one side brings about surrender without the reciprocal injuring of battle) the opponent is often made to believe that your capacity to injure is much greater than or more proximate than it actually is. The crucial place of "cunning" and "deceit" in military strategy is acknowledged throughout Clausewitz's *On War*,[141] and Napoleon specified as the single most important act in achieving victory, the severing of the opponent's lines of communication.

Thus, within war itself, the indisputably physical reality of the mounting wounds has as its verbal counterpart the mounting unreality of language. The two—authentic physical content and unauthentic verbal content—are such inevitable counterparts of one another that they have often been understood as near synonyms. Machiavelli continually coupled "force and fraud" as though they were nearly indistinguishable;[142] and Schopenhauer, too, saw "violence and lies" as inseparable.[143] Thus the fact that the one is brought about by the other, that injuring requires strategy, that wounding requires lying, is not surprising. But the dissolution of language is also present within other areas of language: first, verbal alliances; second, the verbal reports or history of the events within war; third, the everyday conversations of the participants. The immediate circumstances surrounding the outbreak of war so routinely include secret treaties—verbal alliances between the governments of two countries unknown to the populations of those countries, or unknown to the government and population of a third country—that most peace plans have had to include their prohibition as a major provision. Kant's treatise on international government, *Perpetual Peace*, opens with Article I, the stipulation that no secret reservations could be included in treaties; Bentham's plan for peace prohibited secret diplomacy; Saint Pierre's banned treaties formed without the consent of the union; the Fourteen Points preceding the League of Nations also called for the prohibition of secret treaties.[144] Similar provisions exist within the covenants drawn up by the Fabian Society, the Victory Program of the League to Ensure Peace, the Union of Democratic Control in England, and the Dutch Anti-War Council.[145] So, too, the architects of covenants such as the Nuclear Test Ban Treaty have had to so assume the possibility of the deconstruction of the covenant through lying that a provision has had to be designed to preclude that possibility, a "confirmation" provision stipulating ways of certifying that a country's claims about its weapons are true.[146] The recurring presence of prohibitions against "secrets," "withheld meaning," "fictions," "lies" in all these covenants attempting to maximize the possibility of peace suggests how central to war is the licensing of these phenomena.

As fictiveness becomes a major attribute of language which precedes physical injury (the instrumental language of strategy and alliance that brings the wounding into being), so too it becomes a major attribute of language which follows the injury, the language reporting the history of the scenes that took place that day. The habitual discrepancy between the sheer facticity of dead bodies and the fictiveness of "body counts" is familiar from every recent war. So, too, descriptive accounts of the outcome and significance of a particular engagement are often submerged in obscurity: though Stalingrad was early recognized as a turning point in World War II, the deliberate falsification on both sides had been so deep that any coherent description was for a time irrecoverable;[147] so, too, it was for a while almost an act of treason for anyone in Japan's navy to discuss the outcome of Midway;[148] the surrender of the Italian fleet, whose location and timing were determined by the Italian naval authorities, was mythologized on the Allied side as having been directly authored by the Allied radio broadcasts;[149] the fate of British troops at Dunkirk was until the 1970s described as though it were perfectly appropriate and conventional to understand the act of moving away from the opponent as military heroism;[150] Eisenhower's house-by-house movement through the Ruhr Valley at the end of World War II, repeatedly cited in the United States as brilliant strategy, was perceived as an incomprehensible delay by an alarmed British military command as well as by a pleased German military command.[151] Thus in any battle, campaign, or war, more men may be lost than are recorded as lost; a near draw or even a retreat may be remembered as a victory; a "defeat" may be reinterpreted as a "diversionary tactic";[152] and "unconditional surrender" may be renamed "honorable capitulation."[153] As the nature of any specific engagement may be misrepresented, so there may come to exist general structures of explanation or historical interpretation that are fictitious: the "great man" paradigm of military history pervasive during the Napoleonic Wars of the nineteenth century and not wholly absent in the twentieth century has, for example, been described at length as a systematic fiction in Tolstoy's essays on war.[154]

The three categories of language thus far—strategy, alliance, history—may wrongly suggest that only official language becomes unanchored; but the dissolution radiates in all directions and enters the everyday life of civilians at home and soldiers on the front. Freud, in his "Reflections Upon War and Death," attends to the massive depression of the population at home, and attributes it in part to the sudden sanctioning of lies.[155] So, too, the ordinary soldier—as in the portrait provided in Marc Bloch's *Memoirs of War*—dwells every day in the midst of determinate wounds and indeterminate meaning. It is not just every fragment of language but every sound, every noise that is at once resonant with meaning and wholly indeterminate in meaning: the innocent "tap-tap of the raindrops on the foliage" may instead be "the rhythm of distant footsteps"; the "metallic scraping sound of very dry leaves falling on the leaf-strewn forest floor" is repeatedly mistaken for "the click of an automatic loader introduced into a German rifle breech."[156] As the soldier surveys the incontestable bullet-holes in his canteen, a colonel rides by and announces a French victory; the soldier swoons with elation; only later does he realize that the words he has heard do not have the same substance as the pain in his arm, that he did not at that moment really know the distance or direction of the German soldiers whose presumed absence so elated him.[157] Language becomes increasingly severed from material substance and, as Bloch argues in "Les Fausses Nouvelles de la Guerre," tall tales, false reports, rumors, and legends begin to fill the soldier's speech, especially as small groups of men on the move pass and speak for a few minutes with other small groups of men on the move.[158] Like the language of strategy, alliance, and history, the language of daily conversation is not necessarily false, for that would mean it had more clarity than it has: with each sentence one hears, the possibilities are completely open; each verbal utterance has at all times the explosive duality of being at once very possibly true and very possibly false— and of course there will be many moments in which the accuracy of the guess as to its truth or falsity will determine whether one lives or dies.

If the fluidity of language interior to war were described not in terms of the speaker—as it implicitly is in the categories used here of strategy, alliance, history, and daily conversation—but instead in terms of content or subject matter, the list of contents would be very long. What would have come to be unrepresented or misrepresented would include the motives of individuals, of governments, of armies, the direction and location of thousands of men, the location and scale of one's injuring power (submarines drop beneath the visible surface of the water, tanks mime the stodgy innocence of cisterns, guns branch forth as part of the foliage, and missiles float to hidden locations as though they were only birds carried on the wind), the disposition of events within battles, the bodily fate of large populations of soldiers. But of all disappearing content, the most crucial is the one this chapter opened with, the disappearance of the injuries themselves from all spheres of language, whether strategic, historical, or conversational. The radical unanchoredness of the language of war is in nothing so visible as in its separation from that phenomenon (the alteration of hands, heart, lungs, brain) that is, in the midst of so much fictiveness, not only the most indisputably and unalterably real phenomenon but also the phenomenon that is with massive, obsessive, dogged repetitiveness being brought into being often hundreds of times each day, day after day. The eventual transfer of the attributes of injuries to a victorious national fiction requires as prelude the severing of those attributes from their original source, an act of severing and disowning that has a wide, perhaps collective, authorship.

Of primary importance to the analysis here is the unanchoredness of the *framing* issues of war, and the relation of that unanchored verbal frame to the relentless physical content of bodily alteration in war's interior. But it is also relevant to witness and understand for a moment the relation between the unanchoredness of the framing issues and the dissolution of verbal coherence and meaning *within* war. The language in war's interior aspires to the condition of the framing language. The fictions occurring inside war may be—as they traditionally are— understood in the context of immediate motive and need: one (a strategist) deconstructs the bodies of opponents by first separating those opponents from the human community of shared verbal meaning; another (a government leader) misrepresents those events that occurred in the battle because ensuring the continued participation of the population requires exiling that same population from mental participation in what has just occurred; another (a soldier) repeats a wholly false tale of enemy atrocity, or alternatively fails to repeat a wholly true tale of enemy atrocity, because he is now moving gun-in-hand across a terrain that has been emptied of all substantive speech. But however accurate it is to describe these events in terms of their local motives, they should also collectively be seen within the overall occurrence, the overarching frame of what war is.

War is in the massive fact of itself a huge structure for the derealization of cultural constructs and, simultaneously, for their eventual reconstitution. The purpose of the war is to designate as an outcome which of the two competing cultural constructs will by both sides be allowed to become real, which of the two will (after the war) hold sway in the shared space where the two (prior to war) collided. Thus, the declaration of war is the declaration that "reality" is now officially "up for grabs," is now officially not only to be suspended but systematically deconstructed, a deconstruction that will be carried far enough on both sides so that either one, if designated the loser, will have less difficulty reimagining itself as "without" its disputed aspect of self-definition than it would immediately prior to the war. The lies, fictions, falsification, within war, though authored by particular kinds of speakers in any given instance (government officials, journalists, generals, soldiers, factory workers) themselves together collectively objectify and extend the formal fact of what war is, the suspension of the reality of constructs, the systematic retraction of all benign forms of substance from the artifacts of civilization, *and simultaneously*, the mining of the ultimate substance, the ultimate source of substantiation, the extraction of the physical basis of reality from its dark hiding place in the body out into the light of day, the making available of the precious ore of confirmation, the interior content of human bodies, lungs, arteries, blood, brains, the mother lode that will eventually be reconnected to the winning issue, to which it will lend its radical substance, its compelling, heartsickening reality, until benign forms of substantiation come into being.

Thus in war injuring performs two distinct functions: it serves as a basis for a contest, a means that has no advantage over other means of arriving at a winner and a loser; second, it provides a source of substantiation for the issues designated winner as a result of the first function. Because these two profoundly different functions are embedded in a single act, they are usually indistinguishable: at most, injuring is perceived to perform the first function of determining a winner; and sometimes, the referential instability at work in the second, rather than alerting the observer to the presence of that second function, instead brings about the retroactive disappearance of even the first function. Thus injury comes to be understood as "functionless" as in all those descriptions that designate it a useless "by-product" or "accidental outcome" of war. Only when the two very different functions are literally assigned a separate space—as they were in Hitler's Germany where injuring-as-contest could be understood as residing in the external space of confrontation with Allied countries, and injuring-as-substantiation as residing in the internal space of concentration camps—is the structure of war exposed. This double space is not exclusive to Germany (the wars in the Middle East, Vietnam, Korea, for example, sometimes entailed an external space of battle and an internal space of torturing imprisoned soldiers); and much more crucially, even when the two are occurring in the singular space of external conflict, a very similar (though not identical[159]) "second use" of the human body is always occurring there. That is, the collective casualties of World War II—the hurt bodies of Allied soldiers, Allied civilians, Axis soldiers, Axis civilians, neutral civilians—all contribute to the "memorialization function" (objectifying the process that just occurred) or the "as-if function" (certifying the reality of the not-yet enacted disposition of winning issues) and would do so whether, as is the case, victory belonged to the Allied powers or, instead, to the Axis powers. This by no means means that the issues on the two sides are the same, for there may be, as there was, an expanse of justice that separates them; but it does mean that the substantiation process itself, the collective work of the bodies injured in the external space of conflict, is the same regardless of what it substantiates, is the same whether what it substantiates is a construct suffused with beauty and justice or one containing the very antitheses of these attributes. Had the American South won the Civil War, what was substantiated would have been radically different from what, through the victory of the North, was instead substantiated; but the war's contribution to the substantiation process would have been in the two cases identical. So, for example, the American Revolutionary War substantiated the right of secession; the American Civil War substantiated the right of the union to prevent secession.

The conflation of the two functions increases our confusion about and reliance on injuring. If the two were disentwined and replaced by stylized and rationalized versions of themselves, each would require a very different form of substitute. First, a comparatively benign means of arriving at a winner and a loser would be carried out, one that—whether based on talent or instead luck—would be of great duration and intricacy. It could never be a toss of the coin, though it could conceivably be an extraordinarily prolonged and elaborate procedure of forming, forging, weighing, and tossing massive numbers of coins in complicated mathematical combinations in order to engage over time a depth of attention consonant with the depth of imaginative reorientation eventually required of the loser. Second, as the week approached for the final and decisive series of acts, people of the disputing countries and of countries in any way affected by the outcome would not only begin to wear or enact agreed upon signs objectifying that such a contest took place (the "memorialization" component of the substantiation function), but would also (to fulfill the "as if" component of the substantiating function) by prearrangement now gather in large groups throughout their homelands; and at the end, multitudes of individuals would each hold up or simply hold onto an animal organ or entrail in confirmation of the idea of winning and of the issue that was the winner and henceforth "real."

This second form of substitution (though introduced here only for the purpose of structural clarification) is, of course, ghastly to contemplate: there would no doubt be a universally shared species shame at picturing ourselves engaged in so atavistic, so primitive a ritual. But however primitive such a surrogate would be, seeming in that hour to carry us back thousands of years, it would be one very large step closer to the benign and familiar forms of substantiation in civilization than is the fiction-generating process now relied on in war. That is, in almost all arenas of human creation, the work of substantiation originally accomplished by the interior of the human body has undergone a hundred stages of transformation, but the first stage, the first step was the substitution of the human body with an animal body. War is one of the few structures for the derealization and reconstitution of constructs in which this very first form of substitution has never occurred. Thus, the inner voice that protests that the imagined ritual would carry us back thousands of years must be reminded that war carries us back those thousands plus one.

V. Torture and War: A Difference Between Them

Thus war, like torture, is (in its second function) a structure through which the attributes of the hurt body are connected to unanchored verbal constructs: what within torture happens between *two people*—the body of a prisoner deprived of his voice, and the verbal constructs of the interrogating torturer, himself disembodied through his immunity to pain—is a phenomenon of transfer in war happening on a *now vast scale*, between the collective casualties and collective national constructs. The two are structural analogues. Like torture, war claims pain's attributes as its own and disclaims the pain itself. There is, however, a crucial element that differentiates the substantiating function in the two events and partially explains why war has a moral ambiguity torture simply does not have.

Before looking at that differentiating element, it is useful to recall that the fact that torture and war are different is, of course, visible outside the particular vocabulary and structural analysis provided here. Although there has never been an intelligent argument on behalf of torture (and such an argument is a conceptual impossibility[160]), there have been many on behalf of war. Such arguments fall into at least three categories. First, "defense" is widely held to be a justifiable basis for entering a war that has already been started against one's own country: a long philosophic tradition identifies the very basis of political legitimacy as self-help, the ability to defend oneself;[161] this principle was explicitly assumed, for example, by the architects of the United States, as is visible throughout the Federalist Papers.[162] However, even if by "war" one means not just retaliation but initiation—not just defending oneself once attacked but also oneself beginning the attack—at least two additional categories of argument are, if never wholly persuasive, not wholly unpersuasive. The first of these tends to be psychological in its vocabulary, though it is sometimes individual, sometimes class, and sometimes national psychology that is being described. As the psychological argument evolves through each of these spheres it becomes successively more difficult to assess, either to confirm or discredit. So, individuals who refuse to go to war may be described by advocates of war as motivated by "cowardice." On this individual level, the argument seems less an argument than an arbitrary accusation. It is not very persuasive since it is so clear that the courage needed to refuse participation may be as great as what is required to enter war: those who went to Vietnam and those who stayed home in protest were substantiating two different conceptions of a nation, as well as using two different forms of substantiation. Once, however, the framework changes to class, the same argument takes a much more provocative and potentially compelling form. Carl Schmitt's 1928 essay, "On the Concept of the Political," includes an analysis of the pacifist impulse as class privilege: he takes Hegel to mean by the word "bourgeois" the word "pacifist"; a bourgeois is "an individual who does not want to leave the apolitical riskless private sphere," one who believes his property exempts him from conflicts which will, despite his absence, nevertheless take place.[163] The psychological argument, dismissable in its individual form, less dismissable in its class form,[164] is perhaps least dismissable in its national form (where it tends to be sidestepped by opponents rather than argued against). The explanatory framework carried through the term "Scandinavization" is one in which the desirable national traits of "neutrality" and "pacifism" are read as "signs" of the negative national traits they are asserted to accompany, decadence, the demise of self-belief, the exhaustion of a conception of nationhood, propertied self-exemption from conflict and from self-renewal.

The two major categories of argument attended to so far both implicitly acknowledge war as a fiction-generating, or construct-substantiating process (even though they do not worry that it is the most atavistic form of creation). That is, the defense argument, the equation of political "legitimacy" or "sovereignty" with self-help, can be rephrased as the capacity of a people to substantiate—to first "make" and then "make real"—its own beliefs, its own territorial and ideological self-definition. So, too, as the psychological argument is lifted out of the individual idiom of "cowardice" and carried through the class idiom of "self-exemption" to the national idiom of "exhaustion," it becomes clear that what is all along being called into question is the capacity for self-renewal or self-recreation, that it is not so much "moral" attributes but the attributes required for "creation" that are at issue. Tocqueville, for example, argued that the intellectual creativity of a population and its readiness to go to war are related: a reluctance to war or to revolution arising from the condition of equality is accompanied by a reluctance to any revolution in ideas and to the suppression of individual genius; genius will either not arise within those boundaries or will, if it should arise, fail to be recognized.[165] So, too, it has often been noticed that artistic creation frequently occurs in conjunction with an absorption with military matters both in the realm of the individual artist (Leonardo da Vinci is the most familiar example)[166] and in the realm of the nation-state—the history of changing national supremacy in art (belonging now to one country, now to another) tends to follow the same path as the history of changing political supremacy.[167] So, too, major studies of the history of technology, such as those by Siegfried Giedion, Lewis Mumford, and William McNeill,[168] all demonstrate the technological inventiveness occasioned by war. This point will be returned to after looking at the third major form of argument on behalf of war.

This third form is an unselfreferring argument about "justice": it is not the argument that "justice" is an attribute of the ideology on behalf of which one fights ("my country's cause is just") but, rather, the argument that "justice" as an attribute of the international realm (the whole earth) is only present there if that international realm includes the possibility of going to war. That is, what is about to be described here should not be confused with the claim that a particular country's entering war has justice, for it is instead the much less specific and much more radical claim that the world that has permanent peace is "less just" than one in which peace can be interrupted by war. Clausewitz observed that it is always the most powerful country that is "peace loving"; the disposition of issues and boundaries perpetuated by the reigning peace overwhelmingly favor that country. Thus it is the less powerful country (what in some descriptions is called the "continental challenger") that conceives of itself either as directly suppressed by the other country or, perhaps equally drastically, as having a potential for change and growth that is indirectly suppressed by the present disposition of power held in place by the reigning peace. Nothing can be learned about the validity of this argument from its invocation by specific leaders or countries: that Lenin, for example, loved this passage in Clausewitz[169] only shows that the insight may be "useful to" those who wish to go to war rather than that it is "true." Its possible validity is, however, suggested by architects of various peace plans who, originally motivated by the passion to rid the world of war, have sometimes concluded that a world emptied of war may be an unjust world: outlawing war altogether would be, even if possible, possibly morally undesirable because what had up to that moment been a world characterized by "fluidity" of boundaries, trade arrangements, markets, ideological patterns would suddenly be "frozen," and frozen in a way that favored certain sets of people over certain others. So, although some peace plans (for example, Bentham's, Kant's) have required the absolute abolition of standing armies, others, such as that of the Fabian society, explicitly include a provision that allows countries, after waiting a period of time, to go to war.[170] As one moves from general (and unselfinterested) peace plans to specific "peace treaties" between sets of countries, so routine is the inclusion of "exception clauses" (so routine is the guarding of the "privilege" of going to war) that it has been argued that such treaties, far from minimizing the possibility of war, instead specify the next occasion of war; they in effect become predictive models or architectural maps of the next war.[171]

This third form of argument, like the first two, again seems to assume that what is at stake in the question of war is a "fiction-generating" capacity: here, however, it is not so much national self-recreation but, in effect, terrestrial recreation that is being guarded. Central to this argument is the awareness that the overall configuration of political, territorial, and cultural constructs across the entire earth is in a constant condition of being reinvented, remade, and that war—the contest to out-injure—is crucial in this process.

Whether or not any of the three forms of argument on behalf of war has validity is too complicated a question to answer here. What is instead crucial to see is that *even if they were valid*, they would not have shown the necessity of war, but instead would only have shown the necessity of a "contest based on a reciprocal activity that would produce a nonreciprocal outcome abided-by by all."[172] That is, even if the self-recreation of a country, continent, or world as a whole requires a process that allows the periodic derealization of cultural constructs and simultaneously permits their reconstitution by providing forms of substantiation until they can be fully realized, the fiction-generating process need not itself stay frozen in its most elementary form (war), but can potentially itself be derealized and replaced by a reconstituted version of itself. It is precisely because the capacity for invention and self-recreation is so universally held to be precious that the most legitimately seductive arguments on behalf of war are arguments (in different guises) that war contributes to the re-creation process; but in turn the universally shared recognition of the importance of invention ultimately signals the importance of subjecting the substantiation process itself (war) to the same process of recreation. The surrogate for war would be in the international realm the mechanism for ensuring the possibility of periodic change that is the equivalent within the nation-state to the mechanism of elections which ensures the fluidity of constructs interior to national boundaries. Perhaps instead of the next war, the disputing countries could agree to a contest in which each invents a surrogate for war, one that accommodates all war's structural attributes; the country authoring the best substitute would win that international contest, though what it had gained (the issues made real as a result) might well themselves a few generations later be derealized when, in its turn, the country had to enter the process of dispute resolution it had earlier invented. At any rate, however hypothetical (and thus itself "unanchored," not-yet true) is the idea of a substitute for war, evidence that such reinvention is possible is present in the overwhelming fact that civilization has (in areas other than war) repeatedly managed to displace the earliest form of substantiation with forms equally compelling but much more benign. The reinvention of the substantiation process is the subject of the second half of this book.

But before turning to that subject, we must return to the original point of this final section, the articulation of the element that differentiates the structure of torture from the structure of war. Even if war were never replaced by a substitute, it should be recognized that it is not the equivalent of torture. This is, perhaps, intuitively obvious: no one conceives of "soldiers" as "torturers" or of the collective activity as having the absolute moral onus of torture, even though the number and permanence of injuries, as well as the amount of suffering prior to death, are much greater in wars than in torture. The critical difference between torture and war should be articulated, not as an apology for war, but for these three reasons. One, the more accurately the nature of war is described, the more likely the chances that it will one day be displaced by a structural surrogate. Two, insofar as war is already in a state of continuous reinvention, it will always either be progressing in the direction of the benign model of substantiation found in civilization or instead degenerating into a more radical deconstruction, the absolute model for which is torture. Three, once war and torture are differentiated, it will become clear that "nuclear war" is closer to the model of torture than it is to the model of conventional war.

In both war and torture, the normal relation between body and voice is deconstructed and replaced by one in which the extremes of the hurt body and unanchored verbal assertions (pain and interrogation in torture; casualties and verbal issues in war) are laid edge to edge. In each, a fiction is produced, a fiction that is a projected image of the body: the pain's reality is now the regime's reality; the factualness of corpses is now the factualness of an ideology or territorial self-definition. But the nature of the "fictiveness" is in the two very different. This difference does not depend on the content of the fiction, for that content (if considered apart from the human hurt used to confirm it) may itself (in both war and torture) be either good or bad, itself just or unjust. The issue of war might be "human equality" or instead "slavery." Again, if one tortures to certify the reality of Christ (as in the Inquisition), the fact that the idea of Christ is itself a benign and beautiful construct does not change the fact that this is precisely the same event as torturing for the intrinsically ugly construct of Aryan supremacy. That the distinction between the "fictiveness" of torture and the "fictiveness" of war is independent of the content of the fiction is most easily illustrated by the fact that the two forms of substantiation can occur on behalf of the same construct: substantiating a state constitution through war would be different from substantiating that same constitution through torture, even though the content of the thing substantiated is the same.[173] The difference in the "fictiveness" of the two is much larger than is signalled by using the negative idiom of "fraud" to refer to the one and the neutral or equivocal idiom of "fiction" to refer to the other, and turns on the fact that though the constructs of both are deprived of benign sources of substantiation, one is fundamentally "untrue" while the other is "not-yet true": that is, one requires the substantiating power of the hurt body precisely because it does not have consent, while the other has both populations' consent but time is required for them to enact their consent.

It may at first appear that the substantiating function in war has only one half of the collective population's consent, since the casualties of the losing side confirm outcomes antithetical to their original beliefs; but all participants of both sides, though fighting for different issues, have consented to contribute their bodies to a process which assumes *at its very beginning* the nonreciprocity of outcome, contains *from its opening moments* the certainty that only one side's issues will in the end be declared real, and it is the work of every soldier to push forward to this concluding inequality. Thus it is deeply inaccurate to say that in entering war, they agree to substantiate their own and no others' issues. By the very act of entering war, they have agreed to certify *if not their own, then the other side's issues*. After a period has elapsed, they will more actively and specifically assent: after the American War of Independence, the population of England comes not only to accept but even to applaud the United States' act of secession; after the American Civil War, the population of the South comes not only to accept but to take pride in its presence within the larger Union, not only accepts but takes pride in the dissolution of slavery from its territory. The prisoner of torture has at no time consented to contribute his body to the fiction of the regime's power produced through his pain. The mime of power in torture is "fictitious" in that it is "unreal," while the winning issues in war are "fictitious" in that they are "not yet real." In one case "fiction" means "a lie about reality" while in the other, "fiction" means "an anticipation of reality."

The relation of torture and war is clearer if placed within the overall framework of the difference between substantiation in its extreme deconstruction (torture) on the one hand and substantiation in civilization on the other, for war hovers somewhere between the two. As has been suggested earlier,[174] and as will be elaborated at some length in later chapters, in benign forms of creation, a bodily attribute is projected into the artifact (a fiction, a made thing), which essentially takes over the work of the body, thereby freeing the embodied person of discomfort and thus enabling him to enter a larger realm of self-extension. The chair, for example, mimes the spine, takes over its work, freeing the person of the constant distress of moving through many small body postures, empties his mind of absorption with the pain in his back, enabling him instead to attend to the clay bowl he is making or to listen to the conversation of a friend. In torture the opposite takes place. Rather than relieve the body of discomfort, extreme pain is inflicted; rather than enabling the person's movement out of the boundaries of the body into the shared realm of extension, it instead brings about a continual contraction and collapse of the contents of world-consciousness; rather than "making objects that relieve pain" it deconstructs already existing objects in order to inflict pain. It at once reverses and itself apes the benign process of imaginative making, for the pain becomes the intermediate "artifact," the "*produced*" bodily condition whose attributes are themselves projected out onto and become attributes of the regime's power. If these two radically antithetical events are taken as models for the structure of creation on the one hand and as the deconstruction of the structure of creation on the other, then it is clear that war belongs on the same ground as torture, for here, too, what is "produced" is physical distress and bodily alteration rather than an artifact that eliminates pain; war requires the contraction of the contents of world-consciousness (as the prisoner's mind is filled with bodily pain, so the soldier's mind is obsessively filled with the bodily events of dying and killing); it also requires that attributes of the hurt body be separated from the body and projected onto constructs.

The distance separating the model of creation from the model of decreation is vast. Because certain words—such as "produce," "body," "project," "artifact"—are common to the description of both, it may seem that the two are less radically antithetical than they are. But the overlapping vocabulary is itself the sign of how absolute the difference is between them, for they share the same pieces of language only because the one is a deconstruction of the other, a reversal of the path of creation to decreation. That is, torture *begins* at precisely the point where the other has left off: it starts by appropriating and deconstructing the artifacts that are the products of creation—wall, window, door, room, shelter, medicine, law, friend, country, both as they exist in their material form and as the created contents of consciousness.[175] Torture *ends* at what is the other's starting point: it "produces" the pain that has not only been eliminated by the act of creation, but whose very existence had been the condition that originally occasioned the act of creation. In the one, pain is deconstructed and displaced by an artifact; in the other, the artifact is deconstructed to produce pain. Thus torture not only deconstructs the "products" of the imagination, but deconstructs the act of imagining itself. Each of the two entails a structure of action that occurs along a path common to both, but the two move in opposite directions, and only one can be occurring on that path at any given moment. The idiom common to both announces thay they both aspire to inhabit a space which can, at any given moment, be inhabited by only one; thus they are not simply "opposites" but are "mutually exclusive"; they are not simply "deeply antagonistic" to one another—the very existence of each requires the other's elimination.

Civilization and its deconstruction (what for lack of an appropriate term will be covered by the awkward word "de-civilization") exist in these two models as absolutely distinct, an absolute distance separating them because they cannot exist on the same ground. But there is a conceptually (not always chronologically) earlier moment in which the distance separating them is less, a moment in which they are not yet radically distinct but are only in the process of distinguishing themselves. In this earlier moment in creation, the made thing, the artifact, is not yet (as in the chair) given a material form, and thus is not yet self-substantiating. A chair is an "invented thing" yet is also a "real thing." Its reality is confirmed in exactly the same way as the reality of a naturally occurring object, such as a tree, is confirmed: each can be perceptually experienced within the body of the perceiver. Of course, attributes (color, shape, sound) of objects in the physical world (trees, chairs) vary in the degree of reality intuitively accorded them because they vary in the number of bodily senses through which they are experienced. Those attributes confirmed by both vision and touch tend to be felt to have a greater reality than those attributes experienced by only one of the two, or by one of the other senses without these two. So, for example, Locke's famous distinction between primary and secondary qualities may turn on the sensory modes through which the quality is perceived; primary qualities (shape, mass) are attributes accessible to both vision and touch; secondary qualities are accessible to only one of the two (color can be seen but not touched) or by neither of the two (sounds and smells are heard and smelled but neither seen nor touched). But in the process of creation, there is a stage at which the invented thing has *no* material presence in the world, is thus accessible to *none* of the senses, and is therefore judged to have no reality, to be unreal, or to be not-yet real. To have a material form is to have a self-substantiating form; to lack a material form is to lack the autonomous power of self-substantiation. The creation of the chair can be broken down into two phases of "making": first, "imagining" the chair (inventing the idea or image of a chair); second, "making" the chair (inventing the material form of the chair). In both phases the object is "invented," but only in the second is it also "real," self-substantiating. Both will be called "created object," but the near synonym for "created," "fictitious" tends only to be applied to the first, for the second is real. The city in the poet's mind is a fiction; the city on the island of Manhattan is a fact. And though both "fiction" and "fact" are from the words meaning "to make" (*fingere/fictum* and *facere/factum*), the difference between them in everyday use recognizes that in the second, a further stage of invention—material realization—has taken place. Thus, as will be elaborated in a later chapter, the full process of "making" in civilization entails the two conceptually distinct steps of "making-up" and "making-real"; the imagination only considers her work complete when her own activity is no longer recognizable in the object she has produced.

In the distinction between the model of civilization and de-civilization looked at above, the benign model was one in which creation had not only "made" but "made real" the artifact. But at an earlier moment of creation, the "making real" phase has not yet occurred. It is here, where creation has only arrived at the "making-up" phase, that the distinction between the models of creation and de-creation are more confused; they are not yet wholly separate but are in the midst of separating themselves. That is, the "reality" of inventions in this first phase poses a problem, whether what is invented is the image of a chair, the idea of a god, a prophetic intuition about the future appearance of a city. The imagined image, or believed-in belief, may itself be capable of taking a future self-substantiating form (as in the chair) or it may be fundamentally incapable of ever taking that form. But in either event, at the moment when the made thing is only an idea or image or belief (with no verbal or material extension), insofar as it has any reality at all, it will only have reality in the mind of the embodied imaginer who is at that moment in the act of imagining or believing it. It will have no reality for anyone outside the boundaries of the believer's body. It is crucial to notice here that even for the imaginer it has "less reality" than do the contents of his or her perception. The image of the chair, however wonderful, will be less palpable and full than the tree now leaned against, seen and touched; for its trunk bears one's weight, the leaves brush one's face and make sounds as they move. Those who have attempted to describe the experienced difference between "an image" and "a perception"—Hume, Jaspers, Sartre—have all agreed that the perception has a "vivacity" (from *vivere*, to live—the sensorially "alive" experience of seeing, smelling, touching, tasting; thus one's own aliveness is experienced and seems to certify the object's reality, its "vivacity"[176]) that the "image" simply does not have. The image, to use Sartre's description, is by comparison "depthless," "two-dimensional," "impoverished," "dry," "thin."[177] The distinction they observe may be observed by anyone by closing one's eyes, imagining for a few minutes in as much detail as possible another place, then opening them again and looking at the room in which one sits: even if the place imagined is more intimately known than the actual room itself, the actual room will have an immediately recognizable reality that the imagined place simply does not have, a vibrancy of color, form, presence that is so overwhelming that there will not be a moment's confusion about which locale one at that moment actually inhabits.[178] Only when there is no competing perceptual content, as in sleep and dreams, will the imagined object have the "vividness" of the real world, and that "vividness" will quickly dissolve when the dreamer wakes and is at once confronted with much more "vivid" mental content.[179] Thus, even for the imaginer, the image is less real than his or her perceptions, and for anyone outside the imaginer it has no reality at all.

When, as here, the invented object, the fiction, has itself no autonomous source of substantiation, its substantiation may be provided by analogy: for example, the nonselfsubstantiating idea of a god, whose asserted reality is itself inexperienceable, may by analogy be confirmed by the beauty of the trees that can be experienced perceptually—a god's presence is "felt" in the perceptually "feelable" reality of trees, river, wind, mountains. It is not, however, just benign natural occurrences such as landscapes but also the compellingly experienceable sentient presence of pain, or the compellingly experienceable spectacle of visible bodily alteration in injury, that come to be used as a source of analogical substantiation. A dividing line between civilization and decivilization can be drawn between analogical substantiation that requires sensations free of hurt (seeing, hearing, touching) and analogical substantiation that requires instead the sensation of hurt. But even within the realm of hurt alone, a dividing line can be drawn between human hurt and animal hurt; for the displacement of human sacrifice with animal sacrifice (and its implicit designation of the human body as a privileged space that cannot be used in the important process of substantiation) has always been recognized as a special moment in the infancy of civilization. If this is taken as an early line of separation between civilization and its antithesis, it is clear that both torture and war still belong together on the side of decivilization, since each requires *human* hurt in the analogical verification of its fictions.

But now, finally, a third backward step may be taken into the realm of human hurt alone. Even here two separable categories exist, with only one allowing the forward movement up through the successively more benign displacements from human hurt to animal hurt to, finally, no hurt.[180] That is, within this overall negative realm a distinction can be made between a *comparatively* benign situation in which the human body hurt in confirmation of the belief belongs to the believer himself (as when, for example, a tribe chooses to read its own physical sufferings as sentient verification of the existence of a god) and a profoundly different situation in which the belief does not belong to the person whose body is hurt in its confirmation (as when a nonbeliever is sacrificed or tortured to "demonstrate" the authenticity of the fiction, whether the fiction is political, religious, or something else such as the assertion that the gate is impregnable or that the tribe will next year grow wings and be able to fly). The first relation between body and belief—in which both the body and the belief (or, both the pain and the image) belong to the same person—is a primitive structural equivalent for the most benign and familiar form of substantiation that occurs, for example, when an observer confirms the existence of a tree by "seeing a tree." Just as here the entire structure of the perception—both the perceptual act (seeing) and the perceptual object (the tree that is the content of the seeing)— is present within the body of the observer, so in this early form both the sentient experience of the pain (not now "seeing" but "hurting") and the object (not now the tree but the imagined tree, or to make the distinction clearer, the tree-god) occur together in a closed loop of interior mental experience. Further, when the person reads the "reality" of the pain in his hand as an attribute of the idea of tree-god in his imagination, he in effect (as in the later model of creation) deconstructs the pain in the realization of artifice. Depending on how important the tree-god is to him, he may even willfully increase the pain in his hand, so that he can again deconstruct it in transferring its attributes to his imagined tree-god; alternatively, he may experience the pain as so distressful that he chooses to heal his hand and eliminate the pain, and accept the diminished vibrancy of the tree-god. Because both are occurring within the boundaries of his own body, he himself can compare the disadvantage of suffering the aversiveness of the pain with the advantage of having present to his mind the idea of a tree-god, and can modulate their interaction and choose between them. Eventually it may even occur to him that he can have both the vividness of the tree-god *and* the absence of pain; that is, he will move on to discover benign sources of substantiation, such as carving an image of the tree-god which is now perceptually available to his senses as a freestanding object in the external world, and thus he need no longer depend on pain to provide vibrancy.

But the other model, the wholly deconstructed relation between body and belief—where the substantiating body belongs to one person and the substantiated belief belongs to a second person—contains none of these three modestly redeeming advantages. Sentience and self-extension are now wholly severed and work against each other. For there to be hurt occurring in one person's hand and an image of a tree-god in a second person's imagination is not an anticipatory mime of the structure of perception eventually relied on to confirm actualities in the external world. The equivalent here would only exist if there could be one person who had the sentient experience of "seeing" without any content (if, for example, a flood of light suddenly burst upon his eyes) while simultaneously a second person experienced the object of perception without the act of seeing itself (a vibrant image of the tree suddenly presented itself to his mind even though he were himself blind). Second, the pain in the victim's body is not deconstructed in the realization of the artifact since the artifact is not present to him; he disbelieves it. Third, since the hurt and the fictional object reside in different locations, since they are not contained within a single comprehensive consciousness, there is no possibility of judging their relative advantages and disadvantages, modulating their interaction, and choosing between them. Because the believer is separate from the aversiveness of the pain, there is less chance that he will go on (like our first believer above) to invent more benign ways of making his tree-god vibrantly real.

Thus, the difference between the two models is this: in one the belief belongs to the person whose body is used in its confirmation; in the other, the belief belongs to a person other than the person whose body is used to confirm it. It is perhaps unfortunate that so overwhelming a difference is stateable in a single sentence, for the fragility and brevity of utterance cannot accommodate the magnitude of the difference it attempts to articulate: the sentence starts and stops, is spoken, and then is over. Its significance may be held or lost, may be sustained or may, like the sentence itself, slip from view. Part of the difficulty in holding visible its significance arises from the fact that although the twentieth century has engaged in constant acts of invention to a degree never occurring in the world before, and although it has also been preoccupied with identifying the presence of the human hand in constructs that were not always in earlier centuries recognized as human creations (not only is God a fiction, and law a fiction, but childhood is a fiction,[181] sexuality is a fiction,[182] even "wilderness" has been persuasively identified as an invented construct[183]), very little inquiry into the *nature* of fictions has actually occurred, and thus creation—which will eventually come to be understood as having moral and ethical import at least as great as what in earlier centuries was ever perceived to be entailed in questions of "truth"— is at present barely understood in even its most elementary forms. When one day the nature of human creation is fully unfolded, a new language will accommodate a long array of distinctions that are now nearly invisible, and that only begin with the profound difference between a creation and a lie, between a fiction and a fraud.[184] It will be clear that creation is not a unitary event but entails various temporal stages of a realization process, and that new problems arise at each successive stage; it will be clear that the moral and aesthetic value of a given creation does not just depend on the content of the fiction but on the nature of the substantiation used in its confirmation in the transitional period when it is between the states of having been already made-up and not yet made-real; and far, far back in the most atavistic form of substantiation that entails the use of the injured human body, it will become strikingly clear that the most fundamental relation between body and belief turns on the question of whether a person uses his own body in confirmation of a symbolic displacement of that body, or instead uses the body of someone outside the benefits of the invented construct in the confirmation process. Though this distinction will be clearer later, it is needed here now; for torture and war, so deeply alike, are at last different in this critical element. If each requires the most atavistic form of the reality-conferring process, in the one it is the nonbeliever's body and in the other the believer's body that is enlisted into the crisis of substantiation.

The distinction also makes it possible to recognize the size of the moral chasm that separates conventional war and nuclear war. Up to this point, the two have been spoken of as one (and the inclusive term "war" used) because the structural attributes—reciprocal activity of injuring to produce a nonreciprocal outcome abided-by all because of the reality-confirming power of injured bodies[185]— were equally characteristic of each. But now an overwhelming structural difference must be attended to: in conventional war the beliefs substantiated by injuries belong to those who are injured, while in nuclear war the beliefs do not belong to those whose bodies have been used to substantiate the constructs. That is, the very nature of the weapon eliminates the possibility that those injured can have consented to contribute their bodies to the substantiation process; they do not have the option of deciding whether the disadvantages of physical suffering and the advantages of the beliefs to be confirmed exist in such a ratio that a war should be fought (that the aversiveness of the pain is *less* than the benefits to be derived from the constructs) or instead should not be fought (that the aversiveness of the pain is *greater* than the benefits to be derived from the disputed constructs). Because the populations of the disputing countries cannot be consulted when the moment comes to fire the missiles, nuclear war has all the scale of conventional war (rather than the two-person structure of torture) but conforms in all its new massiveness to the model of torture rather than that of conventional war.

Revolutions in the technology of weapons have always tended to modulate the degree of participation of the combatants. When, for example, in Stendhal's *The Charterhouse of Parma*, the young hero, Fabrizio, first wanders onto the Napoleonic battlefield, an old vivandière warns him, "Your grip isn't strong enough yet for the saber fighting that will go on today. If you had a musket I wouldn't say anything, because you could fire your bullet as well as anyone else."[186] The old woman recognizes that one kind of weapon requires years of practice—years devoted to anticipatory participation before the actual participation begins—while the other makes it possible to begin the act of participation, to enter the process of substantiation, beginning on the day when the war itself actually begins. Thus one requires a much longer and deeper level of participation than the other. The significance of this difference is widely recognized outside fiction. In his study of the arms question as it occurred in feudal Japan, Noel Perrin observes the same difference between sword and gun; the nature of the sword made it necessary for the samurai to devote their lives to attaining the skill to use it so that they would have the power to defend their beliefs when the occasion arose; they were thus shocked when the introduction of the gun made it possible for the farmer (himself unskilled in weapons) to kill as easily as the samurai could only after years of practice.[187] As the result of such a technological revolution, fighting is in effect democratized, and the need for a warrior class, in this case the samurai, is eliminated. Similarly, William McNeill shows that in Europe the introduction of the crossbow eliminated the need for a class of knights.[188] Later, the introduction of the gun would further extend the effect of the crossbow; and still later, more and more technically uniform guns would diminish the amount of practice required to shoot them, for the soldier would no longer have to learn the sighting and firing peculiarities of each weapon.[189] In each of these historical moments, more and more of the skill formerly required to use a weapon is built into the new weapon itself, thus increasingly freeing the participant from the amount of his life devoted to that act of participation. It can be argued[190] that with these successive inventions, populations become increasingly liberated from war: rather than devoting an entire lifetime (samurai, knights) to preparation for the actual occasion on which one has to substantiate one's beliefs, a person need only enter those acts when the actual crisis arises. Thus they are freed to devote themselves to world-building in the period when there is no dispute. But it should be noticed that although more and more of the injuring "skill" is becoming built into the weapon itself, thereby freeing the human body from the requirements of learning these skills, some level of skill is always needed of the body: the guns do not move on to the field of battle by themselves—each requires a human body to carry it there, aim it there, and fire it there. This minimal requirement, of course, disappears with the introduction of nuclear missiles.

Nuclear arms may at first appear a radical extension of the freedom: now the population is not only free of the constant preparation of learning how to handle the weapon, but their presence will not even be required in the weapon's firing when the actual war begins. So completely have the formerly embodied skills of weapon use been appropriated into the interior of the weapon itself that no human *skill* is now required; and because the need for human skills is eliminated, the need for a human *presence* to fire it is eliminated; and because the human presence is eliminated, the human act of *consent* is eliminated. The *building-in of skill* thus becomes in its most triumphant form, the *building-out of consent*. It is, of course, only at the "firing" end of the weapon that human presence is eliminated: their bodies' presence at the receiving end is still very much required.

That is, injury will still have the same two intended functions as it had in conventional war: it is by one side being out-injured that a winner and a loser will be arrived at; the injuries will collectively memorialize the fact that such a reciprocal activity took place and will, presumably, certify the reality of the verbal issues. All that has changed is the overwhelming first premise of war and the contest structure, that the beliefs belong to a given population, that the bodies used to substantiate the beliefs belong to a given population, and that this most extreme use of sentient tissue can only be sanctioned by those to whom the tissue belongs.[191] Although the vocabulary of "consent" has its most recent and thus most familiar elaboration in the concept of democracy, it is much older than democracy; what is deconstructed in torture and in nuclear war is not just the democratic impulse but the basic impulse of civilization, the fundamental integrity of the relation between body and belief present in the infancy of symbolic thought. It may be that the elimination of this first premise from war will carry with it the elimination of the second function of the injuries: although nuclear war will bring about an unprecedented level of injury, the injuries may not in the end work to produce an abiding outcome, since they will memorialize that a contest took place, but not one willfully entered into. The injuries may be experienced by the populations as something "done to" them by two governments, and those two governments may be disavowed. What is remembered in the body is well remembered, and that no consent was given in this massive sentient alteration is likely to be part of what is remembered. Thus it is possible that rather than an acceptance of outcome there would instead be an indifference to outcome and a rebellion against the existing governments; of course, if such a disavowal is to occur, it would be preferable that it occur before the weapons were used.

It might be argued that even in conventional war, the "agents" of the war are the "kings and cabinets," rather than the populations, of the disputing countries: Rousseau, for example, is one of many who have argued this; and the existence of compulsory "draft" has been interpreted as eliminating the act of willed participation. By comparison with nuclear war, however, this assertion is simply untrue; for if there theoretically exist a hundred degrees of consent, in nuclear war the level is zero while in conventional war it may fluctuate between, for example, eighty-eight and one hundred. Let us say, that in World War I an American boy has been drafted into the war and that although unsure of himself, he chooses to go, rather than be jailed or go into hiding. Though a constrained act, some level of consent has occurred. He now goes to his job to announce his departure, gets his papers in order, says goodbye to his friends, and in particular, his sister Margaret. With each of these small acts, his level of consent and participation is growing. He walks to the ship, he picks up a gun, he puts on a uniform, not once, but each morning, day after day, he re-puts it on, renewing each time his act of consent. He could have, on the first day, refused to go; or he could now on the three hundredth day, feign injury or appendicitis and so, for a time, exempt himself from further combat; he may stay on the front and fire at the enemy, or stay on the front and fire straight up into the air. His own authoring of his embodied participation did not just occur on the first day but on the third, the fortieth, the five hundredth; and for this reason government leaders, military commanders, and comrades will work to sustain his "consent," keep high his morale, over endless days. His fundamental relation between body and belief takes many degrees and radiates out over thousands of small acts. Perhaps the issue "for which" he risks his body is not the same as the government's issue: he may contribute his body to substantiate an idea of courage he has, or an understanding of ambition, or a world safe for democracy; but he will have some belief for which he performs the work he performs, daily recertifying its importance.

He may have been "misled," and may be fighting for injustice rather than justice; but at least his government had to present him with its "case" against the opponent rather than considering its military actions independent of its own population as jury. (That a jury can be misinformed is one thing and not the same thing as the abolition of the jury system.) The war cannot be executed without his consent: it is authored by—it cannot be carried out without the "authorization" of—the population. One may object that although he has given consent over his own body, he is acting to kill another body over which he does not have the power of consent. Though this is certainly true, it is also true that the opponent facing him across the trenches, by his very presence on this ground, has himself consented. Further, the two ends of the weapon exist in a one-to-one relation: one soldier's body is put at risk while one opponent's body is put at risk. Of course, if he is a good soldier, he will injure more than one opponent and thus the ratio will be one to five, or one to twenty; but there are good soldiers on the other side as well, and thus the fundamental framing ratio of agency to risk is sustained. At any given moment, one body of the opponent may be taken to substantiate the boy's beliefs and, conversely, he makes available to the opponent one body, his own. This one-to-one ratio—beyond seeming merely "fair"—has the critical function of ensuring that the free-floating beliefs cannot be wholly severed from awareness of the aversiveness of injuries, cannot become wholly unanchored from the knowledge of the cost of the beliefs to the believers. The ratio of persons at the two ends of the weapon, the ratio of authorization to risk, changes in nuclear war from something close to one-to-one (depending on the war, one-to-three, one-to-twenty) to one-to-millions, a ratio that can be more precisely determined by counting the number of political and military persons involved in the decision and the number of people who are the act's casualties.

Consent is in nuclear war a structural impossibility. At no point does the population exercise any will over their own participation, as may be seen by contemplating three separate temporal moments: first, the time prior to war; second, the actual moment of the war's outbreak; third, the succession of moments constituting the duration of the war.

First, the time period prior to the war. It might be argued that the existence of a nuclear arsenal, paid for by the population's taxes, contains an implicit act of consent, that a contract for participation was forged and renewed on April 15 of each year (or an equivalent date in other countries). There are two bases on which this argument is wrong. First, taxes are paid to provide for an array of social programs and uses, and there is no provision within the tax procedure that allows the stipulation of money for the salaries of members of Congress and the building of schools on the one hand, and none for weapons on the other. Presumably, however, a tax form could be invented that gave the taxpayer this very power of stipulation. Even then, however, the problem would not be solved. Even were they to agree to finance nuclear weapons, this would not be a sanctioning of the war itself. The equivalent might be an election in which the population voted to forfeit permanently all subsequent elections; that is, they agreed not simply to alter, by means of a provision within the constitution, the constitution itself but agreed to surrender any subsequent power of constitutional self-alteration. A more precise analogue might be the invention of a voting machine that eliminated the need for the population to come to the polls and vote, because the power to guess the will of the people was built into the machine itself. If the invention and manufacture of the machine had been paid for by tax dollars, the sibylline vote would still not be the population's vote. The existence of an arsenal of guns in the United States in 1858 does not mean that President Lincoln (without further consultation of the people) is thereby authorized one day in 1861 to retract all life-support systems from beneath the feet of five-hundred thousand persons; for the existence of weapons and the use of weapons require two separate acts of consent. That is, the existence of weapons and the use of weapons—the "making-up" and the "making-real" of the capacity to injure—are two radically different occurrences. To agree to the existence of weapons is to say, "There could conceivably someday be a use for war," and not to say, "The hypothetical existence of a use sanctions any use."

Second, consider the moment of war's outbreak. Suppose it were possible to consult the world's population in the hours before the nuclear weapons were fired; suppose also the unlikely outcome that the consulted world population voted "yes." Would this make a difference? The certain and correct answer to this question is yes, it emphatically would make a difference. But it is crucial to see that not only is the "yes" vote improbable, the vote itself is impossible. That it is an absolute structural impossibility is not just an unfortunate and accidental attribute of nuclear arms; it is essential to what it is; it is what it is. The missile *is* the building-in of an unprecedented capacity to injure and the building-out of consent, the shattering of the ratio of willed authorization to willed risk by a factor of millions.

Third, *if* "consent" *were* given by the *impossible* world vote on the first day, this event would vastly improve the claim that nuclear weapons have a rightful place within civilization; but there would not *even then* be nearly the degree of consent that exists in conventional war. Estimates about the duration of nuclear war tend to range from hours to months. It is sometimes said that the number of casualties in the first day's exchange might be about fifty million, the approximate number of deaths that occurred in five years of fighting in World War II. How do the two differ? As observed above, the individual soldier in a conventional war does not give or withhold his participation in a single moment of decision: he must continually renew his presence over many successive days. His willingness may dissolve when he first sees the battlefield; instead, he may, seeing the battlefield one day, still be willing until at the end of the second day he finds a bullethole in his backpack; instead, he may be unfazed by the bullethole in his pack the second day but become unwilling on the third day when he receives a nonlethal wound in his shoulder; instead he may be willing on that third day and only become "unwilling" on the fourth day when he receives a second, this time mortal, wound and dies. This same continually open modulation of authorization must be understood in terms of an entire population. Each side works to out-injure the other: that is, as clarified earlier, each side works to bring the other side to its perceived level of intolerable injury faster than it is itself brought to its own level of intolerable injury. Each side works to bring about in the opponent a perceptual reversal: instead of body damage seeming acceptable and loss of territorial-ideological attributes seeming unacceptable, it will come to perceive additional damage as unacceptable and loss of national attributes as acceptable. This means, importantly, that at the point where one side undergoes the reversal and designates itself "loser," the other side stops injuring. That is, each side exercises control over how much it is injured and can on any day bring about the cessation of injuring by announcing itself the loser. A country may decide this after three days and one hundred twenty deaths; instead after fifty days and ninety thousand injuries; instead after three hundred days and eleven million injuries; instead after five years and fifty million deaths. It each day has the renewed option of saying, "This is enough." Thus the fifty million deaths in one day's nuclear exchange and the fifty million in five years of World War II are incomparable. Even if the nuclear exchange had begun with the population's vote (itself an impossible proposition), it would even then only be the same as World War II if in World War II the disputing governments and populations had been given a chance to say "yes" or "no" on a single day in 1939, and were not again given the option of saying "yes" or "no" until one day in 1945. Instead, week by week over three hundred weeks, day by day over almost two thousand days, the collective populations renewed their consent to be more greatly injured (as in other wars—Korea, Vietnam, Falkland Islands— the national participants have consented to a however terrible, much smaller level of injury). Because in nuclear war the collective casualties occur almost simultaneously, the injured populations do not have the power to authorize the degree and level of injury: if in the moment the missile strikes a four-mile radius of a given city, the words "We surrender" are spoken, they will not stop the blast of heat and fire from extending out over another twenty miles; and as the range of that blast reaches that twenty-mile perimeter, the words "We surrender" will not stop the radiation in its outward movement over hundreds of miles more. Thus, just as the act of consent is built-out of the initial firing of the missile, so it is eliminated from the massive interior of its self-amplifying action.

If there should be a nuclear war, when it is looked back upon by populations of future centuries, the differences between it and previous conventional wars will probably appear much greater to them than it does to us now. They may look back upon our present situation the way we now look back upon the slaves building the pyramids of Egypt, as a population who have ceased to exercise political autonomy over their own most intimate property, the human body. Of course, the analogy is not accurate,[192] for the physical distress of forced work in substantiation of a cultural construct like the pyramids is not the absolute surrender of physical autonomy that occurs in bodies maimed and burned in substantiation of national constructs, nor is the elimination of embodied consent occurring in the same degree in the two; for the slave still authorizes the movements of his body as he each day wakes up, walks to the pyramid, puts his hand to the stone, and begins to lift and carry. Perhaps he believes the very beautiful artifact to which he contributes his embodied labor implicitly includes him in its civilizing embrace, that he is its partial author. Perhaps instead he perceives himself as excluded, but chooses (along with the rest of his generation) to devote his lifetime to this aimless project rather than to the shorter life's project of rebellion. Yet from a distance of many centuries, we often ask why they permitted it; for it is a universal fate of those from whom the power to author their own fate has been retracted that later populations reattribute to them the power of authorship and speak of them as "permitting" it. This question is not only asked, retrospectively, of the slaves forty centuries ago, but of the concentration camp prisoners four decades ago. The same question, however unfair, will be asked of us. If there is to be an answer to those future populations, it will only be heard in the words spoken to contemporary military and political leaders, words that will have to be spoken very clearly and soon.

PART TWO

Making

3

PAIN AND IMAGINING

T HE SUBJECT in the second half of this study is the opposite of what it was in the first half, for what will be attended to is no longer the deconstruction of the world but that world's construction and reconstruction. Thus, the particular structure of activity that will be isolated here is now not unmaking but making. As will very gradually become apparent over the next few chapters, the activity of creation has an identifiable structure. A recognition of that structure requires as only a first step the recognition of the relation between physical pain and imagining. If that relation, described immediately below, echoes our earlier subject, it is because the uncovering of the structure of torture and the structure of war is the uncovering of the inverted form of that relation. It is impossible to speak of either torture or war without attending to the destruction of the artifacts of civilization in either their interior and mental or exterior and materialized forms; even more significantly, the infliction of pain in torture is inextricably bound up with the generation of a political "fiction" just as the injuries of war are bound up with a process of conferring facticity on unanchored cultural "constructs." Because these events entail the appropriation, aping, and deconstructing of the territory of creating, they entail some of the very elements that will now be looked at in their benign form. When the relation between physical pain and imagining occurs in its forward form, it has the following shape, a shape necessitated by the exceptional place that each has within the psychic arrangements of intentional states and their objects.

It was noticed at an early point in this book that physical pain is exceptional in the whole fabric of psychic, somatic, and perceptual states for being the only one that has no object. Though the capacity to experience physical pain is as primal a fact about the human being as is the capacity to hear, to touch, to desire, to fear, to hunger, it differs from these events, and from every other bodily and psychic event, by not having an object in the external world. Hearing and touch are of objects outside the boundaries of the body, as desire is desire of x, fear is fear of y, hunger is hunger for z; but pain is not "of" or "for" anything—it is itself alone. This objectlessness, the complete absence of referential content, almost prevents it from being rendered in language: objectless, it cannot easily be objectified in any form, material or verbal. But it is also its objectlessness that may give rise to imagining by first occasioning the process that eventually brings forth the dense sea of artifacts and symbols that we make and move about in. All other states, precisely by taking an object, at first invite one only to enter rather than to supplement the natural world. The man "desiring" can see the rain and know it is its cessation that he is longing for, so that he can go out and find the berries he is hungry for, before the night comes that he fears. Because of the inevitable bonding of his own interior states with companion objects in the outside world, he easily locates himself in that external world and has no need to invent a world to extend himself out into. The object is an extension of, an expression of, the state: the rain expresses his longing, the berries his hunger, and the night his fear. But nothing expresses his physical pain. Any state that was permanently objectless would no doubt begin the process of invention. But it is especially appropriate that the very state in which he is utterly objectless is also of all states the one that, by its aversiveness, makes most pressing the urge to move out and away from the body.

The only state that is as anomalous as pain is the imagination. While pain is a state remarkable for being wholly without objects, the imagination is remarkable for being the only state that is wholly its objects. There is in imagining no activity, no "state," no experienceable condition or felt-occurrence separate from the objects: the only evidence that one is "imagining" is that imaginary objects appear in the mind. Thus, while pain is like seeing or desiring but not like seeing x or desiring y, the opposite but equally extraordinary characteristic belongs to imagining. It is like the x or the y that are the objects of vision or desire, but not like the felt-occurrences of seeing or desiring. While, then, pain is like other forms of sentience but devoid of the self-extension that is ordinarily the counterpart of sentience, the imagination is like other forms of the capacity for self-extension without the experienceable sentience on which it is ordinarily premised. It may well provide an object for other forms of sentience, an imaginary object of hearing (like Ryle's inwardly hummed tune[1]) or an imaginary object of touch (the way the perfume on Annie's letter conjures up before Sartre's mind for a brief instant the palpable near-proximity of Annie herself[2]) but the object it provides is never provided for any experienceable form of sentience unique to itself.[3]

Although the gerund "imagining" assumes an activity, and although in some philosophic contexts it is described as though it were made up of both an intentional act and an intentional object,[4] in fact (as has long been intuitively recognized in the centuries-old game played by children and philosophers alike) it is impossible to imagine without imagining something. The attributes that the invisible and itself (apart from its objects) inexperienceable activity of imagining is understood to have, tend to be derived from the attributes of whatever "imagined object" happened to have been taken as a representable instance of imagining. If, for example, a discussion of imagining takes either Pegasus or a unicorn as its instance of the imagined object, the imagination is then likely to be itself characterized by the qualities of the specified image. Although the unicorn or Pegasus expresses the imagination's freedom from natural occurrence, its ability to rearrange wings and legs into new combinations, and although it expresses as well the imagination's eventual capacity to create beyond "need," since neither the unicorn nor Pegasus is striking for its usefulness to people in the twentieth century, such an image, by its very frivolity, wrongly suggests that the activity itself is trivial or marginal: that is, discussions conjuring up this type of image tend also to underestimate the centrality and significance of imagining in everyday life. Such an object is especially misleading because it wrongly suggests that the imagination is self-announcing—that it is only at work when its object is immediately recognizable as "made-up."

Very different conclusions about the nature of the imagination will be reached if one takes as an indicative object something more richly implicated in everyday life. If, like Sartre, one contemplates not an imagined Pegasus but an imagined Pierre, not a unicorn but an Annie, then the imagination's relation to daily acts of perception as well as its part in the deepest events of loss, love, and friendship are likely to be more immediately recognizable.[5] Even here, however, inferences about the nature of the activity that produced the objects will follow from and be limited by the attributes of the chosen objects. Sartre, for example, draws conclusions from the fact that his imagined Pierre is so impoverished by comparison with his real friend Pierre, that his imagined Annie has none of the vibrancy, spontaneity, and limitless depth of presence of the real Annie. But, of course, had he compared his imagined friends not to his real-friends-when-present but to his wholly absent friends, his conclusions would have been supplemented by other, very different conclusions. That is, the imagined Pierre is shadowy, dry, and barely present compared to the real Pierre, but is much more vibrantly present than the absent Pierre, and it was that absence that had occasioned the introduction of the image both into Sartre's mental life and into his philosophic account of that life. He only began to imagine Pierre because Pierre was away, lost to him, walking in Berlin down streets far beyond Sartre's sensory and perceptual reach; and thus his choice was not between a two-dimensional Pierre and a real Pierre but between a two-dimensional Pierre and a world utterly devoid of, bereft of, that friend's presence. Further, the same generic embodied imaginer capable of picturing, making present, an absent friend, is also capable of inventing both the idea and the materialized form of the telegraph, as well as devising the specific message, "Come home at once," as he is also capable of inventing many other mechanisms for transforming the condition of absence into presence, the telephone, train, airplane, all of which originate as the imagination's object.

The object that is selected as indicative of the imagination's activity, then, may vary from discussion to discussion; and the attributes that are generalized to the activity tend to be derived from the attributes of the model object. The appropriateness of various kinds of model objects will be returned to at a later moment, but for now the point of central importance is that the imagination is only experienced in the images it produces, and that even discussions of its characteristics as a state tend to be instead discussions of the characteristics of the invoked object. Almost never is the imagination "imagined" without an object, though the Hebraic scriptures come very close to requiring that believers do just that, that they apprehend the capacity for creation devoid of any representable content: attributing to God a representable form is explicitly forbidden, though the objects of mental creation may here be contemplated as the objects that God himself produced, the universe and its inhabitants. The immense problems surrounding that requirement, and the gradual modulation out of that requirement in both the Hebraic and Christian scriptures, will be elaborated in a later chapter. For now, whether the object imagined is God, the imagination itself, Pegasus, Pierre, Annie, a unicorn, a wall, a telegram, or an airplane, the activity producing the object tends to be coterminous with and only knowable through that object.

Physical pain, then, is an intentional state without an intentional object; imagining is an intentional object without an experienceable intentional state. Thus, it may be that in some peculiar way it is appropriate to think of pain as the imagination's intentional state, and to identify the imagination as pain's intentional object. Of course, it is probably inaccurate to identify an essentially objectless state as an "intentional state without an object" since only by having an object does it exist as an intentional state: in isolation, pain "intends" nothing; it is wholly passive; it is "suffered" rather than willed or directed. To be more precise, one can say that pain only becomes an intentional state once it is brought into relation with the objectifying power of the imagination: through that relation, pain will be transformed from a wholly passive and helpless occurrence into a self-modifying and, when most successful, self-eliminating one. But to argue that physical pain and imagining belong to one another as each other's missing intentional counterpart seems only to argue that "hurting and an imagined x" may occur together in a closed loop of interior occurrence that is a structural analogue for other intentional acts like "hearing a voice," "touching a windowpane," whereas it seems possible that something much more important is occurring here. What may instead be the case is that "pain" and "imagining" constitute extreme conditions of, on the one hand, intentionality as a state and, on the other, intentionality as self-objectification; and that between these two boundary conditions all the other more familiar, binary acts-and-objects are located. That is, pain and imagining are the "framing events" within whose boundaries all other perceptual, somatic, and emotional events occur; thus, between the two extremes can be mapped the whole terrain of the human psyche.

That this is an appropriate and useful way of understanding the relation between pain and the imagination is suggested by a number of observable phenomena. The more a habitual form of perception is experienced as itself rather than its external object, the closer it lies to pain; conversely, the more completely a state is experienced as its object, the closer it lies to imaginative self-transformation. So, for example, a woman (perhaps her name is Ruth), working in the fields, touching the wheat, feels not only the wheat but her fingers touching. Touch, as is recognized in traditional descriptions of the senses, lies closer to pain than does vision. Looking across the fields, she is filled with images of grain: though to some degree the perceptual event is feelable as occurring in a horizontal band between her cheekbones and her forehead, she tends "to experience" there the sheaves of wheat and barley rather than any self-conscious state of feeling in her eyes. It is because vision and hearing are, under ordinary conditions, so exclusively bound up with their object rather than their bodily location that they are the senses most frequently invoked by poets as the sensory analogues for the imagination. Through them, one seems to become disembodied, either because one seems to have been transported hundreds of feet beyond the edges of the body out into the external world, or instead because the images of objects from the external world have themselves been carried into the interior of the body as perceptual content, and seem to reside there, displacing the dense matter of the body itself.

But while forms of perception, like touch and vision, can be differentiated from one another by the relative degree of emphasis within them on the feeling state or instead on the object, any one of them in isolation contains the potential for being experienced either as state or as object, and thus has within it the fluidity of moving now toward the vicinity of hurting, now toward the vicinity of imagining. Although vision and hearing ordinarily reside close to objectification, if one experiences one's eyes or ears themselves—if the woman working looks up at the sun too suddenly and her eyes fill with blinding light—then vision falls back to the neighborhood of pain. Or if the objects in the external field— the grain, the figures of other workers, the trees off to the side—begin one year to appear distorted or blurred to her (that is, if the objects begin to become lost to her), she will cease to experience vision only as objectified interior content and will begin to become more self-conscious of the event of "seeing" itself: she no longer experiences the images of grain, persons, and trees without also experiencing her own body in the mode of aversiveness and deprivation (a deprivation that in its most extreme form is physical pain). So, too, though touch always has both somatic and external content, it may be experienced now more as one, now more as the other; and the more it is experienced as the first, the closer it lies to pain; the more it is experienced as the other, the closer it lies to self-displacement and transformation. Thus, if a thorn cuts through the skin of the woman's finger, she feels not the thorn but her body hurting her. If instead she experiences across the skin of her fingers not the awareness of the feel of those fingers but the feel of the fine weave of another woman's work, or if she traces the lettering of an engraved message and becomes mindful not of events in her hands but of the form and motivating force of the signs, or if that night she experiences the intense feelings across the skin of her body not as her own body but as the intensely feelable presence of her beloved, she in each of these moments experiences the sensation of "touch" not as bodily sensations but as self-displacing, self-transforming objectification; and so far are these moments from physical pain, that if they are named as bodily occurrences at all, they will be called "pleasure," a word usually reserved either for moments of overt disembodiment or, as here, moments when acute bodily sensations are experienced as something other than one's own body.[6]

The topography of act and object in sensation and perception is even more immediately recognizable as a description of emotional and somatic states. A state of consciousness other than pain—such as hunger or desire—will, if deprived of its object, begin to approach the neighborhood of pain, as in acute, unsatisfied hunger or prolonged, objectless longing; conversely, when such a state is given an object, it is itself experienced as a pleasurable and self-eliminating (or more precisely, pleasurable because self-eliminating) physical occurrence.[7] The interior states of physical hunger and psychological desire have nothing aversive, fearful, or unpleasant about them if the person experiencing them inhabits a world where food is bountiful and a companion is near.

While it may be immediately apparent in these examples why it is appropriate to identify the objectless state of consciousness as approaching the condition of pain,[8] the appropriateness of identifying the objectified state with the opposite framing boundary of imagining may be less immediately apparent since (in the examples given) the sources of objectification (wheat to eat, another human being to love, golden fields to see) originate in the natural world.[9] The appropriateness of the identification, however, arises from the fact that beyond the expansive ground of ordinary, naturally occurring objects is the narrow extra ground of imagined objects, and beyond this ground, there is no other. Imagining is, in effect, the ground of last resort. That is, should it happen that the world fails to provide an object, the imagination is there, almost on an emergency stand-by basis, as a last resource for the generation of objects. Missing, they will be made-up; and though they may sometimes be inferior to naturally occurring objects, they will always be superior to naturally occurring objectlessness. If no food is present, imagining grain or berries will, at least temporarily, allow the hunger to be experienced as potentially positive rather than as wholly aversive; and the imagined image may remind the person to walk over the next hill to find real wheat and berries. The transformation of hunger into eating (a trans-formation in which there is a literal displacement of the passive, aversive sen-sation of hunger by the active incorporation of external objects into the body) no longer requires the self-announcing presence of objects, because if they are not themselves sensorially present to vision, they will present themselves to the imagination, and will motivate either a search (an alteration in the ground of the world) or an act of material invention (an introduction of a new object onto the ground of the world). Similarly, imagining a companion if the world provides none, may—at least temporarily—prevent longing from being a wholly self-experiencing set of physical and emotional events that, emptied of any referential content, exist as merely painful inner disturbances. It may be that "dreaming," too, should be understood in this way, as sustaining the objectifying powers of people during the hours when they are cut off from the natural source of objects, so that they do not during sleep drown in their own corporeal engulfment. That is, the particular content of the dream images (now terrifying, now benign; now full of uncanny secret intelligence about the sleeper, now ignorant, arbitrary, and nonsensical) is itself insignificant beside the overall fact of the dreaming itself, the emergency work of the imagination to provide an object—this object, that object, any object—to sustain and to exercise the capacity for self-objec-tification during the sleep-filled hours of sweet and dangerous bodily absorption.

The appropriateness of identifying imagining as a boundary condition of in-tentionality may, then, be recognized in the fact that imagining provides an extra and extraordinary ground of objects beyond the naturally occurring ground; it actively "intends," "authors," or "sponsors" objects when they are not *pas-sively* available as an already existing "given." But the appropriateness of the identification may also be understood by phrasing the relation in the opposite direction: rather than apprehending the imagination as an extreme miming of the ordinary given condition in other forms of consciousness, other forms of con-sciousness can be understood to entail more moderate and modest acts of the authoring, self-alteration, and self-artifice routinely and dramatically at work in imagining. If one takes "hurting and an imagined object" (for example, "acute thirst and an imagined cup of water") as intentional counterparts, and if one sees the two together as one in a series of intentional events—acute thirst and an imagined cup of water, touching a flower, hearing a baby cry, watching a train, fearing a storm—it may at first seem that its essential relation between state and object (where the object comes into existence specifically *to eliminate* the condition) differs fundamentally from the relation between state and object in all the other examples, in each of which state and object co-exist as ongoing counterparts. But even these other ordinary perceptual and emotional acts entail self-alteration and artifice. Although in "seeing a field" the "field" is not "taken" in order to nullify the act of "seeing" (the way water is imagined in order to eliminate the thirst, or a blanket is imagined in order to eliminate the state of being cold), the person may well change the direction of his gaze, and thus "see the city" to his left in order to displace, eliminate, his "seeing the field." Ordinary perception is self-modifying because, at the very least, it alters and nullifies its own content, continually exchanging one object for another, exercising control over the direction and content of touch, hearing, seeing, smell, and taste. Thus the radical alteration that occurs in the landscape of the natural world in pain-and-imagining—"being on a desert and hallucinating a tree-filled oasis," "dying of thirst and 'seeing' water on the next sand dune"—is habitually mimed in daily acts of shifting one's seeing to the east rather than the west, reaching out to touch the objects on the left rather than those on the right, attending to the sounds coming from the room above rather than the room below. Further, it may be that the cancellation not only of the object but of the interior state itself is occurring in ordinary perception since the person tends to alter the object of touch, vision, or taste precisely at the moment when he becomes self-conscious of the bodily state itself—that is, when the already given object fails to permit the ongoing achievement of self-objectification, throwing his attention back onto the body. If the sun is too bright for a woman's eyes, she moves into the shade, and as she does, her eyes again fill with seeable objects rather than aversive sensation; if in turn the shade grows too dark for her to differentiate, without straining, the seeds she is sorting, she moves back out into the light; she may shift her vision to some nearby children to ease her discomfort at remembering her lost child; or if that only "makes" her more acutely self-conscious of her loss, she may watch some birds instead. Through her relation to these objects, she continually modifies the degree to which she displaces self-consciousness with unselfconscious objectified content. So too in somatic and emotional states like hunger and desire, a person can continually modify the state itself—now minimizing it, now letting it occur, now intensifying and sustaining it, now eliminating it altogether—by continually modifying and adjusting his or her relation to the object.

A third characteristic of the relation between imaginary and real objects that reveals the appropriateness of identifying "imagining" as a framing state of intentionality at the edge of the human world opposite to the boundary formed by physical pain is that the imagination seems to provide a standard for judging the acceptability of objects in the naturally given world. The more exactly the object of desire or hunger or fear fits or expresses the state, the more precise a projection of the state it is, the more will it seem to have been generated by the interior state itself and will it be considered a visionary solution. That is, one of the distinguishing features of the made world, as opposed to the natural world, is that it is brought forth for the precise purpose of being the objects of these states, to be a precise fit; and so when a naturally occurring event seems to have this quality of fit, it seems to belong to the made world. Conversely, the less the object accommodates and expresses the inner requirements of the hunger, desire, or fear, the less there is an object for the state and only the state itself, the more it will approach the condition of pain. The more perfectly Boaz fits the inner shape of Ruth's desire, the more will he seem a kinsman and companion brought forth for her not by a bountiful earth but by a bountiful heaven (by, that is, willed artifice). The less he fits those interior claims, the more will she reside in those interior claims and the more she will suffer. That the imagination is somehow implicated in assessments of objectification does not mean that "made objects" are preferable to natural objects, for one may very much prefer the "given" (raw wheat, raw berries) to the "artificial" (wheat bread, berry jam, berry dumplings, berry pie, whiskey, berry cider, berry wine, and so forth); but it does mean that the very preferability of natural objects at a given moment is itself recognizable because the standard of "perfect fit" has been established by the full array of natural and imagined objects that collectively accommodate hunger in all the varying degrees of its insistence, the nuanced petulance of its claims. This is also true of most other states, and so it is familiar to hear people express their amazement at the natural world by an implicit reference to an imaginary standard: "This stone looks like it was hand-made," "I cannot imagine a more kind person," "The wind in the trees looks like the principle of intelligence itself," "There could not be a planet more physically beautiful than earth."

That pain and the imagination are each other's missing intentional counterpart, and that they together provide a framing identity of man-as-creator within which all other intimate perceptual, psychological, emotional, and somatic events occur, is perhaps most succinctly suggested by the fact that there is one piece of language used—in many different languages—at once as a near synonym for pain, *and* as a near synonym for created object; and that is the word "work." The deep ambivalence of the meaning of "work" in western civilization has often been commented upon, for it has tended to be perceived at once as pain's twin and as its opposite: in its Hebrew and Greek etymological origins, in our spoken myths and unspoken intuitions, and in our tradition of religious and philosophic analysis, it has been repeatedly placed by the side of physical suffering yet has, at the same time and almost as often, been placed in the company of pleasure, art, imagination, civilization—phenomena that in varying degrees express man's expansive possibility, the movement out into the world that is the opposite of pain's contractive potential.[10] Any sense that this duality is arbitrary dissolves when work is seen against the full array of intentional acts and objects; for work (like all the intentional states looked at above but to a much greater degree than was apprehensible there) conforms to this same arrangement. The more it realizes and transforms itself in its object, the closer it is to the imagination, to art, to culture; the more it is unable to bring forth an object or, bringing it forth, is then cut off from its object, the more it approaches the condition of pain. So, as an example of the one extreme, is the fact that the collective artifacts of civilization —its paintings, poems, buildings—are habitually referred to individually as "works." Indicative of the opposite extreme is the fact that historical moments when work has been identified with suffering have been moments in which those persons performing the activity of work have been separated from the benefits of the objects that are the product of that activity. Slavery, whether occurring in ancient Egypt or in the nineteenth-century American South, was an arrangement in which physical work was demanded of a population whose members were themselves cut off from ownership, control, and enjoyment of the products they produced. So, too, the nineteenth-century British factory world is one in which work is described as approaching the condition of pain, not only in the extensive writings of Marx but in the British parliamentary bluebooks on which he relied so heavily. The proximity of work to pain is here specifically attributed to the massive hunger, sores, disease, airlessness, and exhaustion suffered by the industrial population, but all these conditions are in turn attributed to the more fundamental shattering of the essential integrity of act-and-object in the human psyche; for the body at work was separated from the objects of its work; the men, women, and children bringing forth out of their labor a multitude of objects (coal, lace, bricks, shirts, watches, pins, paper, plaited straw), themselves inhabited a space wholly outside the realm on which those objects conferred their benefit, a realm that belonged to a set of people who had not themselves directly participated in the making of the objects.[11]

Far more than any other intentional state, work approximates the framing events of pain and the imagination, for it consists of both an extremely embodied physical act (an act which, even in nonphysical labor, engages the whole psyche) and of an object that was not previously in the world, a fishing net or piece of lace where there had been none, or a mended net or repaired lace curtain where there had been only a torn approximation, or a sentence or a paragraph or a poem where there had been silence. Work and its "work" (or work and its object, its artifact) are the names that are given to the phenomena of pain and the imagination as they begin to move from being a self-contained loop within the body to becoming the equivalent loop now projected into the external world. It is through this movement out into the world that the extreme privacy of the occurrence (both pain and imagining are invisible to anyone outside the boundaries of the person's body) begins to be sharable, that sentience becomes social and thus acquires its distinctly human form.

In this process of externalization, each of the two components is itself diminished, which is in the one case a great benefit and in the other only an apparent loss. Although the activity of work may itself at any given historical moment involve a degree of aversiveness in which it begins to be identical with physical pain,[12] it is by no means an internal requirement of the activity that it have this aversive intensity, and it does not ordinarily do so. It does, however, under all circumstances, and regardless of whether it is primarily physical or mental labor, entail the much more moderate (and now willed, directed, and controlled) embodied aversiveness of exertion, prolonged effort, and exhaustion. It hurts to work. Thus, the wholly passive and acute suffering of physical pain becomes the self-regulated and modest suffering of work. Work is, then, a diminution of pain: the *aversive intensity* of pain becomes in work *controlled discomfort*. So, too, imagining achieves a moderated form in the material and verbal artifacts that are the objects of work. If, for example, a person standing in a field imagines himself to be instead standing by the sea, he has (in imagining) brought about a large alteration in the world, displacing the whole physically "given" context with an invented one. If, in contrast, he fashions out of the clay of the ground a cup and introduces this "new" object into the field, he has again brought about an alteration in the physically "given" world, but a much more modest one than in the first case. While imagining may entail a revolution of the entire order of things, the eclipse of the given by a *total reinvention of the world*, an artifact (a relocated piece of coal, a sentence, a cup, a piece of lace) is *a fragment of world alteration*. Imagining a city, the human being "makes" a house; imagining a political utopia, he or she instead helps to build a country; imagining the elimination of suffering from the world, the person instead nurses a friend back to health. Although, however, artifice is more modest and fragmentary than imagining, its objects have the immense advantage over imagined objects of being real, and because real, sharable; and because the objects are sharable, in the end artifice has a scale as large as that in imagining because its outcome is for the first time collective.

That is, if there were a hundred persons, each of whom imagined himself the inventor of a town, each would find that he could instead in a given week only "make" a few hundred bricks, or construct part of a wagon, or clear a piece of a road. But because each person's made objects now inhabit the sharable external space outside his own body accessible to all, the objects he makes can be coupled with those objects made by the second person, and the third, and so the large imagined town gets made. In imagining the town, each person had to invent and sustain the image individually, and thus the hundred persons continually duplicated each other's efforts. Further, for any one person to make the town continually available to himself, he had to devote each day to sustaining the image, and then the next day (day after day), reinvent it once more. In the collective work of artifice, in contrast, the town becomes a freestanding object; it no longer depends for its existence on the mental labor of daily reinvention. Thus the imaginers may move on to other projects. It may be that in the year the town was being built, more focused and sustained exertion was required than would have been required by a hundred persons daydreaming about a town day after day for three hundred and sixty-five days; but in the long run, the effort required to perpetuate the fantasy city would be much greater, since the act would have to be sustained and renewed over fifty years, while those who built the town will in forty-nine of those years be free of its daily reinvention, except for now and then when it needs to be repaired. The advantage of material culture over a culture of belief is (as will be elaborated in a later chapter) difficult to overstate. In work, then, pain is moderated into sustained discomfort; and the objects of imagining, though individually moderated into fragmentary artifacts, are collectively translated into the structures of civilization that have nothing modest about them.

In the attempt to uncover the structure of creation in later chapters, the assumed starting point will be the framing relation of intentionality, described either as "pain and imagining" or "work and its artifacts." There is one additional word that changes in the movement back and forth between these two sets of companion terms, and that should be briefly clarified before starting. The elementary place occupied by the image of the "weapon" in the first set of terms is the place held by the "tool" in the second set of terms: the projection into intentionality through the mediation of "agency" in the pain-weapon-imagined object arrangement (which is in its deconstructed form, the pain-weapon-power arrangement) becomes now the work-tool-artifact arrangement. The modulation of the weapon into the tool has some of the same characteristics as the modulation of the embodied experience of pain into the embodied activity of work, or as the modulation of an imagined object into a materialized or verbalized artifact, but it also has some additional characteristics.

That the sign of the weapon has an elementary place in the transformation of pain into the projected image was suggested earlier, and was described both in its beneficent and, more elaborately, in its deconstructed form.[13] Although in its benign form the displacement of aversive sentience with the disembodied content of objectified images eventually entails an infinite array of images, and although it is not possible to travel back into the origins of human imagining and chart the chronological sequence in which such images first appeared, there are many outwardly visible indications that the image of the weapon is not just one among thousands of signs but is a sign occupying a primal place in the original moment of transformation. Of such outward indications, perhaps the most important to recall here is the centrality of the image in the language of people in physical pain. Physical pain is not only itself resistant to language but also actively destroys language, deconstructing it into the pre-language of cries and groans. To hear those cries is to witness the shattering of language. Conversely, to be present when the person in pain rediscovers speech and so regains his powers of self-objectification is almost to be present at the birth, or rebirth, of language. That the person in pain very typically moves through a handful of descriptive words to an "as if" construction, and an "as if" construction that has a weapon on the other side, indicates the primacy of the sign in the elementary work of projection into metaphor. To describe one's hurt in an image of agency is to project it into an object which, though at first conceived of as moving toward the body, by its very separability from the body becomes an image that can be lifted away, carrying some of the attributes of pain with it.[14] The primacy of this sign in the projection of human pain into disembodied imagining will, in the next chapter, be more elaborately attended to as it occurs in the Hebraic scriptures, where the relationship between the people and their imagined object (God) is repeatedly represented as a relation between a deeply embodied, suffering human being and a wholly disembodied (i.e., immune from pain) principle of creating, mediated by the recurring image of a colossal weapon that transverses the space between them. The only path connecting the body and the power of creating is apprehended as the vertical line running along the edge of a weapon whose one end is on the ground and whose other end is in heaven; on the other side of the concrete, imaginable image an unimaginable, contentless creator is apprehended to exist. It is also useful to recall that the weapon as a materialized artifact is usually assumed to have been present at the infancy of culture. Long before man extends himself out into the world by making other artifacts, he extends himself out into that world by holding onto a found object (stick, stone) that increases, extends, the length and strength of his arm. This weapon may itself be modified into a tool, or the tool back into a weapon, and it is the identity of the two, as well as the profound mental distance separating them, that must for a moment be held steadily visible.

The weapon and the tool seem at moments indistinguishable, for they may each reside in a single physical object (even the clenched fist of a human hand may be either a weapon or a tool), and may be quickly transformed back and forth, now into the one, now into the other. At the same time, however, a gulf of meaning, intention, connotation, and tone separates them. If one holds the two side by side in front of the mind—a hand (as weapon) and a hand (as tool), a knife (weapon) and a knife (tool), a hammer and a hammer, an ax and an ax— it is then clear that what differentiates them is not the object itself but the surface on which they fall. What we call a "weapon" when it acts on a sentient surface we call a "tool" when it acts on a nonsentient surface. The hand that pounds a human face is a weapon and the hand that pounds the dough for bread or the clay for a bowl is a tool. The knife that enters the cow or the horse is a weapon and the knife that cuts through the no longer alive meat at dinner is a tool. The ax that cuts through the back of a wolf is a weapon and the ax that cuts through a tree is a tool. The hammer that hammers a man to a cross is a weapon and the hammer used to construct the cross itself is a tool.

Although one can conceive of exceptions to this basis of differentiation, the exceptions tend to reaffirm the distinction, as well as to call attention to its complexities. If, for example, someone were to object that the ax that cuts through the tree (in the preceding examples) should be called a weapon rather than a tool, the person making the objection would almost certainly turn out to be one who believes that the vegetable world is sentient and capable of experiencing some form of pain; conversely, if one were to object that the knife that cuts through the cow is a tool, the person would be someone who has retracted the privileges of sentience from the animal world and thinks of cows as already-food and therefore, not-quite-alive (as we more routinely think of trees as not-quite-alive). If an ax strikes the side of a house, that ax may—especially to those whose home it is—be perceived as a weapon even though it acts on a nonsentient surface; but this identification itself exposes the fact that we think of human artifacts as extensions of sentient human beings and as thus themselves protected by the privileges accorded sentience. Again, there are certain instruments (such as those in medicine and dentistry) that we call "tools" even though they enter human tissue; but it should be noticed that this identification is "learned" and that even after it is learned, it requires a conscious mental act to hold steady the perception (which violates all intuition) of the object as a tool. Every child recognizes it as a weapon and responds accordingly. Even adults tend to watch the approach of the knife toward an arm with complete equanimity only if they know that the tissue of the arm has been anesthetized and thus made almost nonsentient. Similarly, contemporary arguments about whether abortion is properly understood as a medical operation (tool) or instead as a murder (weapon) have sometimes turned on the question of whether or not the fetus is capable of experiencing pain.[15]

These and many other concrete instances work to reaffirm the mental and moral distance separating sentient and nonsentient surfaces, a distance so great that the object acting on them is perceived and named as two different objects. If one imagines oneself back at an early moment in culture during which a large knife is suspended above a child (Isaac, Iphigenia, any child), and if before the knife falls, the child is moved out of that space and an animal, goat or lamb, is put in its place, that moment of substitution will be recognizable as one that has always (in the retrospective accounts of the culture which followed) been designated a revolutionary moment in the growth of moral consciousness. But if one now holds steadily visible not two pictures but three pictures—the child and the knife looming above, giving way to the lamb and the knife looming above, and now in turn the lamb is moved out of that location and replaced by a block of wood under the still looming knife—so great in the transition from the second to the third picture is the revolution in consciousness that the object itself is now re-perceived as a wholly different object, a tool rather than a weapon, and the anticipated action of the object is no longer an act of "wounding" but an act of "creating."

This difference will be returned to after looking for a moment at a characteristic that remains common to the two objects. The power of alteration resides equally in weapons and in tools. In each there is a tremendous distance between what is occurring at the two ends, not simply because one end is active and the other passive, but because a fairly inconsequential alteration at one end is magnified into an occurrence full of consequence at the other. A small shift in the body at one end of a gun (so small it is almost imperceptible, only the position of one finger moves) can wholly shatter a body at its other end. The pressure of a hand pushing on the handle of a knife, itself too small to alter (to dent, to scar) even slightly the surface of the handle, will as it begins to be transferred and concentrated across the broad half-inch of the handle to the drastically thin surface edge of the blade, be magnified into a huge power of altering whatever surface it touches at its other end. Thus the object, whether weapon or tool, is a lever across which a comparatively small change in the body at one end is amplified into a very large change in the object, animate or inanimate, at the other end. A person using a weapon or a tool can therefore take credit for, "experience," a large alteration without himself "experiencing" any direct bodily alteration; he experiences alteration without himself risking the aversiveness that ordinarily accompanies self-alteration; he objectifies his presence in the world through the alterability of his world. The difference in the alteration that occurs at each end of the weapon or tool is one not just of scale but of duration. The cut made by a sword or a scythe is not only a much greater change than is that of the motion of the lifted arm that caused it, but it also lasts far longer than that arm motion. In addition to whatever practical benefits are gained by hurting an enemy or harvesting grain, there is a magnification of the actor because he has brought about an alteration not only larger than the one he himself experienced but of much greater duration. Whatever assertion of selfhood might be carried by his performing that lift and swing of his arm over and over again in an unbroken sequence over a week, can instead (if there was a tool or weapon held in his hand) be accomplished by a single movement of the arm, for now the alteration itself, the cut grain or the unhealed wound, is the freestanding sign of that momentary motion that itself endures for a week. The altered object becomes a record that prevents the action from having to be endlessly repeated; presence is registered and need not be continually re-enacted.

Although the primary basis for differentiating a weapon and a tool is that the one acts on a sentient and the other on a nonsentient surface, a more precise account would say that the tool, too, acts on a sentient surface but in a delayed way. The making of an artifact is a social act, for the object (whether an art work or instead an object of everyday use) is intended as something that will both enter into and itself elicit human responsiveness. Though the tool does not, like a weapon, act directly on the human body, it does so at two or three steps removed. Marks on a series of trees register the marker's presence as though he stood in all those places: they allow him to inhabit a space much larger than the small circle of his immediately present body. Those marks are now part of the visual field of anyone else who approaches the grove of trees. Rather than using a weapon on someone's eyes, the world is rebuilt or re-presented (even if only modestly altered) in such a way that it must be reseen. That is, rather than directly altering sentience (as occurs in the use of a weapon on a living body), the tool alters sentience by providing "objects" of sentience. It alters without hurting (often even bringing about the diminution of hurt). Through tools and acts of making, human beings become implicated in each other's sentience. Seeing is seeing of x, and the one who has made the "x" has entered into the interior of the other person's seeing, entered there in the object of perception. The objects of hearing, desire, hunger, touch, are not just passively grasped by the fixed intentional states: the objects themselves act on the state, sometimes initiating the state, sometimes modifying it, increasing, decreasing, or eliminating it. Thus when intentional objects come to include not just the rain, berries, stones, and the night but also bread, bowls, church steeples, and radiators, there comes to be an ongoing interaction at the (once private) center of human sentience; for not only are the interior facts of sentience projected outward into the artifact in the moment of its making, but conversely those artifacts now enter the interior of other persons as the content of perception and emotion. Thus in the transformation of a weapon into a tool, everything is gained and nothing is lost.

One final attribute of the tool in the constellation of "work-tool-artifact" is that it is itself the concrete record of the connection between the worker and the object of his or her work; it is the path from the object back to its sentient source; it is the path that if eclipsed from attention allows the object to be severed from its source. This special position of the tool becomes more apparent when work is seen within the framework of intentionality: work is an intentional act; its object (whether a carved statue or a chunk of coal lifted out of the earth) is an intentional object. The tool, as is visible in its two ends, shares characteristics of each. At once act and object, it can be assimilated in either direction: it belongs to the body and is an extension of the human hand; it is also itself an object (the earliest made object, perhaps preceded only by the weapon) which must itself be made before it can participate in the making of other objects.[16] The tool, then, occupies a remarkable position within the intentional frame. In almost all intentional states other than work, the connection between act and object is invisible and magical, signaled only by a preposition (the "of" and the "for" in fear *of* x, hunger *for* y) that is a "placeholder" and a "zero." In work, the locus of the connection becomes for the first time palpable and concrete in the tool. Across its concrete surface, the interior act and the exterior object become continuous; the "of" and the "for" themselves become subject to direction and control. The benefits to sentience are incalculable.

In the following chapters, there will begin to emerge a more complete account of the interior structure of the act of creating as it is objectified and made knowable in the hidden interior of the created object. As the nature of making becomes visible, the significance of unmaking—looked at in the first half of this book—

will itself be more fully apprehended. One of the central problems in exposing the interior of making (which will be called here "imagining" when the activity and its object are interior, and "making" or "creating" when the activity is extended into the external world and has as its outcome a material or verbal artifact) is, as suggested earlier, the selection of an appropriate "model object." Unmaking resides in and can thus be represented by two relatively self-contained events, torture and war, the first of which is its most complete and therefore most perfect representative. But where is the equivalent representative of "making," which manifests itself everywhere, which seems on the one hand fully present in the most fragile and singular of outcomes (a pencil, a white shirt, a clothespin, room, or curtain), yet seems on the other only to be fully present in the overarching structures of civilization whose scale places them beyond the reach of sensory as well as, perhaps, of intellectual apprehension? Any "model object" will be inadequate, either by being too diminutive and fragmentary, or by being too large for description; and the discussion that follows will minimize the problem only by distributing the error in both directions, now erring on the side of the large, now erring on the side of the small, thus moving back and forth between what is fully representative but not representable and what is easily representable but not truly representative. This alternation is itself made easier by the fact that the structure of making appears to remain constant across such changes in scale.

The logic governing the selection of individual material artifacts as model objects will become apparent as those objects are invoked here and there throughout Chapter 4 and centrally in Chapter 5. Such objects (now a blanket, now an altar, now a chair, a coat, or a lightbulb) will be presented in the context of their own self-evident characteristics as well as characteristics consciously or unconsciously attributed to them in, for example, literature and law (a verbal artifact, such as a story or a legal argument, may itself comment on and expose the nature of a material artifact, such as a city or a pennywhistle or an artificial heart). Thus the specific logic of invoking any one of them need not be anticipated here. The logic underlying the selection of fragments from the overarching structure of civilization may not, in contrast, be self-evident when it is encountered in Chapter 4, and so will be very briefly indicated here.

Because the deconstruction of creation takes a specifically political form (torture, war), it might seem most appropriate to trace the outlines of the opposite event again in a specifically political form, such as the moment when a new country is being conceived and constructed (made-up, made-real), or when an already existing country, having been partially destroyed, is being re-imagined and re-constructed (remade-up). As it happens, the human imagination has given a fairly complete account of itself at both such moments. The "creating" of a country, for example the United States, is knowable, *after the fact*, in the freestanding artifact that was its outcome, the United States itself and more particularly, its constitution; but in addition the actual present-tense activity of creating in this instance produced not only an object (a constitution, a country) but also a record of its own present-tense action. In the pages of *The Federalist Papers*, it is possible to see the outline of the act of creation, in part because Madison, Hamilton, and Jay so self-consciously recognize themselves as engaged in an act of "invention," as when Madison stops to differentiate men from angels on the basis that each is self-governing but the first only achieves this through materialized design,[17] or when Hamilton, calling attention to "the interior structure of the edifice we are here invited to erect,"[18] explicitly refers to the nation as a made object. Similarly, the "recreating" or "reconstructing" of partially destroyed nations has a surviving record of both its initial moment of conception and its successive modifications in the written documents and oral history surrounding the Marshall Plan, the Plan for European Economic Recovery, and the European Common Market. If the period between 1939 and 1945 is conventionally identified as one of the darkest in western civilization, then there can also be taken as one of its most luminous the period of years during which Europe was rebuilt, and, in particular, one forty-eight hour period beginning on 5 June 1947 when a quiet speech given before a small audience at an American university set off throughout the night on a faraway continent a series of phone calls between various heads of state who, in a sudden swell of amazement, disbelief, and visionary trust, found themselves (though still standing in the midst of massive rubble, poverty, hunger, and anticipated cold) "imagining" there in the middle of the night a restored Europe, and imagining as well the still only-imaginary path by which they would get there—found themselves also, perhaps, recognizing in some dim corner of their minds that the United States' Secretary of State may have been only speaking hypothetically, may have been only introducing an "as if" clause in the presence of his listeners, but found themselves realizing too (as though participating in a benign conspiracy with some larger, intercontinental imagination) that Europe could invoke its own "as if" clause, and by acting as if the United States quite simply meant what it said, the hypothetical would become real, regardless of original intention.[19] In any event, within one forty-eight hour period, the United States had "supposed," and Europe had begun to act on the supposition.

In both the *Federalist Papers* and the Plan for European Economic Recovery, the work of the human imagination in constructing large units of civilization (for here the unit of shelter is not a room or a house but a country or a continent) is at moments exposed even in the fundamental framing relation between "pain and imagining," for in the first the United States is being described in its conceptual infancy, and thus both the fragility of union and its anticipated strength are co-present on every page; similarly, the success of European recovery was premised on an almost unprecedented willingness on the part of each of the participants to expose its own inner fragility ("For the first time in modern history," Marshall noted, "representatives of sixteen nations collectively disclosed their internal economic conditions and facilities and undertook, subject to stated conditions, to do certain things for the mutual benefit of all"[20]). Nevertheless, to step into the intricacies of either the *Federalist Papers* or the Plan for European Economic Recovery is to walk into the middle of the civilizing process when many of its fundamental assumptions are already securely in place, and thus in no need of articulation, and thus not themselves self-announcing. Equally important, though the generic events of torture and war are opposed by the civilizing impulse even when that impulse has a specific and small location, manifesting itself in something as fragmentary as a table, and though, therefore, it is also opposed by the making of something more expansive, like a nation or a group of nations, it is nevertheless misleading to focus on a particular country or continent at a given historical moment, since insofar as they have a true opposite, it is "civilization" itself.

If western civilization is characterized by a long list of attributes, two that must occupy a central position in any list are first, its Judeo-Christian framework of belief, and second, its insistent thrust toward material self-expression. These two attributes have in part guided the selection of the two texts that are invoked in Chapter 4, the Judeo-Christian scriptures and the writings of Marx, in each of which the nature of creating—the relation between body and image, body and belief, body and artifact—is endlessly puzzled over, looked at now from one side and now from another, now from below and now from above, held up before the mind and turned over and around so that all its intricacies and edges become visible. Although the logic of invoking the Biblical writings may seem self-evident, the logic of invoking the writings of Marx may seem less so; for Marx is in the United States so often narrowly perceived in his capacity as critic of western economic structures that it is sometimes forgotten that he is our major philosopher on the nature of material objects, that he not only accepts but embraces and applauds the western impulse toward material self-objectification, and that he himself, though an angry opponent of what he perceived to be its injustices, accepted perhaps ninety percent of its materialist premises. Even that part of his work that is revolutionary should probably be seen as what Jacob Talmon has called a "western heresy,"[21] a heresy through which western materialist assumptions have been exported to non-western cultures, a vehicle through which a constellation of premises embedded in materialism has been carried into what were originally less materially centered, or nonmaterial, even in some instances anti-material, societies. If this description is accurate, then his relation to the west is not unlike the relation of Christianity to Judaism; for although there must have been a day long ago when Christianity was perceived as a radical rejection of Jewish belief, at a distance of two thousand years it is self-evident that Christianity accepted ninety percent of Judaic assumptions, and by means of the ten percent by which it strayed, itself became the vehicle through which a stunning artifact invented in a tiny corner of the Mediterranean could be extended out over an entire hemisphere, conferring its benefits not only on the tiny population who were racially related to the original imaginers, but to the populations of several continents, for the relation between the believer and the object of belief no longer depended on the disposition of genetic material residing in the body of the believer.

Although one of these writings appears relatively early in the civilization it helped to sponsor and the other relatively late, in both of them elementary attributes of the nature of creating are made visible because in neither of them is anything silently assumed. In each an extended meditation on the nature of human making is occasioned by the recognition of a problem in the already existing realm of artifice: the largest created object (God in one case; the overarching economic and ideological structures of society in the other) is perceived to be either insufficiently reciprocating or insufficiently self-substantiating, and is thus itself subjected to a process of modification in the course of the meditation. In both works, there is a recognition that the strategies of "wounding" and "creating" have become conflated, and the work of differentiating and holding the two securely separate in the mind comes to depend on a controlling and redirecting of the referential activity of the sign of the weapon or tool, the sign that precisely because of its inherent instability has allowed the partial deconstruction of making into unmaking. Although at the starting point of the analysis that follows, problems analogous to those encountered in earlier chapters will be re-encountered, these texts very quickly carry us to new territory of understanding. The language in which the scriptural and Marxist explorations occur requires no translation into the terms that are of central concern in the present study, for in each of them the overt and undisguised subject matter is that of "creating," "creator," "body," "artifact," "working," "hurting," "making," "maker." The only new term introduced into this otherwise familiar list is the word "believing" which, in its Biblical context, is close to being a synonym for what has been called here "imagining": "to believe" is to perpetuate the imagined object across a succession of days, weeks, and years; "belief" is the capacity to sustain the imagined (or apprehended) object in one's own psyche, even when there is no sensorially available confirmation that that object has any existence independent of one's own interior mental activity. Because both of these writings take "making" as their central subject, and because both so regularly traverse the full expanse of ground that separates the extreme framing condition of the body in physical pain from the opposite framing condition of self-objectification as it occurs in its most extended and ample of artifacts, the Judeo-Christian and Marxist narratives themselves become—perhaps to a degree not equalled by any other two texts in the west—epic explorations of the human imagination.

4

THE STRUCTURE OF BELIEF AND
ITS MODULATION INTO
MATERIAL MAKING
Body and Voice in the
Judeo-Christian Scriptures and the
Writings of Marx

THE HEBREW SCRIPTURES, along with the Christian scriptures they gave rise to, are among all the singular artifacts of western civilization perhaps the single most monumental, for they can be credited with sponsoring a civilization to a degree shared by no other isolated verbal text (*Hamlet*, *War and Peace*, the United States Constitution) or by any isolated material object (the pyramids, the Panama Canal, the Brooklyn Bridge). Although, however, they are themselves a monumental artifact, they are at the same time a monumental description of the nature of artifice. Product of the human imagination, they are also a tireless laying bare of the workings of the imagination, not merely the record of its aspirations toward disembodiment or of its origins in pain but, more concretely, the record of the very sequence of stages and substitutions by which it regulates and promotes its own acts of self-modification.

As will gradually become apparent over the course of the present chapter, the scriptures can be understood as narratives about created objects that enable the major created object, namely God, to describe the interior structure of all making. Thus the overall strategy of these writings is to first make something and then let it, in "its own" words, reveal itself to you: it is (to invoke an analogy that by its very modesty suggests the magnitude of the actual situation) as though one were to make a table that then not only explained its own evolution but the whole history of the pressure within you to make the table, giving some indication of the psychic risks taken in the act, as well as the dangers that would have come with abstaining from the act; but, in addition, this articulate and freestanding table would then, in its ongoing interaction with its human inventor, go on to issue instructions about its own repair and redesign, as well as to require that it be itself supplemented with other made objects, so that it might, through all these successive modifications, come increasingly both to clarify and to fulfill the original obligations it incurred at the initial moment of its making. For made things do incur large responsibilities to their human makers (and their continued existence depends on their ability to fulfill those responsibilities: a useless artifact, whether a failed god or a failed table, will be discarded); just as, of course, human makers also incur very large obligations to the objects they have made.

That this record of the imagination's activity occurs here in the form of narrative event and story would only in error be called allegory, for the act it describes, the reflex and outcome of the action whose design it traces, is overtly presented as the act and action of the imagination. There is, as suggested earlier, no veiled language here; for maker, making, hurting, believing, working, creating, are the wholly undisguised terms of the story. Furthermore, the categories in which its activity is followed, categories encountered throughout the first half of the present study—body and voice (or, in the language of the Christian scriptures, flesh and word)—are not categories read into the text but the categories in which the text announces itself directly. In turn, it is useful to recognize that these are among the most elementary and least metaphorical categories we have. Compared to them, the rush of analytic categories we ordinarily use to enter and accommodate human experience—hierarchical and dialectical, form and content, authoritarian and egalitarian, reason and emotion, for-itself and in-itself, apollonian and dionysian, and so forth—are (however solid, legitimate, and fruitful) remote and fantastical elaborations of distinctions that are only apprehensible once we are already moving about in a richly fictionalized world—that is, their use arises at the point where we are already extended out into a dense sea of constructs and artifacts, deeply immersed in made culture. The concepts of body and voice, in contrast, though not themselves prior to culture and artifice, are perhaps as close to prior as is possible, for they appear to emerge as explanatory rubrics in early moments of creating, or when there is some problem in the relation between maker and made thing that carries us back to the original moment of making. At the same time, they do not, once made culture has been fully entered, cease to be analytically useful, in part because they are at all times immediately recognizable. They continue to be in the end our best, as at the beginning our only, companions.

As "making" has a structure, so too "believing" has a structure, and in the Hebraic scriptures the structure of verbal and material making is only fully exposed because of a problem that arises within the structure of belief. Thus before attending to scenes centering on made artifacts, it is necessary to back up one step to a very different kind of scene, that in which the human body is perceived to be wounded by the primary Artifact, God. The relation between God and human beings is often mediated by the sign of the weapon. At regular intervals, the vast space separating them is transversed by this image. Even when the text does not designate a specific image of a knife or rod, it is implicitly present, hovering in space with its most essential feature, its two impossibly different ends, helping both to account for and to demonstrate the power and perfection of the divine and the imperfection and vulnerability of the human. The invented god and its human inventor (or, in the inverted language of the scriptures, the creator and his creature) are differentiated by the immunity of the one and the woundability of the other; and if the creature is not merely *woundable* but already deeply and permanently *wounded*, handicapped or physically marred in some way (Leviticus 21:16; 22:21; Deuteronomy 17:1), then that individual is asserted to exist at an even greater moral distance from God than does the "normal" person.[1]

Though scenes of wounding by no means constitute the sole content nor even the major content of these writings, they recur so frequently that no reader, Jewish or Christian, will have failed to notice them, and few readers, Jewish or Christian, will have failed to be troubled by them. God's invisible presence is asserted, made visible, in the perceivable alterations He brings about in the human body: in the necessity of human labor and the pains of childbirth, in a flood that drowns, in a plague that descends on a house, in the brimstone and fire falling down on a city, in the transformation of a woman into a pillar of salt, in the leprous sores and rows of boils that alter the surface of the skin, in an invasion of insects and reptiles into the homes of a population, in a massacre of babies, in a ghastly hunger that causes a people to so glut themselves on quail that meat comes out of their nostrils, in a mauling by bears, in an agonizing disease of the bowels, and so on, on and on. There are, of course, many moments in which God explicitly refrains from an act of wounding, retracts even what is presented as a "deserved" punishment; so, too, there are reassurances about Its mercy and the brief duration of Its anger; there are alternative and profoundly benign images of God ("The Lord is my shepherd"); and above all, the largest framing act of the narrative—the creation and growth of a people and their rescue from their human oppressors—establishes the benevolent context in which these other, themselves terrifying, scenes occur. Yet the positioning of God and humanity at the two vertical ends of the weapon itself recurs so regularly that it seems to become a central and fixed locus toward which and away from which the narrative continually moves. At times, this image seems to define the structure of belief itself. The problematic scenes of hurt, as will be shown, tend to occur in the context of disbelief and doubt: the invisible (and hence periodically disbelieved-in) divine power has a visible substantiation in the alterations in body tissue it is able to bring about. Man can only be created once, but once created, he can be endlessly modified; wounding re-enacts the creation because it re-enacts the power of alteration that has its first profound occurrence in creation.

As an understanding of the nature of material and verbal making requires as prelude attention to the problematic scenes of wounding, so in turn an understanding of these scenes of wounding requires as prelude attention to scenes that

center on the growth of the Hebrew population—scenes in which there is a
benign alteration in human tissue that comes with procreation, pregnancy, self-
replication, and multiplication. Although the nature of what is apprehended to
have been the original genesis of man and world is reaffirmed in scenes of hurt,
it is also reaffirmed in the human generational act itself, which in its insistent
repetitions (especially throughout the opening book of Genesis) becomes the
large framing event against which the less frequent but much more problematic
scenes of hurt occur. It will become apparent that the two very different kinds
of scene, reproduction and wounding, each contain an identifiable relation between the human body and an imagined object (God): in each, the experienceable
"reality" of the body is read not as an attribute of the body but as an attribute
of its metaphysical referent. The constancy of this relation across the two scenes
may explain how the Old Testament mind arrived at its otherwise inexplicable
dependence on the rhythmic invocation of scenes of hurt: that is, it may be that
such scenes are introduced in order to perpetuate a relation between body and
belief once the "begat" sequences can no longer be called on to provide the
needed confirmation (as after Genesis, where the next four books of the Torah
have as their subject a single generation of the Hebrew people).

Thus the problematic scenes of wounding will here be framed on either side
by scenes of production—on the one side, bodily reproduction and on the other,
the production of material and verbal artifacts. As this sequencing of material
is meant to suggest, there is in the scenes of human procreation a profound
intuition about the nature of mental creation; then, in the transition from the first
kind of scene to the second, this intuition is obscured and deconstructed, as the
activity of creating becomes conflated with the activity of wounding; in turn, in
the transition from the second to the third subject, the deconstruction is itself
re-constructed, the original intuition is itself rescued, and creating and wounding
are once more held securely in place as separable categories of action.

The overall sequencing of material in this chapter will occur in the following
order. Section I will show that the human body and God's voice are separable
ribbons of occurrence in the scriptural presentation of human reproduction, and
that events occurring in one ribbon are "read" as confirmation of activity occurring in the other. Section II will attend to the same separable bands of physical
and metaphysical occurrence in scenes of wounding, showing the way in which
the Artifact invented to relieve bodily engulfment now Itself requires bodily
engulfment to confirm its "realness." The relation between man and God here
becomes a power relation based on the fact that one has a body and the other
does not, a relation that is itself radically revised in the Christian scripture where
the moral distance between man and God is as great as in the Old Testament
but no longer depends on a discrepancy in embodiedness. Section III will show
that this Christian solution to the problem is itself anticipated in the Old Testament
modulation of "believing" into "making," for the agonizing labor of sustaining

belief (that is, sustaining over endless days the unobjectified mental object of imagining within one's own psyche) is modified by the external existence of material and verbal artifacts that themselves take over the substantiating function of the sentient body while, at the same time, absorbing into their interior the original moral and ethical import of creating. If the human inventors only very gradually receive from their first Artifact permission to author other artifacts, the very slowness with which permission is granted allows the scriptural mind to expose with startling clarity the depth of moral and psychic categories that are eventually invested in material culture. To the question, "What is at stake in material making?", the Hebrew scriptures answer, "Everything." This answer, as will be elaborated in Section IV, is again given in Marx's similar account of the interior attributes of artifice. The essential character of made objects—which will be presented throughout this chapter in the specific vocabularies used in these two monumental texts about making—will in the final chapter be presented in the more immediately familiar language from our everyday interaction with the things we have made.

I. Behold Rebekah: the Human Body and God's Voice in Pregnancy, Reproduction, and Multiplication

The Human Body

The periodic lists of names of offspring—"When Seth had lived a hundred and five years he became the father of Enosh. . . . When Enosh had lived ninety years he became the father of Kenan. . . . When Kenan had lived seventy years he became the father of Mahalalel" and Mahalalel became the father of Jared and Jared the father of Enoch, and Enoch the father of Methuselah, and Methuselah of Lamech, and Lamech of Noah, and Noah of Shem, Ham, and Japheth (Genesis 5:6–28)[2]—are lists of names which we sometimes allude to as though they were flattened recitations of obscure genealogies. But they have when read in the context of the scriptures themselves a tone of triumph and self-assurance that is simply awesome: it is only their formal cadences and their rigidly parallel sequences that control, that just barely contain, the excitement implicit in their assertions. Far from occurring as an intrusion into the text, they appear there as one of its central matters and gradually come to constitute one of the large structures in the five books of Moses. Something of the scale and magnitude and absoluteness of the initial creation itself is gradually implicated in the slow but increasingly inevitable transformation of two people into "a people," a transformation that begins, "Now Adam knew Eve, his wife, and she conceived and bore Cain, saying, 'I have gotten a man with the help of the Lord,' " a passage which without breaking immediately continues, "And again, she bore his brother" (Genesis 4:1,2),[3] for the re-enactment of the initial creation is insistently multiple; it is only its massive and collective multiplicity that allows it to approximate the singularity of the initial creation. As Genesis progresses, the text breaks with increasing ease into a happy enumeration of persons—"The sons of Japheth: Gomer, Magog, Madai, Javan, Tubal, Meshech, and Tiras. The sons of Gomer: Ashkenaz, Riphath, and Togarmah. The sons of Javan: Elishah, Tarshish, Kittim, and Dodanim. From these the coastland peoples spread" (Genesis 10:2–5)—enumerations of persons so extensive that they soon come to be the sole content of the chapters in which they occur, the sole content of the thirty-two verses of Genesis 5, the sole content of the thirty-two verses of Genesis 10, and again of the thirty-two verses of Genesis 11, and once more the major content of Genesis 36 and again, Genesis 49, until across the twelve tribes of Israel and down through the uncountable generations there exist in the years of the exodus out of Egypt the astounding six hundred and one thousand, seven hundred and thirty people of the double census of Numbers (26:51).

Genesis contains both these passages that trace genealogical lines, passages dense with the names of persons, and very different passages so empty of multiplicity, so centered on a single figure or set of figures that their acts seem to occur in an isolation as great as that of the original garden; each of their gestures has something of the absolute quality of the first human gestures— walking, listening, turning, reaching for some fruit. The connection between the two kinds of passage—list and story—is clear and open, requiring no explication, for many of the stories have as their overt subject some aspect of the successful bringing forth of children: *the finding of a spouse* (the servant of Abraham seeks a wife for Isaac—Genesis 24), *the keeping of a spouse* (twice in the presence of foreign admirers, Abram must choose whether to designate Sarai his sister or his wife—Genesis 12:10–19; 20:2), *the losing of one's spouse* (Abraham's Sarah dies; Lot loses his wife at Sodom—Genesis 23 and 19), *the recovery from the loss of a spouse* (Abraham arranges for his son to become a father; in wine-filled nights, Lot sleeps with each of his two daughters, and from these nights are descended the Moabites and Ammonites—Genesis 24 and 19). There is a persistent, usually encumbered, and always unembarrassed concern with the opening and closing of wombs. Even the lap of birth itself is occasionally depicted in startling clarity:

> When her days to be delivered were fulfilled, behold, there were twins in her womb. The first came forth red, all his body like a hairy mantle; so they called his name Esau. Afterward his brother came forth, and his hand had taken hold of Esau's heel; so his name was called Jacob.   (Genesis 25:24–26)

And again—

When the time of her delivery came, there were twins in her womb. And when she was in labor, one put out a hand; and the midwife took and bound on his hand a scarlet thread, saying, "This came out first." But as he drew back his hand, behold, his brother came out; and she said, "What a breach you have made for yourself!" Therefore his name was called Perez. Afterward his brother came out with the scarlet thread upon his hand; and his name was called Zerah.   (Genesis 38:27–30)

The Genesis stories, forever heavy with concern for many human and inhuman matters, are always attentive to creation and procreation, to genesis and to numbers.

In the lists, then, there is a contraction and rapid iteration of the essential event recounted in the stories; and, conversely, the stories are in part an elaboration of a connection between body and belief so folded in on itself in the lists that it is unenterable. Each of the two contains within itself what is withheld from view in the other: the men and women in the stories contain within their bodies not the singular children specified in the narrative, but the tiers and tiers of offspring contained in the lists; so, in turn, the formal tiering in the lists conceals what only becomes manifest in the stories, the extremity of physical and imaginative work required of human continuity and connection.

But even when the overt subject is far from generation, the form of the narrative—the alternation between list and story—is itself mindful of birth. The list contains a crowd of names; the passage is dense, almost overburdened, with names; then the last in the list, as though under the pressure of the weight of ancestors named before him, is pushed out of the list, emerges out of the crowd of successful parents, the remote thicket of disembodied human presences, and moves toward one in discrete actions that make him large and embodied in the reader's field of vision. Out of the list of names from Adam to Noah in Genesis 5 comes Noah and a story that continues for four chapters. Or again, the enumeration of names in the tenth and eleventh chapters—Noah; then Japheth and his fourteen named descendants; then Ham and his thirty-one named descendants; then Shem and his many descendants including Arpachshad, Shelah, Eber, Peleg, Reu, Serug, Nahor, Terah, *Abram*, Nahor, and Haran—yields to, "Now the Lord said to Abram . . ." and with those words Abram is now separated from the sequence of disembodied parental presences and comes before us in physical and verbal acts in the many stories that follow.

That this alternation between list and story constitutes an essential rhythm is suggested by the fact that it occurs in units much larger than, as well as much smaller than, the several chapter sequences. The alternation, for example, describes the relation between what is on the one hand contained in Genesis and what is on the other contained in the remaining four books of the Torah, Exodus, Leviticus, Numbers, and Deuteronomy. Genesis is in its entirety itself a complex list out of whose final terms emerges in the next four books the intimate, palpable, and sustained story of Moses and his people, a single tier of the generational expanse. The alternation between list and story also describes what occurs in a unit much smaller than the several chapter sequence: the story will sometimes suddenly pause, break into a brief spread of names, then continue once more. In some ways, this most contracted instance of the alternation is the most haunting, for it is as though the large figure is suddenly for a moment swallowed into a group of figures on a dusty road, from which a moment later we watch him break away and emerge toward us once more. In the midst of the story of Abram, for example, the story gives way to, then separates from, a brief list: "Then one who had escaped came, and told Abram the Hebrew, who was living by the oaks of Mamre the Amorite, brother of Eshcol and of Aner; these were allies of Abram. When Abram heard that his kinsman had been taken captive . . . " (Genesis 14:13,14; see 26:34,35)—and so the story now continues. Usually the brief cluster of names specifies family identification; but even when, as here, it does not, it once more re-enacts the rhythm of substantiation, the passage into the material world.

There is a second occurrence of this phenomenon of emergence *within* the story itself. Just as the single figure of the story is a materialization of the disembodied figures of the list, so there is often within the story a particular point where there occurs a more extreme materialization of the already materialized figure. This may be a moment in which there is an objectification of an interior attribute that would ordinarily be understood as emotional or psychological: the laughter of Sarah (18:12,13), the weeping of Abraham (23:2), the assertion of Isaac's love for Rebekah (24:67). But despite the range of emotional states (here, joy, mourning, desire), to a large extent there are in the Old Testament writings only two psychological categories, because only two psychological states of searing consequence, belief and disbelief, and of these only the first is permitted to be understood as an interesting category. It is through the human body that belief is substantiated and, as will eventually become clear, it is in its capacity of substantiation that the body, the interior of the body, is often represented in these stories.[4]

The most overt instances of this occur in those passages describing the actual passage of children out of the mother's body. Within the mental structures of the Hebraic scriptures, a baby not only emerges out of the interior of the body but itself represents the interior of the parental body. In the recounting of the birth of Esau and Jacob and again in the recounting of the birth of Perez and Zerah (passages quoted above), the struggle between the twins for the privileged sibling position provides the occasion for a graphic presentation of the event itself. While it would be both indelicate and inaccurate to say that in these passages the visible interior of the maternal body is itself depicted, it should at least be noticed that the attributes of the infants singled out (the color red, the condition of hairiness, the act of extending, then pulling back, a small arm from the encircling parent) at least do not deflect attention away from the image of the woman's vagina in a way that other attributes (blue eyes, the particular expression on the infant's face) certainly would.

More often, the immediate fact of birth is only asserted rather than described, and the representation of the mother's interior is less direct. One recurring instance is the image of the well. Both its shape and its containing of a life-giving substance make it an appropriate and inevitable objectification. Its alliance to the womb and its implication of abundance are usually unmistakable, as in that passage where Abraham's line through Hagar and Ishmael is first announced (16:7–11), or again in that passage where that line is about to end with Ishmael's death in the wilderness. An angel of the Lord appears to Hagar and says, " 'Arise, lift up the lad, and hold him fast with your hand; for I will make him a great nation.' Then God opened her eyes, and she saw a well of water; and she went, and filled the skin with water, and gave the lad a drink" (21:18,19). The reading of the well as a discreet representation of the interior of Hagar, the implied place of the well within her body and the assertion of her material extension into the future, is made imagistically unambiguous: the promise of abundance accompanies the instruction to re-enact the original unity of mother and infant (closing his hand within hers, as earlier he was himself enclosed within her), which is immediately followed by the appearance of the well. The three retrace in reverse the path of birth: as the womb leads forward to a separation from the child and his freestanding perpetuation in the world, so in this sequence the joining of the two bodies leads back to the womb, to its externalized substitute, the well. He lives because he is given water: thus the rescue requires that the interior content of the well now also become the interior content of Ishmael's body; that is, rescue occurs by reaffirming the interior bodily continuity of the two generations, their sameness. The mediating detail of placing the water within another once-alive container, "a skin," works to underscore the kinship between the living (mother) and nonliving (well) containers of life.

A second artifact through which the inside of the human body is projected out into the world is the altar. It is sometimes explicitly merged with the object of the well and with the generative affirmation:

And Isaac dug again the wells of water which had been dug in the days of Abraham his father; for the Philistines had stopped them after the death of Abraham; and he gave them the names which his father had given them. But when Isaac's servants dug in the valley and found there a well of springing water, the herdsmen of Gerar quarreled. . . . And he moved from there and dug another well, and over that they did not quarrel; so he called its name Rehoboth, saying, "For now the Lord has made room for us, and we shall be fruitful in the land."

From there he went up to Beersheba. And the Lord appeared to him the same night and said, "I am the God of Abraham your father; fear not, for I am with you and will bless you and multiply your descendants for my servant Abraham's sake." So he built an altar there and called upon the name of the Lord, and pitched his tent there. And there Isaac's servants dug a well. (26:18,19, 22–25)

The building of the altar by Noah after the flood immediately follows God's command, "Breed abundantly on the earth, and be fruitful and multiply upon the earth" (8:17). So, too, Abram builds two altars in the land of Canaan following God's announcement of His intention to make him a great nation (12:1–7); and again Abram's building of the altar by the oaks of Mamre immediately follows God's prediction of man's incomprehensible plenitude (13:6). The relation between the three shapes of womb, well, and altar should for a moment here be placed side by side. When they are, it becomes evident that in the transition from the living womb to the externalized artifact of the well, the shape of the containing vessel, as well as the relation between outside and inside, is held constant: in each it is not the outside surface but the inside surface that functions "to contain," that holds what is precious. But in the transition from the first and second to the third, the containing shape of womb and well is turned inside-out, for what had been the interior lining now becomes the exterior, tablelike surface: it is now what is perceived to be in contact with this outward surface (God) that is most precious. That the altar's surface is the reversed lining of the body is made more imagistically immediate in all those places where blood is poured across the altar. The significance of the altar will be returned to at a later point, but for now it is important to recognize that it is a turning of the body inside-out; for, as will become increasingly clear, belief in the scriptures is literally the act of turning one's own body inside-out—imagining, creating, the capacity for symbolic and religious thought begin with the capacity to endow interior physical events with an external, nonphysical referent.

The progression from the list of names to a story centered on a single embodied figure, and then once more from the story of this figure to a passage containing a representation of the interior of the body itself, are two stages of a process of substantiation that, though described here at some length and with some labor, actually occur with great tact and ease in Genesis itself. At that point, for example, where the servant of Abraham goes with ten of his master's camels and many gifts to the city of Nahor in Mesopotamia to find a wife for Isaac, Rebekah appears before us in these eight verses. The double movement into materialization occurs in the first sentence, and is only elaborated in the next seven:

> Before he had done speaking, behold, Rebekah, who was born to Bethuel the son of Milcah, the wife of Nahor, Abraham's brother, came out with her water jar upon her shoulder. The maiden was very fair to look upon, a virgin, whom no man had known. She went down to the spring, and filled her jar, and came up. Then the servant ran to meet her, and said, "Pray give me a little water to drink from your jar." She said, "Drink, my lord"; and she quickly let down her jar upon her hand, and gave him a drink. When she had finished giving him a drink, she said, "I will draw for your camels also, until they have done drinking." So she quickly emptied her jar into the trough and ran again to the well to draw, and she drew for all his

camels. The man gazed at her in silence to learn whether the Lord had prospered his journey or not.   (24:15–21)[5]

Out of the net of names (Bethuel, Milcah, Nahor, Abraham), Rebekah materializes, and then (still within the first sentence) out of Rebekah materializes in turn an object, a jar,[6] an image of abundance and generosity whose scale—the gigantic thirst of camels is renowned and there are ten of them; the servant looks on in stunned silence—leaves no question about the part she can play in bringing forth a great nation. Her dignity is enhanced rather than compromised by this revelation of her eager hospitality and largesse, which in its strength and sweet concern falls immediately into place within what in the very small number of lines of this chapter has become a dense texture of human promises and commitments, for in this story a servant is acting on behalf of his master, who is in turn a father acting on behalf of his son. In the chapter immediately preceding this one, Sarah has died and Abraham has buried her in a cave in a field in the land of the Hittites. It is against this background that Rebekah is sought and found. Abraham is old, but Isaac is young; Sarah is dead, but Rebekah has come—the generations will continue.

There is of course in the story of the finding of Rebekah a crucial element, the presence of God, that has so far been unattended to, as it has also been unattended to in the foregoing account of the other scenes of reproduction. The story of Rebekah will be returned to, after first returning to the general framing event of generation.

God's Voice

Genesis is filled not only with the emphatic material reality of the forever multiplying human body, but with God's voice which takes two different forms, a command ("Be fruitful and multiply") and a promise ("You will be fruitful and multiply"). Whether a promise, a prediction, or an instruction, it is one that is iterative in its occurrence, for God or his angel many times repeats it to Abraham or to his descendants, or Abraham himself repeats it to his descendants, or those descendants repeat it back to God as a reminder of His pledge (for example, 32:12). It is, of course, iterative not only in its form but in its content, its images and metaphors of unimaginable multiplicity. "I will make your descendants as the dust of the earth; so that if one can count the dust of the earth, your descendants also can be counted" (13:16); and again, "Look toward heaven, and number the stars, if you are able to number them. . . . So shall your descendants be" (15:5); and again, "Behold . . . you shall be the father of a multitude of nations. . . . I have made you the father of a multitude of nations" (17:4,5); and again, "I will multiply your descendants as the stars of heaven and as the sand which is on the seashore" (22:17); and as the narrative passes from Abraham to Isaac, the promise is still recurring, "I will multiply your descendants as the stars of heaven" (26:4). It is directly through the promise that the multiplication of humanity comes to be understood as a re-enactment of the original creation, for the power "to make" bodily tissue at the original appearance of Adam and Eve is also the power to alter, magnify, multiply the amount of that bodily tissue. Thus to the words "I have *made* you" are added the words "And I will *make* your descendants"; to the words "I have *formed* you" are added the words "And now I will *multiply* you," "I will *make* you multiple"; to the words "You are what I have *made*" are added the words"I will *make* of you a father," "I will *make* you the father of multitudes." The language of the promise strengthens the sense of the re-enactment, for the dust of the original creation is now the luminous dust of the stars, dust, sand, and soil that surround the promise with images of boundless multiplicity.[7]

The place of man and the place of God in the human generation that so dominates Genesis are easy to separate from one another: the place of man is in the body; the place of God is in the voice. The narrative records momentous alteration in the human body. The list makes emphatic the sheer repetitive nature of matter, for to move outside the body and immediately find another presence and another and another is the method of the list. The story emphasizes the factual density not by the repetition of matter as one moves outward but by the same solidness and self-repeating quality as one moves inside, or when what is inside emerges out, as in the double rhythm of intensified materialization. Once inside any single figure, one is back with the list—the crowd of future humanity resides within the parental body. Both story and list are ways of making the awe-inspiring alterability of matter visible. The change that occurs to the individual body in pregnancy is, of course, an overwhelming one: a single human body becomes a much bigger body and then breaks into two. So, too, the overall multiplication, the capacity of something (whether human tissue, or a blue flower, or a microbe) not simply to persist in its existence, to stay, to itself endure, but to re-assert and replicate its aliveness, and soon occupy more and more space, to achieve greater and greater presence, is again a visible alteration whose power to overwhelm can only in the twentieth century be missed because we now live at a time when the growth of populations is simply assumed.[8] The scriptural preoccupation with the list carries us back to a time when the duration of a people was more precarious, and when its survival and rapid growth were thus occasions for triumphant recitation. The act of counting (here and in many other contexts) has a fixed place in the landscape of emergency;[9] and when the count is favorable, counting and recounting are also one of the deepest sources of pleasure.

In contrast, God's presence within the story of human generation is exclusively verbal. A large change is occurring in the scale of the collective human body. No change is occurring in God himself; for he is unalterable and unchanging;

that is, he has no body (it is precisely the condition of having a body that makes one susceptible to the original alteration of having been created, and that makes one then susceptible to subsequent alterations, such as multiplication and growth). The statements of God are almost wholly separable from the material reality they describe: they accompany human events without entering them. The physical and the verbal run side by side, one above the other, as two distinct or at least distinguishable horizontal ribbons of occurrence. The first only participates in the second by anticipating it: that is, it is as though the upper ribbon has been pulled back one interval so that its content will always immediately precede the content of the lower ribbon. The verbal enters the human phenomenon of generation by being placed before it and so coming to be perceived as its cause or agent.[10] The descriptions attributed to the primary Artifact take (as was noted earlier) two different forms—a command or instruction ("Be fruitful and multiply") and a promise or oath ("I will make you fruitful and multiply you")—both of which are anticipatory. Hence the actual fact of the magnification of the human body, the literal event of procreation and multiplication, is never simply an event in and of itself but becomes in the first form an obedient acting out of the thing that had come before, and in the second form a divine fulfillment of the thing that had come before. The fact that in the first case the transition from voice to body requires human work while in the second it requires God's work is unimportant beside the fact that in both instances an alteration in the sheer quantity of human matter is given a referent that is profoundly immaterial.

What is to be noticed here is that however more powerful the Word of God is than the Body of man, it is within these stories always the case that the Word is never self-substantiating: it seeks its confirmation in a visible change in the realm of matter. The body of man is self-substantiating: iteration and repetition (the material re-assertion of the fact of their own existence) is the most elemental form of substantiating the thing (existence, presence, aliveness, realness) that is repeated. But the body is able not only to substantiate itself but to substantiate something beyond itself as well: it is able not only to make more amply evident its own existence, presence, aliveness, realness but to make ever more amply evident the existence, presence, aliveness, realness of God. With each successive increase (yesterday there were five hundred people and today six hundred: here are their names, the space they fill, the presence that can be seen, the garments that can be touched, the laughter and conversation that are audible), they reassert not only the sensorially confirmable realness of their own existence (We are. We are. We are. We are.) but the sensorially confirmable realness of God's existence (He is. He is. He is. He is.), or in the voice that is attributed to him, "I am. I am. I am. I am." Although the invented Artifact allows a people to experience itself in the capacity of spirit rather than matter, it is here not a diminution of the body but its amplification that sponsors the increased apprehensibility of the spirit. An extreme change in the visible world now has a referent in the invisible world; the body in its most intense presence becomes the substantiation of the most disembodied reality.

This separation of the material and the verbal, and the intensification of the material required by the substantiation of the verbal, are apparent not only in the overall event of generation in Genesis—the growing of two people into a people and the verbal descriptions, the prophetic commands and promises accompanying that growth—but also within specific stories. Within any given story, the body of man and the voice of God are still separable, though more difficult to describe separately because God is here directly credited with the physical act of opening and closing the womb. Here it is again apparent that the attribution of an occurrence within the human body to God entails not a diminution but an intensification of the bodily occurrence. The ordinary alternation in the female body between the conditions of being sometimes "not with child" and sometimes "with child" becomes within these stories the intensified and absolute conditions of "barrenness" and "fertility": "not with child" has become the much more extreme "not with the capacity to conceive a child" and the capacity to conceive a child is now coterminous with (and exclusively reserved for) the phase of pregnancy. Sarah is not "without child" and then "with child" but first barren and then able to conceive (17); Rebekah, in turn, is barren and then, by God's action, a mother (25:21); God closes, then opens, all the wombs of the house of Abimelech, preventing, then permitting, the power to bring forth children (20:17,18); the childbearing phases within the lives of Leah and Rachel are similarly described in the language of these absolute categories (29 and 30). Everything is at stake in the alterability of the body, for this attribute is at once intensified and lifted away from the body and attributed to God. First, the self-alterability of the body is denied, for to be barren is not just to be without child but to be unalterable, unable to change from the state of without child to with child: barrenness is absolute because it means "unalterable" except by the most radical means, unalterable except by divine intervention. This, then, is the second step, a doubling and lifting away of the power just denied in the woman: God in changing the body from barren to fertile is not simply changing it from being unpregnant to pregnant but changing it from being "unchangeable" to both changed and pregnant. Onto the humanly created Artifact is projected in the opening story of creation the very power of creation that brought It into being (God, not man, is the Creator); and now again here, the authored object is Itself understood as the Author; the produced objectification of the human capacity for self-replication and self-extension (God) is itself made responsible for all acts of human reproduction. This intensification of the body and lifting of its central attribute away from it and assignment of that attribute to the immaterial and spiritual are (as will become increasingly clear) crucial in the overall conversion of body into belief.

The categories of material and verbal, or body and voice, or sentience and self-extension, are for the most part throughout these stories kept securely separate. It is true that though man is primarily a body, he is (at least at the edges of experience) not wholly without a voice. The degree of his verbal participation is, in fact, the one fluid element in the stories of generation: it is human prayer, the prayer of Abraham, that leads God to re-endow the house of Abimelech with the powers of conception (20:17,18); it is again the prayer of Isaac that brings God to Rebekah's womb (25:21).[11] But if at the boundary that separates the human body and God's voice the degree of man's verbal participation is negotiable, the degree of God's bodily participation is emphatically unnegotiable. God confirms his existence before humanity in the bodies of the human beings themselves rather than in any materialization of Himself separate from their bodies. It is surely in fact in part the longing for a separate materialization of God that makes "the land" such a resonant category in the scriptures: it is almost God's body, a separate form of substantiation, a form of material amplification (a large and solid fact to match the large and solid fact of their physical amplitude) intimately bound up with and almost offered in exchange for their own material magnification. Human magnitude and God's material magnitude will occur simultaneously: "To your descendants I give this land" (15:18); "I will give to you and to your descendants . . . " (17:8); "To your descendants I will give this land . . . "(24:7); "I will give to your descendants all these lands . . . " (26:4). But the land is not, of course, an exception to the purity of the verbal category, for it is itself a verbal construct: it is never in the Pentateuch "the land," but instead "the promise of land," the promised land.

Between the extremes of the physical body and the voice of God, there is no transition. They remain separate bands of occurrence. Although the scriptural mind labors to keep the two separate, at the same time precisely how one then gets from one to the other becomes a source of great concern. Though alterations in the material world make real (believed-in) the existence of an Alterer, the path from the realm of matter to the Referent must itself be materialized, made real, believable. Even after the invisible path from the voice describing generation to the material fact of generation is identified as causal, the path itself remains inaccessible to the senses. There remains an anxiety about the nature of the crossing that is apparent in the search for models that periodically surfaces in the stories.

The way in which a disembodied cause influences or "instructs" (from "*struere*," to build; instruction is thus itself a form of creation) may have one such model in the story of Jacob's mating rods:

> Then Jacob took fresh rods of poplar and almond and plane, and peeled white streaks in them, exposing the white of the rods. He set the rods which he had peeled in front of the flocks in the runnels, that is, the watering troughs, where the flocks came to drink. And since they bred when they came to drink, the flocks bred in

front of the rods and so the flocks brought forth striped, speckled, and spotted.   (30:37–39)

Just as human reproduction entails the incorporation of the disembodied word of God into the human body, so there is here an analogous phenomenon: the absorption of the disembodied "word" of man, the abstract pattern of the artifact, into the animal body. The simplest piece of human culture (one in many diverse contexts invoked to exemplify the elemental cultural impulse, the cutting of the mark on the tree[12]) becomes in its relation to the animal body the equivalent of the force of the divine voice on the human body. The disembodied stripes of the artifact are now not just the abstract analogue to the embodied stripes of the flocks but the cause of them: the animal stripes, whatever their merit in and of themselves, become the record of the action of the spirit on matter, the record of the power of voice on matter, evidence that fragments of self-extension circle back and themselves act on the sentient source out of which they arose. Although this passage occurs in the midst of a story clarifying the nature of the relation between Jacob and his earthly father-in-law, Laban, it works more widely to clarify the relation between Jacob and his heavenly father-in-law; for Jacob appropriates the gifts of both fathers, the flocks of Laban and (in his animal husbandry) the cultural authorship and authority of God. The passage provides a model for the connection that exists between the two too easily separable ribbons of voice and body: it does not explain *how* the voice acts on the body but simply asserts that that can in fact occur, re-enacts it, providing a visible image.

A second very different, and much more representative, model again depends on re-enactment, and carries us back to the very beautiful story of the finding of Rebekah. The short distance between the start of Chapter 24 and the passage in which Rebekah appears contains a sequence of four verbal acts: first, Abraham describes to the servant the oath he wants him to make; second, Abraham describes to the servant the promise made to Abraham by God about the future of his people; third, the servant (placing his hand under Abraham's thigh) takes the oath; fourth, the servant, now in Mesopotamia, prays a prayer to God in which he describes his present surroundings ("I am standing by the spring of water, and the daughters of the men of the city are coming out to draw water") and then describes a hypothetical future verbal event which if it then happens (and it does so happen a few lines later) will be the sign that the woman found is the woman appointed for Isaac by God, thereby fulfilling not only the servant's verbal prediction but also the three preceding verbal acts. The verbal acts of man and God are, in the courting of Rebekah, braided together, as are the series of representational acts: the act of a father on behalf of a son (as, too, the interior of Abraham's thigh and the interior of Rebekah's womb are allied), the act of a servant on behalf of a master, the act of a people on behalf of their God, and the act of a God on behalf of His people. These representational contexts, the continuous series of "acts on behalf of," establish within a handful of sentences a rhythmic habit of "impersonation" of and commitment to something beyond one's own person. It is a deeply serious form of theatre in which humanity so acts on the assumption of God's personal (impersonated) participation that his participation is enacted. The passage from the verbal and wholly disembodied realm of God to the wholly embodied realm of man occurs through the half-embodied states of moral theatre and mime in which one person devotes himself to being someone other than who he is, at first lending his body to one who has no body because he is not at that moment physically present (as Isaac is not present when Abraham *becomes him* in planning for a wife; as again Abraham is not present in Nahor when the servant *becomes Abraham*, and by extension Isaac, in locating a wife) and ultimately lending his body to one who has no body because he is permanently bodiless—that is, because he is God.

The second model is a more representative solution to the anxiety of separation than the first (and than other models not introduced here) because in its work of braiding together human and divine body and voice, it achieves in an abbreviated space a form of transition between the two realms that, as will be returned to at a much later point, the scriptures as a whole achieve. For now, it is important to stress that in the inner content, as opposed to the formal fact, of the scriptures, such moments are exceptions to the separation of body and voice: they do not undercut, or even threaten, the categorical separation but instead reassure by making present (from "*prae-sens*" that which stands before the senses) an imaginable path from the upper ribbon to the one below.

In the narratives of human generation, there begins to be exposed the mental structure present in the activity of "believing," the intensification of the body and the projection of its attributes outward onto a disembodied referent. Here the inner reflex of "faith" or "sustained imagining" is almost wholly benign. But the same mental structure will also be visible in the much more disturbing scenes of wounding attended to below: the two kinds of scene—generation and wounding—have so much in common that it seems possible that the persistent and troubling occurrence of human hurt is the result of its conflation with generation in the Old Testament habit of mind, a conflation of the sources of pain and creation that the Judeo-Christian scriptures can be read as slowly and at great cost laboring to disassociate so that eventually the two would come to occupy two securely separate spaces in the civilization sponsored by those scriptures (the very difficulty we may now have in understanding how the two could ever have been conflated is itself the sign of how successfully they were eventually separated). Although the two kinds of scene occur together in Genesis, the second seems to become especially pronounced in the books that follow, suggesting that it is here more routinely invoked precisely because these books are preoccupied with a single generation and thus the tracing of the tiers of offspring and the awe-inspiring spectacle of growth is not now itself available for confirmation. The reliance on the scenes of wounding may also arise because they contain an element absent in the other kind that makes them easier to invoke. In the scenes of generation, there is no fixed path imaging the passage from the upper to the lower ribbon: insofar as there is one, it must be improvised with each new instance of generative affirmation and thus it changes from instance to instance (as in the stories of Jacob's mating rods and the finding of Rebekah). But in the scenes of wounding a single, easily available form of conceptualization can be invoked in every instance: the image of the weapon. It provides the human mind with something that is a singular vertical line yet so radically different at its two ends that it can in the same moment be pictured as connecting the two realms *and* preserving their absolute difference.

II. Scenes of Wounding and the Problem of Doubt

Sometimes as one reads through the Hebraic scriptures, God's existence seems so absolute and human belief in that existence so assumed and widely shared that doubt within the story of any one individual's life or any one epoch seems like only a small tear in the page, a tiny fold in an almost invisible shred of tissue in the heart, the dropping of a single stitch in the endless rounds of a woven cloth. God's realness, His presence, seems so steady, so immediately available for apprehension, that the individual person or group that fails to apprehend Him seems only an idiosyncratic exception, perversely denying of what is obvious. Yet at other readings—perhaps even almost simultaneously— it seems as though what is on every page described in these writings is the incredible difficulty, the feat of the imagination and agony of labor required in generating an idea of God and holding it steadily in place (hour by hour, day by day) without any graphic image to assist the would-be believer.

Body, Weapon, Voice in Individual Scenes

In scenes of both reproduction and wounding, the graphic image of the human body substitutes for the object of belief that itself has no content and thus itself cannot be represented. But what is almost wholly benign in the first form of analogical confirmation is not wholly benign in the second. The extreme purity of the categories of material and immaterial in genesis is uncompromising and untroubling because believer and object of belief are in the same occurrence magnified: the growth of the population, their increasingly prominent physical presence, not only confers the attribute of largeness and largesse on the meta-physical referent (a benign outcome) but also, even considered apart from that referential activity, is itself a benign event—physical survival and growth them-selves benefit the people. But in the scenes of wounding, men and God are no longer magnified in the same occurrence: the magnification of God is here instead located in the contraction of his people, either in the diminution of the race as a whole as in the thousands who in one moment die in Numbers, or in the individual contraction of physical pain. The category of embodiedness is here, as in the earlier kind of scene, intensified, for to be oneself in pain is to be more acutely aware of having a body, as so also to see from the outside the wound in another person is to become more intensely aware of human embodiedness; but while pregnancy and multiplication are a happy form of physical increase, the alternative form of increase is aversive, experienced as a diminution of personhood. Here, then, the same purity of body and voice becomes deeply problematic both to the figures within the narratives and to centuries of readers who have sometimes recorded their response in troubled meditation.

Just as the overwhelming power of alteration in the original creation brought forth both the physical earth and the species man, so the land and man continue to be two canvases on which God's power of alteration continually re-manifests itself. The subsequent alterations of the land recur not only in Genesis (the flood that for a time obliterates the earth's surface, making it when it again appears a "new" thing) but periodically throughout later books: "The uneven ground shall become level, and the rough places a plain. And the glory of the Lord shall be revealed" (Isaiah 40:4,5); "What are you, O great mountain? Before Zerub-babel you shall become a plain; and he shall bring forward the top stone amid shouts of 'Grace, grace to it' " (Zechariah 4:7). So, too, this power of alteration repeatedly manifests itself across the surface of the human body:

> Their flesh shall rot while they are still on their feet, their eyes shall rot in their sockets, and their tongues shall rot in their mouths. And on that day a great panic from the Lord shall fall on them, so that each will lay hold on the hand of his fellow, and the hand of the one will be raised against the hand of the other; even Judah will fight against Jerusalem.   (Zechariah 14:12–14)

Perhaps the continual movement of the people across the land is itself a conflation of the two locations of alteration, land and body. Many of the Genesis chapters open with the instruction to "Arise and go" from this place to another (12, 26, 28, 31, 35, 46). This physical movement, which continues in later books, re-enacts the idea of original world creation because there is a sense of erasing, wiping away, the ground of events, beginning again, starting over. When Moses records the history of the people, he does so by writing down their "starting places" (Numbers 33:1,2); and the repeated language of "starting places" also calls attention to the kinship between this event and bodily reproduction, the connection between the movement of extending-outward (or going forth) and the movement of extending outward from the "lap of birth," a different kind of "starting place."[13]

It is, however, not the surface of the land but the surface of the body that will be attended to in this section. Throughout the writings of the patriarchs and the prophets, we again and again and again return to a scene of wounding. It is a scene that carries emphatic assurance about the "realness" of God, but one that (for the participants inside) contains nothing that makes his "realness" visible except the wounded human body. The powerful God does not have the power of self-substantiation. The body is not simply an element in a scene of confirmation; it is the confirmation. Apart from the human body, God himself has no material reality except for the countless weapons that he exists on the invisible and disembodied side of. Whether the material object is itself physically present in the story (as in the flaming torch and pot of Genesis 15, or the burning bush of Exodus) or instead verbally invoked (as in the song given by God to Moses in Deuteronomy 32—"I will spend my arrows upon them; / they shall be wasted with hunger, / and devoured with burning heat / and poisonous pestilence. . . . If I whet my glittering sword . . . I will make my arrows drunk with blood"), the weapon is a material sign of Him separate from the human body and explaining the path of connection. In this respect, as was suggested earlier, the stories of hurt differ from the stories of reproduction, where there is no immediately available, mediating sign. Here there are many—fire, storm, whirlwind, plague, rod, arrow, knife, sword—and on the other side of these objects, our brooding and terrible heaven. Only the last four objects in this list have the actual physical structure alluded to earlier, a vertical line connecting two radically different ends. But this same structure is implicitly present in all the others as well: a fire, too, has two ends, for at one terminus it is ignited and at the other it burns; just as plague, storm, and wind each has a passive terminus from which it is brought into being and pointed in a certain direction, and an active terminus where it mutilates and imperils.

Across the image of the weapon, the material and the immaterial are severed. As God in the scene of hurt is a bodiless voice, so men and women are voiceless bodies.[14] God is their voice; they have none separate from him. Repeatedly, any capacity for self-transformation into a separate verbal or material form is shattered, as God shatters the building of the tower of Babel by shattering the language of the workers into multiple and mutually uncomprehending tongues (Genesis 11:1–9). The book of Numbers is a book of constant murmuring and complaint:

And the people complained in the hearing of the Lord about their misfortunes; and when the Lord heard it, His anger was kindled. (11:1)

Moses heard the people weeping through their families, every man at the door of his tent; and the anger of the Lord blazed hotly. (11:10)

> Then all the congregation raised a loud cry; and the people wept that night. And all the people of Israel murmured against Moses and Aaron . . . "Would that we had died." (14:1,2)

> "How long shall this wicked congregation murmur against me? I have heard the murmurings of the people of Israel, which they murmur against me . . . [You] who have murmured against me, not one shall come into the land." (14:27,29,30)[15]

Their voices whine and murmur. Devoid of any content other than complaint, their utterances are self-trivializing and dissolute, a form of inarticulate pre-language that carries no power to legitimize their suffering, their hunger, their fear, their doubt, their exhaustion, or to legitimize our notice of these things. If their voices were able to form and express these things, the story would be a different story. As it is, their voices deprive them of God's sympathy and of ours: they seem debased and dismissible even if now and then for a moment— as in their expressed longing for meat, fish, melons, cucumbers, leeks, onions, garlic and the longing for a fixed ground that is implicit in this simple enumeration of root fruits (11:5)—one catches a glimpse of the edge of that other story. Physical suffering destroys language, and moral rightness (in the Old Testament as in most other human contexts) tends to lie with the most articulate. So we linger with the people only a moment; then continue on in the hope that their cries will end.

Although an occasion of wounding is often described (within the narrative itself as well as in commentaries on the narrative) as a scene of disobedience and punishment, it is in many ways more comprehensible and accurate to recognize it as a scene of doubt, for it is a failure of belief that continually reoccasions the infliction of hurt. Unable to apprehend God with conviction, they will—after the arrival of the plague or the disease-laden quail or the fire or the sword or the storm—apprehend him in the intensity of the pain in their own bodies, or in the visible alteration in the bodies of their fellows or in the bodies (in only slightly different circumstances) of their enemies. The vocabulary of punishment describes the event only from the divine perspective, obscures the use of the body to make experienceable the metaphysical abstraction whose remoteness has occasioned disbelief.

Moments in which the people have performed an immoral act (other than doubting) and where the idiom of punishment may therefore seem appropriate, must be seen within the frame of the many other moments where the infliction of hurt is explicitly presented as a "sign" of God's realness and therefore a solution to the problem of his unreality, his fictiveness. Such passages openly identify the human body as a source of analogical verification. Specified forms of hurt are overtly presented as demonstrations of His existence:

> And Moses said, "Hereby you shall know that the Lord has sent me to do all these works, and that it has not been of my own accord. If these men die the common

death of all men, or if they are visited by the fate of all men, then the Lord has not
sent me. But if the Lord creates something new, and the ground opens its mouth,
and swallows them up, with all that belongs to them, and they go down alive into
Sheol, then you shall know that these men have despised the Lord.'' (Numbers
16:28–30)

The spectacular and self-consciously innovative form of devastation immediately
follows (16:31–35). Hurt here, as in many other moments, becomes the vehicle
of verification; doubt is eliminated; the incontestable reality of the sensory world
becomes the incontestable reality of a world invisible and unable to be touched.
Centuries of men and women ready, even longing, to deepen their faith may
feel that a small and benign sign, the blossoming of Aaron's rod, would suffice;
but belief is not within the mental structures of the Hebraic scriptures an exercise
in open choice. As barrenness in the scenes of reproduction may be changed by
God to a living presence in the womb, so in the infliction of pain the anterior
condition of unfeeling emptiness (doubt) is replaced by an amplification of
sentience about which no uncertainty is possible and no decisions need be made.

This is true whether the hurt actually takes place, as in the narratives of
Genesis, Exodus, and Numbers, or whether it takes place in a warning and
ritualized form, as in the repeated refrain, ''Lest you die,'' ''Lest he die,'' ''Lest
I turn my face against you'' which follows the enumeration of rules in Leviticus.
Moments in the Old Testament where punishment is rendered because of doubt
are to a large extent paradigmatic of all other moments of punishment: that is,
immoral behavior or disobedience or cruelty are extreme forms of doubt, extreme
failures of belief, the failure to absorb into oneself and to embody in one's acts
and attitudes a concept of God. To offer incense at a time when this act has
been prohibited, to question the authority of the human leader appointed by God,
to break a specified rule of ritualized cleansing, to devote one's body to a dance
around a golden calf, to turn back for one last look at the cities of Sodom and
Gomorrah and their godless behavior, to complain of present hunger and anticipated hunger when one should perceive oneself as under the protection of a
bountiful Overseer, are all demonstrations of an inability to incorporate into
everyday patterns of gesture and speech a depth and totality of certitude, and
thus all demonstrations of doubt. So too to be a foreigner—a second explanation
for God's many moments of inflicting hurt—is also an extreme form of disbelief,
a state of existing wholly outside the circle of faith.

The failure of belief is, in its many forms, a failure to remake one's own
interior in the image of God, to allow God to enter and to alter one's self. Or
to phrase it in a slightly different form, it is the refusal or inability to turn oneself
inside-out, devoting one's physical interior to something outside itself, calling
it by another name. Disobedience or disbelief or doubt in the scriptures is
habitually described as a withholding of the body, which in its resistance to an
external referent is perceived as covered, or hard, or stiff:

But they refused to hearken, and turned a stubborn shoulder, and stopped their ears
that they might not hear. They made their hearts like adamant lest they should hear
the law and the words . . .   (Zechariah 7:11,12)

They refused to take correction.
They have made their faces harder than rock;
They have refused to repent.   (Jeremiah 5:3)

Blessed is the man who fears the Lord always;
But he who hardens his heart will fall into
calamity.   (Proverbs 28:14)

He who is often reproved, yet stiffens his neck
will suddenly be broken beyond healing.   (Proverbs 29:1)

O that today you would hearken to his voice!
Harden not your hearts, as at Meribah . . .   (Psalms 95:7,8)

Yet they acted presumptuously and did not obey thy
commandments, but sinned against thy ordinances, by
the observance of which a man shall live, and
turned a stubborn shoulder and stiffened their neck
and would not obey.   (Nehemiah 9:29)

Because I know you are obstinate and your neck is
an iron sinew and your forehead brass.   (Isaiah 48:4)

Perhaps most familiar are the descriptions of the ''stiff-necked'' Israelites who
trouble God while Moses is receiving the ten commandments (Exodus 32:9,
33:3, 33:5, 34:9) and earlier, the opening passages of Exodus in which the
Egyptian Pharaoh's refusal to listen, to be moved to belief by the turning of the
Nile to blood, the plague of frogs, the invading swarms of gnats and flies, are
described in these same terms: ''Pharaoh's heart remained hardened and he would
not listen'' (7:22); ''He hardened his heart and would not listen to them'' (8:15);
''Pharaoh hardened his heart this time also, and did not let the people go'' (8:32);
''But the heart of Pharaoh was hardened, and he did not let the people go''
(9:7).[16] The processes of belief are openly revealed in this story, for the physical
hurt is here explicitly designated a ''sign'' of God's realness, and even more
significantly, the Pharaoh's disbelief in God is itself caused by God in order to
occasion the wounding of the Egyptians and thus provide to the watching Israelites the ''signs'' of God's realness: ''I have hardened his heart and the hearts
of his servants, that I may show these signs of mine among them, and that you
may tell in the hearing of your son and of your son's son how I have made sport
of the Egyptians and what signs I have done among them; that you may know
that I am the Lord'' (10:1,2; see also 10:20; 9:12; 9:35).

In all of the passages cited above, the withholding of the body—the stiffening
of the neck, the turning of the shoulder, the closing of the ears, the hardening
of the heart, the making of the face like stone—necessitates God's forceful

shattering of the reluctant human surface and repossession of the interior. Perhaps the most overt acting out of this occurs in the final plague on the house of the Pharaoh, the final entry into his hard heart, the massacre of the innocents in which the interior of the body as it emerges in the first-born infant is taken by God (Exodus 12). The fragility of the human interior and the absolute surrender of that interior that does not simply accompany belief, that is not simply required by belief, but that *is itself belief*—the endowing of the most concrete and intimate parts of oneself with an objectified referent, the willing rereading of events within the realm of sentience as themselves attributes of the realm of self-extension and artifice—are in this history acted out with terrible force and unequivocal meaning.

The taking of the Egyptian infants is a few lines later followed by the willing consecration of the Israelite infants, the willing consecration by the Israelites of their own interiors: "The Lord said to Moses, 'Consecrate to me all the first-born; whatever is the first to open the womb among the people of Israel, both of man and of beast is mine' " (13:1,2). This, in turn, is itself followed by a second way in which God's entry into the human interior is ensured, the prohibition against leavened bread: "And it shall be to you as a sign on your hand and as a memorial between your eyes, that the Law of the Lord may be in your mouth" (13:9; see also Deuteronomy 11:18). The human child, the human womb, the human hand, the face, the stomach, the mouth, the genitals (themselves circumcised, marked)—it is in the body that God's presence is recorded. Either, as in the case of the Egyptian massacre, the body is unwillingly given and violently entered or, as in the case of the Israelite consecration that follows, it is willingly surrendered.

Although any part of the body may become the focus of this conversion, the essential habit of mind is caught and dramatized in its full force in the stories of the sacrifice of offspring, for here the two extreme forms of physical alteration, self-replication and wounding, converge. It may be in part for this reason that the story of Abraham and Isaac is returned to again and again as containing the central mystery of the Judeo-Christian God. Again, the story does not merely describe the rigors of belief, what is required of belief, but the structure of belief itself, the taking of one's insides and giving them over to something wholly outside oneself, as Abraham agrees to sacrifice the interior of his and Sarah's bodies, and to participate actively in that surrender. The building of the altar externalizes and makes visible the shape of belief; the hidden interior of sentience is lifted out through work into the visible world. Itself the body turned inside-out, it in turn becomes the table on which the body will once more be turned inside-out. Here the stages of materialization, heightened forms of self-presentation, are successively increasing powers of one another. Abraham's three times repeated, "Here I am" is a simple statement of amplified self-presentation. The interior of himself and Sarah—Isaac—is the once heightened and externalized form. Then Isaac himself will be cut open, his own interior exposed. As in the stories of reproduction, we encounter the rhythm of substantiation, more and more emphatic presentation and re-presentation of the body to confer the force and power of the material world on the noumenal and unselfsubstantiating. Who can be ignorant of the risk he takes and the cost he incurs in agreeing *to re-create* God, *to make* an already existing God more immediately apprehensible, *to remake* of an unapprehensible God an apprehensible One. To be alone on a mountain in the wind and afraid—one might feel only an overwhelming alone-ness, the fragility of one's own existence, the distance, dimness, and unreality of God; but the anticipated sight of the interior of the body then makes the dimly apprehended incontestably present; for the object of conviction acquires a com-pelling and vibrant presence from the compelling and vibrant sequence of actions with which the human willingness to believe, to be convinced, is enacted.

Belief is the act of imagining. It is what the act of imagining is called when the object created is credited with more reality (and all that is entailed in greater "realness," more power, more authority) than oneself. It is when the object created is in fact described as though it instead created you. It ceases to be the "offspring" of the human being and becomes the thing from which the human being himself sprung forth. It is in this act that Isaac yields against all phenomenal assessment to Abraham, that Abraham yields to God, and that the reader yields to the narrative:[17] it is not simply the willingness to give one's interior to some-thing outside oneself but the willingness to become the created offspring of the thing in whose presence one now stands, as Isaac at that moment is not the many things Isaac is but only Isaac-son-of-Abraham, as again Abraham the patriarch, Abraham the husband, Abraham the father of Isaac, Abraham the father of the twelve tribes of Israel all now converge into Abraham-the-created-offspring-of-God, and as the reader in his or her many capacities ceases to be the many things that he or she is and becomes in the stunned and exhausted silence of Genesis 22:1–19 the created offspring of the text, of this text and of the many stories through which the framework of belief is set in place.

That God in this story requires Abraham only to anticipate rather than to carry out the wounding—requires him only to enact imaginatively the physical action (just as the reader is required only to follow the event imaginatively rather than to perform an equivalent physical sacrifice in his own family)—calls attention to the fact that the structure of belief outlined here is (as will eventually be elaborated in Section III) itself deeply modified. But before turning to that modification, both the pervasiveness and continual re-assertion of the original structure must be clarified.

Body, Weapon, Voice in the Larger Framework

The extreme separation into body and voice that characterizes what almost seem randomly recurring scenes of hurt (as well as the scenes of reproduction) is also recognizable in larger and more central units of thought. The ten commandments are an explicit articulation of the separation. They are themselves divided into an initial three that describe God and seven that describe man, a division insisting on the categories of voice and body. In the first God asserts that He and only He is the Lord. In the second, he forbids all forms of materially representing him, prohibits all attempts to endow him (and other aspects of the invisible world) with a body. In the third he specifies that he is to be represented only by a single verbal construct, a name; and that must be used with extreme reluctance and care. His province is absolute precisely because he is unlimited by any specifying acts of representation, except in the human beings who come to represent him in the commandments that follow. In effect, the first three describe God by prohibiting description. The next seven command that man embody God, that man's actions describe God. If God is materialized, he is so only in the actions and prohibited actions of men and women, beginning with the fourth commandment, the mimetic re-enactment every seven days of the initial Creation: the temporal rhythm of work and rest is absorbed into, enacted by, the body of the believer.

This fourth commandment is a crucial transition between the three devoted to God and the seven devoted to man, for it is explicitly about creation, and the purity of categories insisted on in the overarching, three-seven division is itself a way of maintaining the original structure of creation. To insist that God be only verbally indicated is to ensure that the initial division between creator and created is maintained. That is, to have a body is to be describable, creatable, alterable, and woundable. To have no body, to have only a voice, is to be none of these things: it is to be the wounder but not oneself woundable, to be the creator or the one who alters but oneself neither creatable nor alterable. It is therefore appropriate that the fourth and transitional commandment has as its explicit subject the instruction that humanity abstain from the act of creating; for it is the "rest" part and not the "work" part of God's original act of creation that we are here required to imitate. Although mankind is instructed to re-enact both the six days of work and the one of rest, the whole force of the commandment is focused on the second part: in all the iterations of this commandment throughout the Old Testament scriptures, men and women are never threatened, reprimanded, or punished for not working on the six days of work, while the failure to rest on the seventh repeatedly brings the most severe punishment.[18] Here as so often elsewhere in the Hebraic scriptures, it is not acts of making through which man becomes god-like, for these are either neutral or, as in the present instance, full of the possibility of despicable godlessness. The emphasis on rest in the fourth commandment is intimately bound up with the prohibition against graven images in the second, with the stress on one's position as a created offspring in the fifth, and with the general semantics of negation, passivity, and prohibition in the series of "thou shalt not" constructions that follow in the remaining commandments—all of which inhibit man's perception of himself as initiating actions.[19] God is all things; man becomes godly by only doing some of those things. God both kills and refrains from killing; it is by imitating only his not killing that humanity achieves the condition of the spiritual. Again, God both creates (six days of making) and abstains from creating (the one day of rest); it is by imitating his abstaining from creating that man finds the godlike within him. It is possible that the prohibition against work and material culture arises as a result of the conflation of wounding and creating: that is, the very benign emphasis in Mosaic law on teaching a people to be free of the desire to kill others may—because of the confusion of hurting and making—simultaneously require that they be made free of the desire to create. In any event, regardless of the path by which the prohibition arises, it does arise, and is present not only in all those places where it is the overt content of an instruction but, more significantly, in the ongoing insistence on the absolute separation of verbal and physical categories.

In discussions of power, it is conventionally the case that those with power are said to be "represented" whereas those without power are "without representation." It may therefore seem contradictory to discover that the scriptures systematically ensure that the Omnipotent will be materially unrepresented and that the comparatively powerless humanity will be materially represented by their own deep embodiment. But to have no body is to have no limits on one's extension out into the world; conversely, to have a body, a body made emphatic by being continually altered through various forms of creation, instruction (e.g., bodily cleansing), and wounding, is to have one's sphere of extension contracted down to the small circle of one's immediate physical presence. Consequently, to be intensely embodied is the equivalent of being unrepresented and (here as in many secular contexts) is almost always the condition of those without power.[20]

The narrative frame out of which the ten commandments emerge is one that also stresses the prohibition against bodily representation of God, since it is the place in which while Moses is on Mount Sinai, the people—afraid and longing for the presence of God among them—take their gold and silver and jewelry and make the golden calf, which prompts Moses to shatter the tablets containing the ten commandments and return again to the mountain to receive them once more. The basic categories insisted upon with the three-seven division in the commandments are in the story preserved in the contrast between the embodied representation of God in the golden calf and the wholly verbal representation of Him in the engraved tablets. When this particular period in the desert is recalled in Deuteronomy, God's presence as a voice is insisted on. Of crucial importance to the present discussion is the fact that what permits the recovery of the immaterial voice of God back out of the materialized image in the calf is the intermediate phenomenon of the weapon, in this case, the fire:

Then the Lord spoke to you out of the midst of the fire; you heard the sound of
words, but saw no form; there was only a voice. And he declared to you his covenant,
which he commanded you to perform, that is, the ten commandments . . .

Therefore take good heed to yourselves. Since you saw no form on the day that the
Lord spoke to you at Horeb out of the midst of the fire, beware lest you act corruptly
by making a graven image for yourselves, in the form of any figure, the likeness
of male or female, the likeness of any beast . . .

Did any people ever hear the voice of a god speaking out of the midst of the fire,
as you have heard, and still live?   (Deuteronomy 4:12,13,15,16,17,33)

Passages throughout the Old Testament that deal with the continually troubling
problem of graven images frequently remind the offenders that God is a disembodied voice, a voice hovering on the other side of a weapon. One passage in
Isaiah, for example, that rages against the impulse to make images is soon
followed by another passage which begins by asserting that God is a voice, and
then takes an embodied image of God (significantly, an embodied equivalent
for the voice, a gigantic mouth) and feature by feature reconverts it into a weapon:

> Behold, the name of the Lord comes from far,
>   burning with his anger, and in thick rising smoke;
> his lips are full of indignation,
>   and his tongue is like a devouring fire;
> his breath is like an overflowing stream
>   that reaches up to the neck;
> to sift the nations with the sieve of
>   destruction,
>   and to place on the jaws of the peoples
>   a bridle that leads astray.   (Isaiah 30:27,28)

The lips are deconstructed into the thick smoke of indignation, the tongue into
a devouring fire, the breath into a flood that drowns. By the close of the passage,
not only has the bodily image of God disappeared but it has been transferred
back to mankind: what began as the mouth of God becomes by the end the
bridled jaws of the people.

In the story in Exodus, the categorical division of body and voice is by the
act of casting the molten calf for a moment greatly threatened, and it is in part
to reestablish and reemphasize those categories that the people of Israel are now
made to experience the physical ground of their existence more intensely. At
this point in the narrative a new priestly class emerges and is instructed by Moses:

"Thus says the Lord God of Israel, 'Put every man his sword on his side, and go
to and fro from gate to gate throughout the camp, and slay every man his brother,
and every man his companion, and every man his neighbor.' " And the sons of
Levi did according to the word of Moses; and there fell of the people that day about
three thousand men. And Moses said, "Today you have ordained yourselves for

the service of the Lord, each one at the cost of his son and of his brother, that he may bestow a blessing upon you this day.'' (Exodus 32:27–29)

This first form of massacre is followed by a second, ''a plague upon the people because they made the calf which Aaron made'' (32:35). That God cannot be materialized in a graven image is a truth materialized, verified, in their own bodies. The distinction between the categories of voice and body, blurred by the making of the calf that endows the verbal category with bodily representation, is reaffirmed by a great intensification of the category of the body, so that the distance between the two remains constant: that which should have been a purely verbal category has for a moment been given a material form and, consequently, that which belongs to the material category must be pushed into a much more extreme materialization of itself. Across sword and plague, as first across the fire on Horeb, God's verbal purity is reclaimed; the categories are safely separated.

This same rhythm of events—the humanly sponsored blurring of body and voice followed by a reaffirmation and intensification of their categorical separation—occurs in many other stories.[21] Although in passages about graven images the boundary between material and immaterial is compromised by attempts to give material representation to the wholly disembodied, the threat may also come from the opposite side, from attempts to disembody the human body by increasing the power of the human voice or the capacity for cultural self-transformation through artifice: this kind of blurring of categories is, like the other kind, immediately followed by God's swift demonstration of their absolute distinguishability. The fall of Adam and Eve follows this sequence in such stark outline that a recitation of the narrative events themselves makes manifest the two-part rhythm of conflation and intensified separation. Part of the knowledge that comes with eating of the tree of good and evil is that they stand, without protest, as creatures with bodies in the presence of one who has no body. It is crucial that these two be said together: the problematic knowledge is not that man has a body; the problematic knowledge is not that God has no body; the problematic knowledge is that man has a body and God has no body—that is, that the unfathomable difference in power between them in part depends on this difference in embodiedness. In response, they perform their first cultural act wholly independent of God,[22] the weaving of leaves into aprons. With this act of covering their own sentient tissue, the nature of the relation has suddenly changed. No longer is the One wholly hidden and the other wholly revealed; now instead the One is wholly hidden and the other is only *almost* wholly revealed, very slightly hidden by the fragile intervention of the artifact, as they again more drastically hide themselves behind a larger web of leaves, a grove, when God now enters the garden.[23] God accepts their act and Himself clothes them with skins. But the rhythmic reassertion of categories follows, for while they are slightly disembodied by their clothing, their awareness of the body will soon be correspondingly heightened: the body is made a permanently preoccupying category in the pain of childbirth, the pain of work required to bring forth food, and the ongoing unease in relation to any fixed shelter. God accepts their woven refusal to walk naked in His presence and, simultaneously, makes the physical acts of eating and generation, work and rest, themselves complex and cutting nets of difficulty.

Absolute verbal purity is eternal life: the projected voice is the power of sentience separated from the fragility and vulnerability that attend sentience at its unprojected site and source. The eviction from the garden is in the Genesis story described not as a direct punishment for their eating from the tree of good and evil but as a prevention of their eating from the second tree, the tree of eternal life: "Behold, the man has become like one of us, knowing good and evil; and now, lest he put forth his hand and take also of the tree of life, and eat, and live forever . . . " (3:22). Man and woman move outside the garden to the realm of the troublesome body, permanently blocked from the verbal category. Significantly, in the final sentence of the eviction from the garden (as in the stories described earlier) there emerges the image of the weapon to keep the categories separate, for permanently placed between Eve and Adam and every approach to the tree of life is "a flaming sword which turned every way" (3: 24).

The Shattering of the Categories of Body and Voice in the Christian Scriptures

Throughout the Old Testament, God's power and authority are in part extreme and continually amplified elaborations of the fact that people have bodies and He has no body. It is primarily this that is changed in the Christian revision, for though the difference between man and God continues to be as immense as it was in the Hebraic scriptures, the basis of the difference is no longer the fact that one has a body and the Other has not. The change that occurs is a change less in the object of belief than in the very structure of belief, a change in the nature of the religious imagination. The consequences are, of course, very great; for in speaking now not of Judaism but of Judeo-Christianity one is speaking no longer of narratives that belonged to a small, racially pure group of people but of narratives that crossed continents, were adopted by many races, and became the framework for a major civilization.

Throughout the Hebraic scriptures, the question of having or not having a body arises not only in man's relation to God but also in man's relation to man. Even among human beings, the one with authority and power has no body for his inferiors. He cannot, for example, be seen without clothing: the family and people of Ham are enslaved by Noah after Ham accidentally sees Noah naked (Genesis 9:20–27). As in the scene in the garden of Genesis or in the desert of Exodus, the violation of the hierarchies necessitates a reaffirmation by intensifying the bodily experience of the inferior—in this instance, by the subjugation to physical labor. No one, we are often told throughout the scriptures, sees God face to face and lives. The recurring warning "See God and die" contains the by now familiar rhythm: to see God is to endow him with a body; to die is not simply to be punished for that act of sentience with the loss of sentience but to reaffirm the threatened categories of body and voice by experiencing the most extreme requirement of the body. Moses is permitted to see not God's face but his "back" or, as it is elsewhere described, his "aftermath" (e.g., Exodus 33:23);[24] and in turn, Moses withholds his own face from his people, covering it with a veil, exposing it only when in the presence of one greater than himself (34:33), as the people unquestioningly expose their own faces to their human leader and teacher.

Even in those places in the scriptures where God permits himself to be materialized—(passages that will be attended to in detail in Section III)—the aspect of God most prominently represented is his unrepresentability, his hiddenness, his absence. Exodus ends with several chapters elaborating God's instructions for the building of his tabernacle, instructions laden with thick sequences of precise requirements that stun the mind with their confident sweep of beautiful detail. Again and again, it is the curtain, covering, or veil that folds and unfolds before our eyes. The tabernacle itself is made of ten curtains of "fine twined linen and blue and purple and scarlet stuff," sewn five and five together with loops of blue and clasps of gold to permit and prevent entry. Beyond this first layer there appears a second: covering the tabernacle is a tent made of eleven curtains of woven goats' hair. Beyond this second layer appears a third: covering the tent is another tent made of tanned rams' skins and goatskins. Beyond this third layer appears a fourth, a veiled courtyard of "fine twined linen" hangings on bases of bronze with fillets of silver (36:8–19; 38:9–20). As the principle of the veil repeats itself in the outward movement, so it is again repeatedly encountered as one moves inward to the smaller objects residing within the fourfold sequence of walls. At the doorway of each of the four layers hangs a woven or embroidered gate (38:18; 36:37); inside the tabernacle is the mercy seat hidden by the unfurled "overshadowing" wings of the cherubim hovering at its edges, their faces turned inward to the seat, themselves eclipsed by their wings (37:7–9); before the altar stands a veil of blue and purple and scarlet stuff and fine twined linen (36:35); hanging suspended from the ledge of the altar, extending halfway down, is "a grating, a network of bronze" (38:4). Though there gradually comes before us in these endless tiers of tissue something that seems the magnificent and monumental tissue of the body of God, what at the same time comes before us is the veil, the materialization of the refusal to be materialized, the incarnation of absence. It is a realm of exclusion, entered only by the priests (whose bodies are, like the altar that is the symbolic representation of the human body, themselves surrounded by layers and layers of woven garments). Of the

priests, none except Aaron and his line may approach the altar or see behind
the veil, "lest they die" (Numbers 18:3). Of Aaron's line, no one may enter
who is blemished, handicapped, or hurt: these, like all those not of Aaron's line,
may not "come near" (Leviticus 21:16–24).

Even, then, at this moment of materialization the profound distance between
humanity and God entails at its very heart an absolute inequality in bodily
representation. It is against this background in which not only divinity but even
much more minor forms of authority and privilege carry with them a withholding
or hiding or nonexistence of the body, that the act of "witnessing" becomes so
important in the Christian supplement to the Hebraic scriptures. It is the opening
emphasis of Luke, who at once designates his story as the "narrative of the
things . . . delivered to us by those who from the beginning were eyewitnesses"
(1:1,2), as it is again stressed in the words of Jesus with which Luke's narrative
ends, "You are witnesses of these things" (24:48), as again the narrative that
comes between these first and final lines is preoccupied with the astonishing act
of witnessing participated in not only by twelve men alone but by the crowds
of people who follow and watch. Again in the Gospel of John, the repeated
invitation "come and see," "come and see" (1:39,46) contains the excited
recognition of a privilege that continues to be acknowledged even in the dread-
filled invitation to the final watch, "The hour is coming," "The hour is at hand"
(5:25,28). The centrality of this act of witnessing cannot be overstated. It is
there not primarily to assure the distant reader that the story is true, that it
happened; rather, the story *is* itself the story of the permission to witness, the
story of the sentient body of God being seen and touched by the sentient body
of man. John the Baptist, for example, is described as being sent to earth not
to be the one seen but to be among those who see the one sent to be seen (John
1:6–8); and it is not a misreading of his story to recognize its essential subject
in abbreviated lines such as these, "The next day he saw Jesus coming toward
him . . ." (1:29), "The next day again John was standing with two of his
disciples; and he looked at Jesus as he walked" (1:35), as this is also the essential
event happening to the crowds of people who the next day "saw," and the next
day again "looked." As in the Old Testament scriptures, in the New the human
body substantiates the existence of God, but rather than that bodily verification
occurring in the bodily alteration of pain, it occurs in the bodily alteration of
sensory apprehension, not the hand of a woman turning leprous but the hand of
a woman touching Jesus, not the thousands cut and killed in the desert beneath
Horeb but the thousands who "watch, hear," and then follow in order to sustain
this perceptual privilege.

Although in the Christian scripture the most crucial form of physical alteration
is perceptual, there is also a second form of alteration, recovery. The rhythmic
return in the Hebraic scripture to a scene of wounding here becomes instead a
rhythmic return to a scene of healing, a scene that occupies the same fixed and

central locus in the strategies of belief: epilepsy, disease, and pain are eliminated (Matthew 4:23), a paralytic walks (9:4; 8:5), a man's withered hand becomes whole (12:9), a dumb man speaks (12:22), a dead child is restored to life (Luke 8:41–56; Mark 5:22–43), a woman who has bled for twelve years no longer bleeds (Luke 8:43), a man frail with thirty-eight years of sickness is cured (John 5:2), fever and demonic possession disappear (Matthew 8:14), the young disciples are prevented from drowning by the quieting of the winds and the sea (8:25). Just as the Old Testament act of wounding is explicitly presented as a "sign," so the New Testament act of healing is explicitly presented as a "sign": the human body is in each the site for the analogical verification of the existence and authority of God, but the alterations are almost always now in the direction of recovery.[25]

God's most intimate contact with humanity, His sensory contact with the human body, is in the Hebraic scriptures mediated by the weapon and in the Christian addition is mediated by Jesus. Hence within the Judeo-Christian tradition it is the weapon (not Jehovah) that Jesus displaces. But the inner logic of this change requires more than the simple elimination of the weapon: absence is too laden with ambiguity to complete the act of perceptual restructuring that is taking place. Although now gone from those parts of the narrative describing God's immediate interactions with people, it is of course not gone from the entire story. In the end, the weapon returns, more concrete and cruel than it ever had been in the older writing: it no longer hovers in hypothetical outline on the horizon, appearing and disappearing, re-emerging in multiple forms; it becomes tangible, singular, and still. The descent of God to earth is the descent of God to the underside of the knife, plague, or rain of fire. The elimination of the weapon from the human assessment of divine power requires not just that the weapon be subtracted out of the relation, or that the human being be lifted away from the pain end and God from the power end of the image that bonds them, but that God's position be explicitly changed to the sentient end. The cross is unusual among weapons: its hurt of the body does not occur in one explosive moment of contact; it is not there and gone but there against the body for a long time. The identification is steady. More important, the cross, unlike many weapons, has only one end: there is not a handle and a blade but only the blade, not a handle and lash but only the lash. Though like any other weapon it requires an executioner, the executioner's position is not recorded in the structure of the weapon itself. The fact of bodily pain is not here memorialized as the projected facticity of another's power, for there is no second terminus to assist such a translation.

Thus the Old Testament weapon disappears and is replaced by Jesus in the Christian story, then reemerges in a heightened and deeply altered form at the end of the narrative. To say that Jesus takes the place previously held by the weapon sounds awkward and unfamiliar; yet the very ease with which we accommodate the displacement in the opposite direction—the substitution of Christ by the cross in Christian ritual and symbol—at the very least signals the plausibility of the earlier substitution and may instead be understood as itself the uncovering of that earlier event. It is not an accident that the image of the cross comes to have such a central place. The weapon becomes the primary sign and summary of the entire religion precisely because the entire religion is at its very heart an alteration in the reading of this sign. The alteration insists that omnipotence, as well as more modest forms of power, be reconnected to the facts of sentience. It is not that the concept of power is eliminated, and it is certainly not that the idea of suffering is eliminated: it is that the earlier relation between them is eliminated. They are no longer manifestations of each other: one person's pain is not the sign of another's power. The greatness of human vulnerability is not the greatness of divine invulnerability. They are unrelated and therefore can occur together: God is both omnipotent and in pain.

That the central work of the Christian story is the fundamental restructuring and redirecting of the referential activity of the generic sign of the weapon can be understood without attending directly to the particular sign of the cross (which is only itself the summary of that change; the cross is not obsessively present in the scriptures themselves but is instead in the following centuries distilled out and made a singular focus). The change from God's disembodied relation to man's body in the older writings to God's embodied relation to man's body in the younger is itself a transformation that conforms to the description given earlier of the transformation of a weapon into a tool.[26] The weapon acts directly on sentience while the tool only acts on sentience by providing it with an object; the weapon, for example, enters the field of vision by cutting the eyes; the tool enters the field of vision by making the loaf of bread, the statue, or the mark on a stone that can be seen. In the Old Testament scenes of hurt, Jehovah enters sentience by producing pain; Jesus instead enters sentience by healing and, even more important, by himself becoming the object of touch, the object of vision, the direct object of hearing. In both the human body is the vehicle of belief, but the vividness of pain in the one becomes the vividness of touch and vision in the other. The fear and doubt and loneliness that motivated the casting of the calf in the wilderness—an act so far from unmindfulness of God, so full of longing for His presence that it is difficult to comprehend why their act, however deserving of punishment, should carry the particular scar of lewd godlessness— are in the embodiment of Jesus eliminated.

The story of Doubting Thomas provides a stark summary of this revised form of verification. The doubt of Thomas after the death and resurrection of Jesus is one of a constellation of parallel failures that can be understood as intensifications of one another: the *sleep* of Simon Peter, James, and John in the garden; the *doubt* of Thomas; the *denial* of Peter; the *betrayal* of Judas. The perceptual lapse in sleep becomes the more active perceptual lapse of doubt, the still more active (because now acted on) failure of denial, all of which climax in the most active form of perceptual lapse implicit in them all, betrayal. Three of the four—sleep, denial, and betrayal—occur together in the fourteenth chapter of Mark, the twenty-sixth chapter of Matthew, and the twenty-second chapter of Luke; but it is the story of Thomas, told at the end of John, that makes most apparent the transformed relation between body and belief:

> Now Thomas, one of the twelve, called the Twin, was not with them when Jesus came. So the other disciples told him, "We have seen the Lord." But he said to them "Unless I see in his hands the print of the nails, and place my finger in the mark of the nails, and place my hand in his side, I will not believe."
>
> Eight days later, his disciples were again in the house, and Thomas was with them. The doors were shut, but Jesus came and stood among them, and said, "Peace be with you." Then he said to Thomas, "Put your finger here, and see my hands; and put out your hand, and place it in my side; do not be faithless, but believing." Thomas answered him, "My Lord and my God!" Jesus said to him "Have you believed because you have seen me? Blessed are those who have not seen and yet believe." (John 20:24–29)

Belief comes not, as so often in the Old Testament, by being oneself wounded but by having the wound become the object of touch. If in the Hebraic scripture we repeatedly move from the human body to a more extreme materialization of that body in the exposure of the interior (whether benignly as in reproduction or not benignly as in scenes of hurt), so now in the Gospels we begin with the body of God and move relentlessly toward the more extreme materialization, the exposure of the interior of Jesus in the final wounds of the crucifixion. The relation between body and belief remains constant in the two scriptures: in each the interior of the body carries the force of confirmation. The difference resides in whose body it is that is required to confer the factualness of the material world on the immaterial realm: in the one it is the body of man that substantiates God; in the other, the body of God that substantiates God. The Artifact becomes for the first time self-substantiating.

The fact of Christ having a body is so central a premise of the narratives that it is unnecessary to follow its elaborations at any length here. His proximity to the poor, to prostitutes, to robbers, to the rich, and, perhaps most important, to the sick entails the risk of contamination that is only a re-enactment within the realm of ordinary action of the incomprehensibly large risk of the initial taking on of a body. Even after the resurrection, he remains embodied: at the very end of Luke, he invites the apostles to look at his hands and to "handle" him to assure themselves he is not a spirit (24:36–40). While they are still "disbelieving for joy," he asks the question that is the universal signal among sentient creatures that they are sentient: "Have you anything here to eat?" (24:41). The disciples then give him a piece of broiled fish which he eats in front of them; the largesse of their hospitality is quietly outdone by the largesse of his resurrected willingness to be hungry. He himself, of course, in the Last Supper and in the communion ceremony enters the food chain, allowing himself to be taken in, now not just as the object of perception but as an object of sustenance. It may be that the second is always implicit in the first—that is, the fact of being eatable may be inevitably contained in the fact of bodily representation, as the calf cast in the wilderness is eaten by the people: "And [Moses] took the calf which they had made, and burnt it with fire, and ground it to powder, and scattered it upon the water, and made the people of Israel drink it" (Exodus 32:20).

The centrality of the body is emphatic not only in the original narratives themselves but also in the choices that are unconsciously and communally made about representations of Christ, the "graven images" of Christianity. An act of representation is an act of embodiment. Christ is himself embodied in the scriptures, long before any visual depictions of him. But it is also interesting that centuries of visual representations have made Christ's embodiment more prominent, have made it their central content. In western art and culture, by far the two most endlessly visualized moments in Christ's life are his infancy and his hours of dying, the two periods in any life when the body is most prominent in asserting its claim. Infancy is the time *prior* to acquired language, selfhood, and world content that accompanies the growth of "a person" within the body, the time when all energy is absorbed in holding a head steady on a neck, or in making an erratic little arm arrive in space at a point already reached by the eyes, the time when the body is all-engulfing because not yet mastered. Christ in the manger close to the animals who feed there, Christ in his mother's lap close to the belly that once contained him—whatever the details of gaze or smile, it is Christ as baby uncompromised by babyhood. So, too, the depicted hours on the cross make visible the inside of the body, the body in its nearly unchallengeable demonstration of its final requirement. These two phases of life are obsessively pictured in work by centuries of known artists, centuries of unknown carvers of ikons, and recent centuries of machines generating massive numbers of images for books and household walls.[27] That it should have come to be these two periods seems in retrospect as inevitable and self-evident as it is actually unanticipateable in the scriptures themselves which objectify the successive phases of his life with equal emphasis. It is perhaps also inevitable that the single image people again and again name as the most overwhelming, the Pietà, should be a conflation of the other two, for it at once pictures Jesus crucified and Jesus in the infant world of his mother's lap.

To have a body is, finally, to permit oneself to be described. God in the Old Testament is seldom the subject of his own utterances. His most essential self-description either has no predicate ("I am") or has a circular predicate ("I am the Lord," "I am your God").[28] The New Testament text is laden with statements that specify a predicate—I am the resurrection, I am the way, I am the light of the world, I am the door of the sheep, I am the good shepherd, I am the bread of life—a proliferation of predicates each of which can be held and turned in the mind. As striking or meaningful as any one of these self-descriptions may be, it is not their specific content but the act of definition itself, a kind of *predicative generosity*, that is of greatest consequence. To consent to be predicated is to consent to be at a given moment confined by some attribute (to be ... like a gate, or instead like a light), just as to consent to have a body is to consent to be, though everywhere, apprehensible at every given moment as only somewhere rather than everywhere. The predicability of Jesus enables him to enter and be held in the human mind, just as the larger fact of his having a body enables him to enter and be held within the narrow perceptual field of human vision, the slender realm of touch (just as iconographically, he is more emphatically confined by the framing manger, lap, or cross against which he is almost always held). The gospels themselves are sustained acts of predication: He is ... the one who walked down the street, the one who was seen by John, the one who was born in a stable, the one who died on a hill, the one who spoke in the temple, the one who healed the leper, the one who fed people on a mountain, and so forth. The consent to have a body is the consent to be perceived and the consent to be perceived is the consent to be described. The fourfold retelling of the stories in the four gospels itself luxuriates in and celebrates the describability of Jesus.

Centuries of commentary on the relation between the Hebraic and Christian scriptures have long attended to the intricate parallels between the stories contained in the two, the life and acts of various Old Testament figures, Abraham, Joseph, David, Jonah, thereby becoming, retrospectively, prophetic emblems of the younger figure who becomes their apotheosis. But it is simultaneously crucial to recognize within the repetition the most central difference. The entire story of Christ's life can be placed against the entirety of what has come before and recognized as a retelling not just of isolated incidents but of the largest framing events of the Pentateuch—first, creation; second, exile; third, rescue—*but a retelling from the point of view of sentience.* The same framing story is told but now it is God himself, not man, who is described as created, exiled, and rescued: hence human sentience is deeply legitimized by its having become God's sentience as well. The Old Testament opens with the genesis, the creation, of man and world; the New Testament opens with the genesis, the creation, the birth of God himself. Each portrays conception as the Word and its material realization (or as it is called in the younger writing, the Annunciation and its material realization); but in the first, the Word of God materializes itself in the body of man and world, while in the second the Word of God materializes itself in the body of God, thereby locating voice and body, creator and created, in the same site, no longer stranded from one another as separate categories, thus also inviting humanity to recognize themselves as, although created, simultaneously creators. Similarly, the essential Old Testament condition of homelessness, which is after the eviction from the Garden the permanent situation of generations of landless men and women, becomes now in turn the essential New Testament condition of God's homelessness, exile, and exclusion from shelter. It accompanies him from the opening moments of his birth at an Inn in which there is no room for him, and continues on into the final moments of his manhood that culminate in the most extreme form of exile from the land—exile from the entire earth, execution. In the interim, it is a condition reasserted in many smaller events, such as the exclusion from the temple (John 12:42; Luke 9:22); it is a condition actively advocated, as when the homeless Christ urges his disciples to leave their homes; eventually it is absorbed into the proverbial texture—

Foxes have holes and the birds of the air have nests; but the son of man has nowhere to lay his head. (Matthew 8:20; Luke 9:58)

and is lifted into the prophetic framework—

He was in the world, and the world was made through him, yet the world knew him not. He came to his own home, and his own people received him not. (John 1:10,11)

For Jesus himself testified that a prophet has no honor in his own country. (John 4:44; see Luke 4:24)

Like the women and men of the tribes of Israel, he is created into a world that does not easily accommodate him: createdness and exile are inevitable counterparts because the overt limitations of the second merely expose the not yet visible limitations already contained in the first. The unaccommodating natural world must be "re-created," and it is the invitation to material recreation that becomes overt in the New Testament shattering of the categorical separation of body and voice, the revised strategy of self-rescue.

One may accurately say that in the Pentateuch God, too, is being described in the recorded histories of His covenant people; but it is His complete separation from the conditions of createdness and exile that are there insisted on—Jehovah Himself created, Himself homeless, is unthinkable. To retell the story in terms of God's own experience of these events is to endow the interior facts of sentience, the facts of being alterable, creatable, woundable, with absolute authority.

In turn, to place sentience and authority together, to make sentience authoritative, is to place pain and power on the same side of the weapon, and so restructure the object itself. One of the peculiar characteristics of pain is that something which is its opposite, power, can reside in a different location yet come to be perceived as increasing or decreasing as pain itself increases or decreases. Usually, pairs of opposites expand or contract not in correspondence to but in inverse relation to one another: the increasingly wet does not become the increasingly dry; to move farther east is to move farther from the west; an increase in light lessens the darkness, rather than carrying a growing darkness with it. But the phenomenon of pain repeatedly occurs in human contexts that allow its increase to be attended by an increase in the power accruing elsewhere. The altered relation in the Christian scripture between the body of the believer and the object of belief subverts this severed relation between pain and power, assuring that sentience and authority reside at a single location and thus cannot be achieved at each other's expense.

The conferring of the authority of the spirit on the fact of sentience has as a second consequence the dissolution of the boundary between body and voice, permitting a translation back and forth. The body in the Old Testament belongs only to man, and the voice, in its extreme and unqualified form, belongs only to God. Across the cross, each of these retains its original place but simultaneously enters the realm from which it had earlier been excluded. Although the discussion here has described only one half of this exchange, God's acquisition of a body, this is accompanied by its counterpart, man's acquisition of a voice, a voice as potentially free of material qualification as if it were, like the God of the Old Testament, wholly disembodied, a voice that is immortal. The Christian narratives, unlike the Hebrew, are eschatological in their emphasis: the "after-life" of western civilization is widely described as a Christian rather than a Jewish invention. Although the body as well as the soul is resurrected, the promise of immortality is the privilege of the verbal category now bestowed on the body. However immeasurably great the distance between man and God, it is no longer based on the fact that man has a body and God has not, nor on the fact that God is immortal and man is not. The authorization of sentience becomes, simultaneously, the license to remake sentience to be free of the original limitations of mortality.

This double exchange between formerly separate categories thus entails a third consequence, the invitation to making and material culture. The deep meanings of body and voice are no longer enclosed within the rigid outline of their own possibilities, stranded from one another. The continual exchange between the categories is not simply permitted but *required in* the very act of belief; thus as believing becomes an essential habit of mind, so the believer's perception of himself as a maker becomes an essential habit of mind. It is a commonplace, though a disputed commonplace, that material culture is an inevitable accompaniment of Christianity. One basis for this association is immediately apparent in the overt content of the narratives themselves where there is a great alertness to the work people do and live through: Joseph the carpenter, Peter, James, John, the fishermen, Paul the tentmaker. Jesus himself repeatedly breaks the fourth commandment to heal on the Sabbath (John 5:16,17; Luke 4:31; 6:5; 13:10; 14:1). After his death and return, this same emphasis continues: Mary for a moment supposes him to be the gardener (John 20:11–18), as Simon Peter,

Thomas, and Nathanael, casting their nets in the Sea of Tiberias, suppose him to be a fisherman on the coast (John 21:1–8). They are to recognize him within the world of work, just as Christ's invitation to Thomas to place his fingertips in the print of the nails (which occurs between the two mistaken identifications of Jesus as gardener and as fisherman—20:24–29) is itself an invitation to a mimetic entry into the world of work.

But the connection between "believing" and "making," recognizable in these overt conflations of Jesus and laborers, is much more deeply carried by the more fundamental and pervasive interchange of physical and verbal categories. The scriptural attribution to God of a body confers spiritual authority on human sentience but conversely requires that human sentience become authoritative: the Jesus who cries to his disciples in the garden, "Could you not watch one hour" (Mark 14:37) and who in these final scenes asks them to "see better" and to know what it is they "touch," is a God asking for heightened, more acute, more responsible acts of perception. If the exquisite acuity now required of the senses has something in common with hurt, it is not in the passivity of hurt that they are to discover him, for he continually heals their hurt. Sentience is no longer the passive surface on which the weapon's power of alteration inscribes itself but is instead relocated to the other, active end of the object and becomes responsible for controlling and directing that power of alteration. In this shift, weapon becomes tool, sentience becomes active, pain is replaced by the willed capacity for self-transformation and recreation, and the structure of belief or sustained imagining is modified into the realization of belief in material making.

The shift from "believing" to "making" has so far here been presented in terms of the contrast between the Hebraic and Christian scriptures: in the older writings, the Artifact is a projection of human capacities but itself requires the intensification of the human body to endow the imagined object with realness; in the younger writings, the Artifact is again a projection of human capacities, but the Artifact itself has undergone the modification of material realization, and thus it need no longer borrow the attribute of "realness" from the human body since it is itself self-substantiating. (Thus a god may, like the mental picture of a table, first exist as an imagined object but then, like a real, sensorially confirmable wooden table, undergo a second stage of invention where it acquires a material form.) In both of the two stages of invention—making-up (mental imagining) and making-real (material realization)—the invented object requires some form of "confirmation," for it cannot be a successful fiction if it is recognizable as fictitious, unreal; but in the transition from the first to the second stage, the attribute of "realness" has been built into the artifact itself. A return to the Hebraic scriptures, however, will make it clear that this shift from "believing" to "making" is itself anticipated in the Old Testament, and in fact anticipated there to such an extent that the Christian story itself seems necessitated by and generated out of a pressure never long absent from the pages of the older writings.

III. The Interior Structure of Made Objects

It can accurately be said—and has often been said—that the Hebraic scriptures are deeply hostile to material culture. The second commandment prohibition of graven images, though most familiar in the story of the casting of the calf beneath Horeb and the story of the adoption by the tribe of Dan of gods crafted by Micah (as well as the complicated record of the uniting of Bethel and Dan, and the succession of sometimes image-worshipping kings that followed), is not confined to these two moments but regularly appears and disappears as a persistent problem. Nor is the aversion to made objects confined to the specific class of objects that are graven images, for the aura of contamination that clings to them also surrounds ordinary objects: even the offensiveness of the molten calf is inseparable from the offensiveness of the jewelry out of which it is made, both of which are seen by Moses and by God as objects of self-adornment; similarly, throughout Jeremiah, the prohibition against graven images is inseparable from the prohibition against a people's act of building for itself beautiful buildings paneled in cedar and painted with vermilion (22:14; see also chapters 11, 12, 14, 32 as well as this conflation in Haggai, Zechariah, and Malachi). The anti-cultural, anti-urban, anti-craftsmen emphasis of the Hebrew texts has been persuasively and elaborately explored by Herbert Schneidau, who shows the way the initial identification of "Cain the murderer" with "Cain the original city-founder" continues in the rejection of the "old urban tradition of architectural and astronomical wisdom" in the story of Babel, in the attribution of unequivocally negative connotations to city-dwelling in the story of Sodom and Gomorrah, in the equation of the Israelites' final entry into the promised land with the necessity of destroying the world's oldest city (Jericho), as well as in an extensive list of parallel moments that only begin with these four.[29] Human acts of building, making, creating, working are throughout the Old Testament surrounded with layer upon layer of prohibition from above and inhibition from below.

At the same time, however, the scriptural attitude toward human acts of creation and culture cannot be simply and summarily identified as dismissive because it is deeply complicated by three very large facts. First, the Hebrews are themselves engaged in a sustained act of inventing an Artifact so monumental and majestic (however problematic) that it perhaps has no peer in any other single artifact invented by another people: the Old Testament prophecy that this Artifact will eventually compel and absorb the attention of the rest of the world (e.g., Exodus 33:16; Zechariah 8:21–23) has not proved inaccurate, and the prohibition against other acts of making can in part be understood as an attempt to prevent the energy of completing this one act of invention from being deflected into more modest outcomes. Second, this Artifact, God, is itself the pure principle of creating: thus "making" is set apart and honored as the most holy, most privileged, and most morally authoritative of acts. Invention is recognized as the thing on which nothing less than the fate of humanity depends. Third, God's very first act is designated to be the creation of the world: that is, in a single stroke, the Old Testament mind has effectively subverted the entire "natural world" and reconceived the whole cosmos as the proper territory for acts of artifice and intelligence. Thus the dense texture of prohibition against human making must be seen within the larger framework of these three facts that together establish "making" as *at once the most morally resonant of acts* (as now in the late twentieth century theories of distributive justice identify as the most urgent locus of ethical action "making" and "distributing what is made") *and the most extensive*, stopping not at the doorway of the house, the gateway of the city, the edge of the shore, not at the seas or the earth itself but out through the stars and the galaxies (so that the Voyager space*craft*, hurtling out of the solar system with its recorded message, "Hello from the children of the earth," only makes concrete the *reach* of inventiveness that was first assumed in the opening lines of Genesis).

Further, as this section will suggest, the human acts of material making are in the Old Testament—even in the very midst of the prohibition—very gradually made compatible with the generous impulses in the larger framework they themselves modify. Material objects in the Old Testament can be grouped into two categories. The first consists of those made by man, either independent of God's permission or explicitly against his prohibition: this category includes graven images as well as an array of domestic objects. The second consists of objects authored by God, or authored by man at God's explicit invitation and instruction; to the list of divine imperatives, "Go...," "Arise...," "Behold...," "Hearken...," "Do...," "Do not...," is periodically added the striking imperative, "Make..." (e.g., Make an ark of gopher wood). As will gradually become clear, the shattering of the categorical separation of body and voice already looked at in the New Testament is in the Old Testament accomplished through these two genres of material objects; for the objects authored by man *successfully* confer on God a body, while the objects authored by God divest man of his body by taking over the work of substantiation earlier required of the human body, thereby placing humanity closer to the verbal category. Each of the two genres of objects will be attended to separately below. Together they expose the interior structure common to all made things—whether an ark, altar, well, written song, robe, recorded law, house—as a conflation of God's body and man's body, that is, as a materialization of the principle of creating and a dematerialization of the human creator.

Objects Authored by Men and Women—Graven Images and the Conferring on God of a Body

As suggested earlier, the making of graven images is in the Hebraic scriptures a blurring of the categorical integrity of body and voice. Consequently, it is almost inevitably followed by events intended to reestablish the extremity of the two categories: either the physical ground of man's existence is intensified or God's verbal purity is reasserted. Both kinds of event are mediated by the weapon either as material object (sword) or material phenomenon (fire), which is singular in being a concrete object or phenomenon other than the human body identified with and allowed to represent God. The making of graven images, however, eventually works to shatter this singular representational object into a multitude of objects. It is as though the colossal mediating sword, spanning the distance from the ground to heaven, is itself fractured, and on the ground beneath the place where it once stood are cups, blankets, songs, and a multitude of other objects.

To say that the act of representing a god is an act of endowing him with a body may be self-evident, perhaps even tautological. Certainly the narratives themselves stress this connection either by the form of the image which, like the molten calf, is often animal-like (for an animal is the phenomenon of sentience uncomplicated by personhood and self-extension) or in its surrounding context, as in the story of Rachel's usurpation of her father's household gods, which are almost absorbed into her own flesh as though they were her own children not yet born:

> Now Rachel had taken the household gods and put them in the camel's saddle, and sat upon them. Laban felt all about the tent, but did not find them. And she said to her father, "Let not my lord be angry that I cannot rise before you, for the way of women is upon me." So he searched but did not find the household gods. (Genesis 31:34,35)

But the seriousness of the claim only becomes clear if the act of representation, the act of embodiment, is itself recognized as an act of description, description not only unasked for but forbidden. The full threat and real significance of the graven images is *not merely that they describe God but that they give rise to a situation that requires God to describe himself.* That is, whether in fury or in sadness, his explanations of the error or injustice or godlessness of the images inevitably engage him in acts of self-description. Even in the moment when the graven images are being themselves destroyed, they are provoking God into acts of predication through which he materializes himself.

Two classes of predicates, the familiar and the unfamiliar, occur here, the first of which will be looked at before attending to the second, more significant class. Some of the acts of self-description occasioned by the graven images are ones that have, in other scriptural contexts, already been granted, never withheld, such as his existence as a voice, a name. So, for example, the fact that in Leviticus the announcement of the prohibition is repeatedly followed by "I am the Lord"—

> You shall make for yourselves no idols and erect no graven image or pillar, and you shall not set up a figured stone in your land, to bow down to them; for I am the Lord your God. You shall keep my sabbaths and reverence my sanctuary: I am the Lord.   (26:1,2 see also Jeremiah 16:18,20,21)

is not startling, for this act of predication is the one about which there is the least hesitation elsewhere in the scriptures, however carefully that name must be spoken. But even this most minimal predicate takes on in the present context a special tone; for a slight suggestion of vulnerability and acknowledged ineffectiveness attends the need to place by the side of the rejected physical image the reasserted verbal image, as in these deeply moving passages from Isaiah:

> Remember this and consider,
> recall it to mind, you transgressors,
> remember the former things of old;
> for I am God, and there is no other;
> I am God, and there is none like me . . .
>   (Isaiah 46:8,9)
>
> I am the Lord, that is my name;
> my glory I give to no other,
> Nor my praise to graven images.
>   (Isaiah 42:8)

"I am the Lord, that is my name"—the repeated announcement almost sounds like that of an orphan protecting himself against mistreatment by introducing his sign of personhood ("I'm Jo") or that of an ineffective bureaucrat reminding his subordinates of his position ("You cannot do that; I'm the president").[30] If the predicate itself is familiar from many other scriptural contexts, there is the presence of something not wholly familiar in its iteration here.

The same is true of a second predicate, God's identification as Creator. This predicate, which originates with the opening of Genesis and is of course assumed throughout the scriptures, becomes in the graven image passages explicit, extensive, possibly even obsessive; for the crafting of idols inverts the primary relation between maker and made, and thus the restoration of the original direction of creation requires the continual reminder that it is God who has created everything, that "Before me no god was formed" (Isaiah 43:10), that it is He who "stretched out the heavens" and brought about all geographical alteration (Jeremiah 10:12,13), that, most important, it is he who has created humanity, brought them forth from the womb (Isaiah 43:7,21; 44:2). Throughout each of these passages and similar passages in Micah, Nahum, and Habakkuk, the graven images outrage the logic of divine creation because they are made by those who, themselves made, cannot be makers. (One cannot be unmindful here of the very large risks taken by the scriptural mind in allowing a made thing, a fiction, God, to itself denounce graven images on the basis that they are made things, fictions, and hence "lies" as in Isaiah 44:12–20 and Habakkuk 2:18–20.) The violation of this prohibition brings forth the most severe punishment—

> And in this place I will make void the plans of Judah and Jerusalem, and will cause their people to fall by the sword before their enemies. . . . I will give their dead bodies for food to the birds of the air, and to the beasts of the earth. And I will make this city a horror, a thing to be hissed at; every one who passes by it will be horrified and will hiss because of all its disasters.   (Jeremiah 19:7,8)[31]

—punishment which, as in the passage cited, sometimes takes the form normally found in other scenes of punishment unconnected to graven images, but at other times does not.

What is most striking about the graven image passages is that in them the threat of punishment often departs from the conventional scene of punishment: the sentient human body that must be shattered because it has made an artifact is here often described as though it were itself an artifact or object of craft:

> Because Manasseh king of Judah has committed these abominations, . . . and has made Judah also to sin with his idols; therefore thus says the Lord, the God of Israel, Behold, I am bringing upon Jerusalem and Judah such evil that the ears of every one who hears of it will tingle. And I will stretch over Jerusalem the measuring line of Samaria, and the plummet of the house of Ahab; and I will wipe Jerusalem as one wipes a dish, wiping it and turning it upside down.   (2 Kings 21:11–13)

> Thus says the Lord God [to the rebellious house]: Set on the pot, set it on, pour in water also . . . . Therefore thus says the Lord God: Woe to the bloody city: . . . Then set it empty upon the coals, that it may become hot, and its copper may burn, that its filthiness may be melted in it, its rust consumed.   (Ezekiel 24:3,9,11 and see 23:30,35,37–39,42 for prelude)

> Lo I am about to make Jerusalem a cup of reeling . . . On that day I will make Jerusalem a heavy stone for all the peoples; all who lift it shall grievously hurt themselves . . . On that day I will make the clans of Judah like a blazing pot in the midst of wood, like a flaming torch among sheaves . . . . And on that day, says the

Lord of hosts, I will cut off the names of the idols from the land, so that they shall
be remembered no more.   (Zechariah 12:2,3,6, and 13:2)

So will I break this people and this city, as one breaks a potter's vessel, so that it
can never be mended.   (Jeremiah 19:11, see 18:15ff)

As I live, says the Lord, though Coniah the son of Jehoiakim, king of Judah, were
the signet ring on my right hand, yet I would tear you off . . .   (Jeremiah 22:24;
see 22:9–14)

In their breaking of the second commandment, humanity itself becomes a dish,
a cup, a pot, a potter's vessel, a piece of jewelry and, in similar passages, a
clay vessel complaining to its fashioner that it lacks handles (Isaiah 45:9), pieces
of silver and gold not yet refined (Malachi 3:3; Zechariah 13:9); elsewhere Israel
becomes a threshing sledge to be used against enemy kingdoms worshipping
idols (Isaiah 41:15), as she in turn in her transgression is consumed by her
enemies like a garment eaten by moths (Isaiah 50:9). These passages are striking
in part because in them God's powers of creating and of wounding are so vividly
conflated.[32] But they are also startling for a second reason: the transformation
of the covenant people into an unwiped dish or a rusty pot is so resonant with
divine contempt that its harsh tone for a moment reestablishes and underscores
the distance between the creator and his creatures; but the continual linking of
the remote fact of cosmic creation with the humanly accessible fashioning of
cup and shirt ultimately entails the collapsing of the distance between the very
categories it seeks to reestablish.

It becomes clear why the prohibition against idols must be absolute, for once
they are made, they create a situation from which there can be no return to the
original position. Once man, undirected by God, creates an object, God has then
one of two choices. He can disown the object, thereby acknowledging human
authority over that one sphere of creation and thus contracting the territory of
creating that belongs to him. Alternatively he can (as in the passages just cited)
extend the boundary of his authorship to include that object as well, thereby
restoring to himself the entire ground of creation but now including within that
province an instance of artifice so diminutive that it discredits the entire enterprise. Either response involves the recognition of a limit within divine creation.

The problem occasioned by the making of an ordinary object is, of course,
greatly compounded if that object is an idol. Once an idol is placed in the space
between man and God, there are from the divine position only four possible
interpretations. Just as in the case of the ordinary object divine disowning and
owning are the same self-limiting act, so the four objections to the idol consist
of two in which it is disowned (rejected because God interprets it as not referring
to Himself) and two in which it is owned (rejected because God interprets it as
indeed referring to Himself). One, the idol refers to another god (e.g., Baal).
Two, the idol refers to nothing beyond itself; it is itself (in the eyes of the people)

another god. Three, the idol does refer to Me and is a misrepresentation. Four, the idol refers to Me and is a representation. These are the four categories of response into which most of the Old Testament objections to graven images fall. Although each provides its own compelling basis for God's wrath, each is to a large extent a restatement of the other three and of the single essential problem contained in the image-making.

The graven image passages in Deuteronomy, Isaiah, and Kings, for example, express deep anxiety about the possibility of God's replacement by or conflation with Baal, Molech, or the self-referring Asherim. This protection of the purity of God's position is also a protection of the racial purity of the Hebrew people, as in Jeremiah where the people are warned against mixing with the image-making Chaldeans or Babylonians. There is both an ontological and a sociological basis for being scandalized by idols that refer to other gods. But often in these passages the objection to the idol's reference to other gods slides over into and becomes indistinguishable from anger over its misrepresentation of God. Is a molten calf more scandalous if it is understood to refer to Baal than if it is understood to refer to God? The psalmist writes in dismay of the event at Horeb:

> They exchanged the glory of God
> for the image of an ox that eats grass.   (Psalm 106:20)

The terrible disloyalty of worshipping Baal is not worse than the searing insult and deep indignity of portraying God Himself as a young ox. Infidelity and misrepresentation come to be almost indistinguishable, even changing places so that attention to Baal is described in terms of its lewd indignity, and visualizing God in material form is described in terms of infidelity, disloyalty, and harlotry. So, in turn, the error of misrepresentation is almost indistinguishable from the error of representation, since any act of representation, any material analogue, is a diminution and distortion of the immaterial, as is quietly articulated in the astonished and pained incomprehension of Isaiah:

> To whom then will you liken God,
> or what likeness compare with him?
>
> To whom then will you compare me,
> That I should be like him? says the Holy one.   (Isaiah 40:18,25–26)

One may say that here and in similar passages, God or the prophet intentionally misrepresents the nature of the act of representation, for the idol need not be seen as suggesting an equivalent, let alone (as in the passage just cited) as indicating the lines of aspiration, an image of what God should seek to become. But ultimately, of course, the idols are just that, a human suggestion to God about what he should try to become, and what he should try to become is embodied. It is precisely because any act of material representation is an act of embodiment that again and again the graven representations in the Hebraic scriptures are so close to what God in the Christian narratives willingly becomes by taking a body. It is difficult to read that "men kiss calves!" (Hosea 13:2) or that they make an image "of an ox that eats grass" (Psalm 106:20) without seeing floating behind the asserted ludicrousness of these animal images the image of Jesus at birth surrounded by cow and donkey, sleeping in their shelter, held in the container out of which they eat, himself almost one of their infant offspring. Again who can read the following passages from Jeremiah without being deeply moved by its bewilderment at the horrible impropriety of representation—

> A tree from the forest is cut down,
> and worked with an axe by the hands of a craftsman.
> Men deck it with silver and gold;
> they fasten it with hammer and nails
> so that it cannot move.
> Their idols are like scarecrows in a cucumber field,
> and they cannot speak;
> They have to be carried,
> for they cannot walk.   (Jeremiah 10:3–6)

—yet not at the same time be haunted by another pair of outstretched arms, another nailed and speechless scarecrow in a cucumber field.

It is the fact of embodiment itself out of which these coincidences arise, and it is the fourth of the interpretive categories—the idols refer to God and represent him—that is the essential violation implicit in them all.

As the foregoing discussion has tried to suggest, God responds to instances of representation by trying to clarify who he is and who he is not, by acts of self-clarification and self-definition through which he materializes himself. Thus the human act of embodying him, though itself interrupted, does in fact end by leading to his embodiment. Even those attributes that are already a given in the scriptures—his existence as a verbal category, his name, "the Lord," and his position as creator—acquire a more coherent because no longer total shape as he now specifies which *parts* (rather than all) of the power of creating are his, and which *parts* (rather than all) of creation have Him as their ultimate referent. He also acquires a much richer emotional predication in this context: as was implicit in the descriptions above, the ordinary tones of anger or vengeful authority that surface elsewhere in the scriptures are now regularly accompanied by psychological events such as bewilderment, incomprehension, vulnerability, jealousy, shame, or embarrassment, attributes that sound awkward and inapplicable to God yet are unquestionably present in the complex emotional tone of the graven image passages. This tone signals the power of images to endow God with a body: they do not simply visualize him in material form but, more important, give rise to responses in him that can only accompany the bodily condition. But there is a second, far more startling signal of this same fact that becomes evident when one attends not to the class of familiar predicates (I am the Lord; I am the Creator) but to the class of unfamiliar predicates.

As one moves through the graven image passages, one encounters a wide assortment of attributes that God directly or indirectly claims in the process of rejecting a specific statue, idol, or pillar. For instance, He may in one instance reject it because it is a representation; in a second, because it is an *animal* representation (implicitly accepting the fact of representation itself); and in a third, he may reject some *particular quality* of the specific animal representation (implicitly accepting not only the general phenomenon of representation but even the whole subphenomenon of animal representation), as in this passage from Hosea where it is only the offensive docility of the calf imagery that he angrily dismisses:

> Men kiss calves! . . . .
> So I will be to them like a lion,
> like a leopard I will lurk beside the way.
> I will fall upon them like a bear robbed of
>   her cubs,
> I will tear open their breast . . .   (13:2,7,8)

But while one could make a long catalogue of the predicates he gradually acquires through this process, one particular ground of dismissal, one particular attribute of the idols, recurs again and again until it becomes perhaps the single most prominent basis on which he differentiates himself from them. It is visible in the following passages:

> And there you will serve gods of wood and stone,
> the work of men's hands, that neither see, nor
> hear, nor eat, nor smell.   (Deuteronomy 4:28)

> Every man is stupid and without knowledge;
> every goldsmith is put to shame by his idols;
> for his images are false,
> and there is no breath in them.   (Jeremiah 51:17; see 10:8,9,14)

> > What profit is an idol
> > when its maker has shaped it,
> > a metal image, a teacher of lies?
> > For the workman trusts in his own creation
> > when he makes dumb idols!
> > Woe to him who says to a wooden thing,
> > Awake;
> > to a dumb stone, Arise!

> Can this give revelation?
> Behold, it is overlaid with gold and silver,
> and there is no breath at all in it.   (Habakkuk 2:18,19)

Divine contempt for these statues is based on the fact that they do not see, do not hear, do not eat, do not breathe, do not move—based, that is, on the startling fact that, despite their material form, they do not have the attributes of bodily sentience. On this remarkable basis of sentience God continually differentiates himself from the idols. It is this that they lack. It is this that He has. Thus the graven images give rise to a situation in which God begins to claim as his own sensory attributes he almost nowhere else has occasion to own as a defining characteristic, sensory attributes that he in other contexts sometimes explicitly rejects (e.g., ''and though they cry in my ears with a loud voice, I will not hear them'' Ezekiel 8:18). Their failing is a failing he has elsewhere taken pride in.

He now not only reminds his people that he is alive, that he sees, moves, hears, breathes (and by implication, even eats, for the absence of this attribute is included in his denunciation of wooden and stone objects), but even that he experiences the most passive, extreme, and unselfobjectifying form of sentience, physical pain. This acknowledgment occurs, for example, in chapters 8 through 12 of Jeremiah where, amid lines condemning the worship of Baal or other gods and idols, comes the repeated cry, indistinguishably that of the prophet and of God, I am wounded:

> For the wound of the daughter of my
> people is my heart wounded,
> I mourn, and dismay has taken hold on me.   (8:21)

> O that my head were waters,
> and my eyes a fountain of tears,
> that I might weep day and night
> for the slain of the daughter of my people!   (9:1)

> Thus says the Lord of hosts:
> ''Consider, and call for the mourning
> women to come;
> send for the skilful women to come;
> let them make haste and raise a wailing over us,
> that our eyes may run down with tears,
> and our eyelids gush with water.''   (9:17,18)

> . . . for his images are false,
> and there is no breath in them.
> . . . .
> Woe is me because of my hurt!
> My wound is grievous.
> But I said, ''Truly this is an affliction,
> and I must bear it.''

> My tent is destroyed,
> and all my cords are broken;
> my children have gone from me . . .   (10:14,19,20)

The same cry of pain, the same admission of a capacity to suffer, occurs in Isaiah, again in a voice that hovers between that of the prophet and that of God:

> For a long time I have held my peace,
> I have kept still and restrained myself;
> *now I will cry out like a woman in travail,*
> *I will gasp and pant.*
> I will lay waste mountains and hills,
> and dry up all their herbage . . . .
> They shall be turned back and utterly
> put to shame,
> who trust in graven images,
> who say to molten images,
> "You are our gods."   (42:14,15,17, italics added)[33]

Again the cry of Godly hurt becomes audible in his confession of need for the Hebrew people:

> How can I give you up, O Ephraim!
> How can I hand you over, O Israel!   (Hosea 11:8)

Thus the idol provides the external image of a body while lacking the interior requirement, sentience, yet simultaneously provides the situation in which God now claims that interior requirement, acknowledging his sentience not only in its aggressive and powerful outcome, as in the authoritative overseeing of men's actions, but in the passive, objectless, self-experiencing fact of itself whose most persuasive sign is physical pain. While in other scriptural contexts He has the objectified content of vision, hearing, and touch, now He has the sentient eyes, ears, and embodied surface that can themselves be wounded, produce the sensation of hurt, shed tears, require comforting, and so forth.

His sentience is in turn accompanied by greater attention to and accommodation of human sentience, for the failure of the statues to hear is designated their inability to answer the human cry for help, their failure to walk is their failure to lead men out of trouble:

> When you cry out, let your collection
>   of idols deliver you!
> The wind will carry them off,
> a breath will take them away.   (Isaiah 57:13)

> [Their idols] cannot speak;
> they have to be carried,
>   for they cannot walk.

> Be not afraid of them,
>   for they cannot do evil,
>   neither is it in them to do good.   (Jeremiah 10:5)

> "And we will say no more, 'Our God,'
>   to the work of our hands.
> In thee the orphan finds mercy."
>   I will heal their faithlessness;
>   I will love them freely,
>   for my anger has turned from them.
>   . . .
> O Ephraim, what have I to do with idols?
> It is I who answer and look after you.   (Hosea 14:3,4,8)

And there you will serve gods of wood and stone, the work of men's hands, that neither see, nor hear, nor eat, nor smell. . . . When you are in tribulation, and all these things come upon you in the latter days, you will return to the Lord your God and obey his voice, for the Lord your God is a merciful God; he will not fail you or destroy you or forget the covenant with your fathers which he swore to them.   (Deuteronomy 4:28,30,31)

It is in this context that God repeatedly identifies himself as rescuer—

I brought them out of the land of Egypt, from the iron furnace, saying, Listen to my voice.   (Jeremiah 11:4)

> You know no God but me,
>   and besides me there is no savior.
> It was I who knew you in the wilderness,
>   in the land of drought.   (Hosea 13:4,5)

and as healer—

You shall not bow down to their gods, nor serve them, nor do according to their works, but you shall utterly overthrow them and break their pillars in pieces. You shall serve the Lord your God, and I will bless your bread and your water; and I will take sickness away from the midst of you.   (Exodus 23:24,25)

> We would have healed Babylon,
>   but she was not healed.   (Jeremiah 51:9)

Again in Isaiah, he contrasts the weight of lifeless idols carried on the stooped and weary backs of men and donkeys with his own willingness to carry man: "Even to your old age I am He, and to grey hairs I will carry you" (46:4). This inclusion of the old and sick is almost a retraction of the previous rule banishing the blemished and handicapped from his presence (Leviticus 21:16f; 22:4,17f; Deuteronomy 17:1). He asks that those who know him know that "I am the Lord who practices kindness" (Jeremiah 9:24), and describes destruction as human rather than divine activity: "My compassion grows warm and tender / I will not execute my fierce anger, / I will not again destroy Ephraim; / for I am God and not man" (Hosea 11:8,9). Reciprocal sentience is at once reciprocating sentience and reciprocal need: "They shall be my people, and I will be their God" (Ezekiel 11:20; see Zechariah 8:4–8).

The graven images elicit from Yahweh a constellation of attributes characteristic of Jesus and that seem to follow from the fact of embodiment itself. Having a body means having sentience and the capacity to sense the sentience of others, reciprocity, compassion. Having a body means being not everywhere but somewhere, no longer hidden: "He will surely be gracious to you at the sound of your cry. . . . your Teacher will not hide himself any more, but your eyes shall see your Teacher" (Isaiah 30:19,20). The idols themselves are repeatedly denounced for their attribute of hiddenness (Habakkuk 2:18–20; Deuteronomy 27:15), now by extension no longer an attribute of God. In turn, standing revealed requires—both in the New Testament and in the graven image passages in the Old Testament—that the believers themselves become responsible for a more accurate sentience, one that is healed and whole and able to differentiate between true and untrue gods: "Hear, you deaf; / and look, you blind, that you may see!" (Isaiah 42:18). Thus, though the people's making of idols always fails, their remaking of God and of themselves does not fail: the material statue is a tool or lever across whose surface they reach, repair, and refashion the primary Artifact.

All this is not to say that such passages are without cruelty, for as was stressed earlier, God's dominant response to idols is to reassert the separation of man's body and His voice, usually by making emphatically visible the precarious fact of man's body. But the bodily sentience God here begins to claim as his own is a frail but persistent theme that challenges the mental structures of the Old Testament, though it only itself becomes dominant once God in the Christian scriptures enters the body directly, bringing with him attributes that make the kinship between Christ and the realm of material culture (only implicit in the graven images) unmistakably clear. Conversely, all this is not to say that passages about idols are the only parts of Hebrew narrative and prophecy that contain this more benign, accommodating (more consistently reciprocating) God. It is only to say that the idols are one very resonant locus of the transformation.

Objects Authored by God Divest Man of His Body

The discussion has so far looked at one class of artifacts in the Old Testament, those made by man: they are independent of God's sanction and are usually made explicitly against his prohibition. There is a second class of artifacts, those made by God, or made by man according to God's overt and often detailed instructions. This second group of objects, like the first, explodes the categorical separations by conferring on God a body and relieving man of his own body. The structure of exchange is once again directly described by the narratives themselves.

The artifacts God authors are, especially by the measure of our own dense material culture, very small in number. Each, however, tends to be a compelling focus of attention, in part because each stands in vibrant isolation against the desert background: in the midst of an unbroken mineral expanse, the blue of the tabernacle curtains breaks across eyes eager for an object. Further, though small in number, they are relatively extensive in the categories of invention they bring forth. The tabernacle is perhaps the most familiar: God's instructions to Moses for its design and construction are intricate enough to fill thirteen chapters of Exodus (25–31, much of which is then repeated in 35–40), for he specifies the pattern, length, height, width, and material of section after section of the large edifice, including precise accounts of its hinges, clasps, and pegs; he specifies also the design of the small objects within its walls (ark, altar, mercy seat, incense altar, table of acacia wood, lampstand of pure gold, snuffers, trays, oils, clothing); and he designates as well the particular craftsmen whose hands will bring this about. Instructions for additional objects enter elsewhere, as in Numbers where He requires "two silver trumpets of hammered work" (10:1) and a lampstand with seven candles, "from its base to its flowers, it was hammered work; according to the pattern which He had shown Moses" (8:4). His direction also extends to objects much larger than the tabernacle, as in his requirements for the number and size of the cities of refuge and cities of the Levites, "And you shall measure, outside the city, for the east side two thousand cubits, and for the south side two thousand cubits, and for the west side . . . " and so forth (Numbers 35:5f). It also includes surprising categories of objects such as the model, map, or miniature theatre that Ezekiel is directed to build: "Take a brick and lay it before you, and portray upon it a city . . . and build a siege wall against it, and cast up a mound against it; set camps also against it. . . . And take an iron plate, and place it as an iron wall between you and the city" (Ezekiel 4:1–3).

Language, too, is an artifact, and when it is written down, the verbal artifact becomes a material artifact. The law given to Moses on Mount Sinai is only the most widely recognized instance of God's authorship of the expansive realm of the materialized voice that extends from the simplest deed of possession inscribed on the object itself—

> There shall be inscribed on the bells of
> the horses, "Holy to the Lord."   (Zechariah 14:20)
>
> This one will say, 'I am the Lord's,'
> another will call himself by the name of Jacob,
> and another will write on his hand, 'The Lord's.'   (Isaiah 44:5)

to the entire history of a people—

Moses wrote down their starting places, stage by stage, by command of the Lord; and these are their stages according to their starting places.   (Numbers 33:2)

and includes the self-conscious category of song, sometimes only in His simple instruction to sing (Isaiah 42:10) but elsewhere, as at the end of Deuteronomy, in His dictation of the words of the songs themselves (31:30; 32:1–44). The isolated phrases explicitly attributed to God within the scriptures are only heightened and self-conscious acknowledgments of an authorship that extends to all the words of the scriptures, for in the Hebrew tradition, it is not just the ten commandments but the Pentateuch and the books of the prophets, and even the oral law eventually written down in the Talmud that are understood to have been given to Moses at Horeb.

As was earlier suggested, God ordinarily permits himself to be materialized in one of two places, either in the bodies of men and women or in the weapon. These two forms of substantiation reappear in the recurring scenes of hurt. These two are also often lifted out of the narrative sequence of events and self-consciously designated as signs. In all of Genesis, for example, there are only two signs: one is the weapon, the bow, "I set my bow in the clouds, and it shall be a sign of the covenant between me and the earth" (9:13) [34] the other is the mark in the human body, "You shall be circumcised in the flesh of your foreskins, and it shall be a sign of the covenant" (17:11). Such signs are themselves, of course, simple artifacts, artifacts not identical with the thing they indicate (since they do not have within them the angry intention and outcome of an actual wound or weapon) but elementary enough to memorialize the literal event of wound and weapon for which they are substitutes. That they are at once "substitutes" for bodily hurt and "signs" of God's peace with man (two ways of saying the same thing) is indicative of the benign consequence of artifice that becomes more pronounced in more freestanding artifacts.

When God authors a particular material or verbal artifact, the object is often presented in the narrative either as his own material form or as a material substitute for the human body. Sometimes the narrative is not explicit and it is instead the surrounding context that specifies this meaning. For example, the suggestion introduced earlier that the layers and layers of woven tissue in the tabernacle are almost the visible tissue of the body of God is not directly stipulated by the text but is instead indicated by the larger framework. The crisis-laden departure of the Hebrews from the culturally rich, graven-image filled kingdom of Egypt into the desert wilderness (the opening of Exodus), followed by the deep crisis centering on the contrast between the molten calf and the voice speaking out of the fire (the middle of Exodus), makes the building of the tabernacle (the end of Exodus) impossible to sever from the issue of material representation that is so emphatic in the first two of the three major subjects. Calf and tabernacle, two lonely objects on an otherwise unbroken expanse of sand and stone, are connected by much more than the jewelry that goes into the fashioning of each (the jewelry taken out of Egypt in the opening; put into, then retrieved back out of, the calf in the second part; then put into the tabernacle). Thus, while the scriptures overtly call for the absolute annihilation of material form and the reestablishment of God's verbal purity, the structure of events instead insists on the more modest distinction between man's version of God's material form (calf) and God's version of God's material form (tabernacle). The tabernacle, like the calf, is an object around which the people gather, a spatially locatable object they can move toward or away from, an object which satisfies their longing for His observable presence among them that had earlier given rise to the incident of idol-making. Even here, however, the image of the weapon is not entirely gone, for it is instead hovering nearby in hypothetical outline: the tabernacle is surrounded by the brooding storm-cloud by day and the fire by night. The perceptual object that permits the translation back out of the material base into the wholly verbal source is held in readiness.

Sometimes, however, the projection of the sentient body into the artifact is explicitly announced by the narrative itself, as in the following passage describing God's consuming punishment of the Hebrews for their disbelief:

> And Moses said, "Hereby you shall know that the Lord has sent me to do all these works, and that it has not been of my own accord. If these men die the common death of all men, or if they are visited by the fate of all men, then the Lord has not sent me. But if the Lord creates something new, and the ground opens its mouth, and swallows them up, with all that belongs to them, and they go down alive into Sheol, then you shall know that these men have despised the Lord."
>
> And as he finished speaking all these words, the ground under them split asunder; and the earth opened its mouth and swallowed them up. . . . And all Israel that were round about them fled at their cry; for they said, "Lest the earth swallow us up!" And fire came forth from the Lord, and consumed the two hundred and fifty men offering the incense.
>
> Then the Lord said to Moses, "Tell Eleazar the son of Aaron the priest to take up the censers out of the blaze; then scatter the fire far and wide. For they are holy, the censers of these men who have sinned at the cost of their lives; so let them be made into hammered plates as a covering for the altar, for they offered them before the Lord; therefore they are holy. Thus they shall be signs to the people of Israel." (Numbers 16:28–38)

"But on the morrow," the story continues, the people again disbelieve; God sends a plague that consumes fourteen thousand seven hundred people, until its action is stopped when the hammered plates are carried into the midst of the assembled people. In the first part of the passage, man's body substantiates God. In the second, the artifact becomes the sign of the human body on whose surface God's presence earlier recorded itself, and now itself performs the work of substantiation. Just as a person may assert his or her presence by continually fluid, continually disappearing, then renewed actions or instead assert that presence by objects which alter the scene they enter and permanently record the person's presence,[35] so God may continually reexpress Himself by repeated moments of actually inflicted hurt, or may instead have the enduring record of a single instance of that bodily alteration in the artifact.

It is not clear whether it is more true to say that the censers are a sign of (substitute for) man or a sign of (substitute for) God; for although it is in this passage explicitly man's body that is substituted for, spared, and therefore "represented," it is also God's body that is represented since man's body was itself hurt to confer the attribute of "realness" on an unapprehensible God—it was itself "a sign," a material verification of His existence. That is, the censers are a substitute for man's body which was originally itself a substitute for God's body. Once this process begins, it can be spread out onto more and more of the realm of artifice: as God has his sign in the human body and the human body its sign in the object, so this first object in turn may have its sign in a second object, and the second in a third. This occurs in Zechariah 6, for example, where all that is to be substantiated by the temple is, until the completion of that temple, substantiated by the more quickly fashioned crown which, after the completion of the larger artifact, will be enshrined there. As the process becomes more richly mediated, any given part of it can acquire a more specific function, as here where the crown serves first in the capacity of prophecy and then commemoration. What is clear from this sequence of substitutions is that the process of the imagination (and the transformation out of the realm of the sensuously immediate made possible by the imagination) begins not by man conceiving of signs separable from himself but by man conceiving of his own body as a sign of (and substitute for) something beyond itself, and then bringing forth other signs that can perform the work of representation in his dear body's place.

Those objects God introduces into Hebrew life are usually accompanied by the meanings described above. When Zechariah sees in a vision an elaborate branching lamp, the angel tells him that the seven lamps are "the eyes of the Lord" and the seven lamps have "seven lips,"[36] and the lamp is God's speech announcing first his power of altering the land through cosmic disturbance and second, his power of altering the land by introducing into its midst a new building (4:5–10). The cataclysmic disturbance need not occur if the thing it was intended to make apprehensible, God's power of alteration, can be made apprehensible in the landscape-altering construction of the temple, and if the thing that power of alteration was in its turn intended to make apprehensible (for God surely has no interest per se in changing land or landscape), the sheer fact of God's existence, the pure force of that fact, can be made apprehensible in an elaborate lamp, memorializing the verbal announcement and thus pre-empting the enactment of that force. The same is true of the seven-faceted, engraved stone placed before Joshua which "will remove the guilt of this land in a single day" (Zechariah 3:6–10), thereby eliminating the need for the usual ghastly procedures for transforming disbelief into belief through material verification. More simply and more powerfully, the altars and wells that God calls for throughout Genesis are, as noticed earlier, themselves representations of the interior of the human body, repeatedly occupying its position in the narrative sequence. Such artifacts are also, as the text periodically reminds us, material commemorations of God having spoken.

Embedded in God's authorship of material culture is the scriptural recognition that in scenes of verification the specific content of God's speech is unimportant. God's speech must have content, and must be detailed, only so that the material enactment of the specified content of his speech will make incontestable the fact of his having spoken; there must be content, and the content must be detailed, but what the content is, what the details are, does not matter. The scriptural attribution to Him of the authorship of material objects contains the recognition that the statement "Let this man be cut in two," "Let this woman's body become leprous," "Let the earth swallow them up" (followed in each case by the enactment) has—not only from man's point of view but from God's—no superiority over the statement, "Let the censers be hammered into altar plates," "Let there come into your midst a three-foot by five-foot altar," "Let the tabernacle be built according to this plan" (followed in each case by the enactment). In seeing that it is the fact of God's voice and not its particular content that matters, the human imagination behind the scriptures gains control of the content of God's speech and in doing so regains its own voice. Thus the artifacts are not only God's body but man's speech.

In any culture, the simplest artifact, the simplest sign, is the single mark on wood, sand, rock, or any surface that will take the imprint. This simplest of all artifacts provides the essential model for all God's artifacts in the Old Testament. It is the red mark on the doorposts and lintels, the passover mark, the mark (itself made of blood, and thus a projection of the body) that substitutes for the infant Hebrew bodies during the slaughter of the first-born (Exodus 12:7,23). The invention of this mark occurs on the eve of and makes possible their exodus out of Egypt. When, many years later, they finally enter the promised land, their crossing of the Jordan into Jericho is obsessively described as a "passing over" (Joshua 1:14; 3:11,14,17; 4:1,7,10,11,22,23),[37] and the sparing of the twelve tribes during that passage is commemorated in twelve stones. It is linked imagistically to the original red sign on the doorposts by the scarlet cord by which Rahab the harlot first lets them climb down into the walled city, the scarlet cord that she later hangs in her window as a sign to the Hebrews that all her kin within (as always in the Old Testament, the unit of rescue is the family) be passed over in the ensuing massacre of the city's people. Here, as on the earlier night, the possession of the fragile artifact depossesses one of the consequences of having a body.

The introduction of this sign is by no means an endorsement of material culture in general, for in both cases that culture is explicitly rejected, left behind in the exodus out of Egypt, destroyed in the entry into Jericho.[38] But in its simplicity, it defines quite openly and unambiguously the extraordinary design residing in each of the artifacts God authors. What, for example, are God's detailed instructions to Noah about the construction of the ark—

> Make yourself an ark of gopher wood; make rooms in the ark, and cover it inside and out with pitch. This is how you are to make it: the length of the ark three hundred cubits, its breadth fifty cubits, and its height thirty cubits. Make a roof for the ark, and finish it to a cubit above; and set the door of the ark in its side; make it with lower, second, and third decks. For behold . . .   (Genesis 6:14–17)

but the simple passover mark elaborated into an intricate blueprint of rescue. With it, Noah is dispossessed of his body and "floats on the face of the waters" (7:18) as his bodiless God once "moved over the face of the waters" (1:2). When the flood comes, the ark is for him the red mark on the frame of his door, the scarlet cord in his window.

Just as all the material objects independently authored by man may be collectively designated by the rubric, "graven images," so all the material objects authored by God (altar, well, tabernacle, censers, lamp, seven-faceted stone, red mark, ark) may be collectively designated by the rubric, "passover artifacts." So, too, the materialized (written, memorized, recorded) verbal artifacts are passover objects. In some instances, this rhythm of substitution and sparing is explicit:

> Then those who feared the Lord spoke with one another; the Lord heeded and heard them, and a book of remembrance was written before him of those who feared the Lord and thought on his name. "They shall be mine, says the Lord of hosts, my special possession on the day when I act, and I will spare them as a man spares his son who serves him."   (Malachi 3:16,17)

Again, the long song God gives to Moses at the end of their wanderings in the desert (Deuteronomy 32:1–43) is explicitly prefaced with God's explication of its function:

> Now therefore write this song, and teach it to the people of Israel; *put it in their mouths*, that this song may be a witness for me against the people of Israel. . . . And when many evils and troubles have come upon them, this song shall confront them as a witness (for it will live unforgotten *in the mouths* of their descendents).   (31:19,21, italics added)

The words of the song themselves describe God's vengeance and jealousy, his hatred of other gods and of image-making, his "glittering sword" and "arrows drunk with blood'' (32:41,42); but the song is itself intended as an almost free-standing artifact which by being ''in their mouths''[39] witnesses Him in their bodies, records His presence that would otherwise be recorded more deeply in their bodies in the hunger, burning heat, poisonous pestilence, and venom of crawling things that the content of the song describes. The form of the song is a substitute of its own content: the logic of the relation between form and content is described by the sentence, ''Because of this, not this.''

The passover function of the Deuteronomy song extends to other materialized verbal artifacts as well, for it is in part introduced to clarify the purpose of the ''commandments and his statutes which are written in this book of the law'' (30:10) that Moses has just completed writing. The song is a coda or epilogue to the law commenting on its inner design. One long chapter of Deuteronomy is devoted to the consequences of disbelief; the body is blanketed with every conceivable sore, sickness, and humiliation accessible to the Old Testament mind. It includes such passages as the following:

> The most tender and delicately bred woman among you ... will grudge to the husband of her bosom, to her son and to her daughter, her afterbirth that comes out from between her feet and her children whom she bears, because she will eat them secretly, for want of all things, in the siege and in the distress. ...   (28:56,57)

and it ends with the progressively deep humilations of objectless work:

> And the Lord will bring you back in ships to Egypt, a journey which I promised that you should never make again; and there you shall offer yourselves for sale to your enemies as male and female slaves, but no man will buy you.   (28:68)

But again, the formal fact of the scriptures is pre-emptive of scriptural content. The law itself is meant to displace the bodily verification of the primary Artifact, as one brief passage makes clear, and as the introduction of the song then reaffirms in expanded form:

> For this commandment which I command you this day is not too hard for you, neither is it far off. It is not in heaven, that you should say, ''Who will go up for us to heaven and bring it to us, that we may hear it and do it?'' neither is it beyond the sea, that you should say, ''Who will go over the sea for us, and bring it to us, that we may hear it and do it?'' But the word is very near you; it is in your mouth and in your heart, so that you can do it.   (30:11–14)

Like the song, the law is an artifact intended to eliminate human hurt. One may say, ''The cruelty is itself occasioned by the law, for bodily hurt occurs if the law is broken.'' But to say this is to forget the overarching sequence of substitutions: bodily hurt (analogical verification of God's realness) is replaced by the law, which may then become the broken law, in which case the bodily cruelty no longer has a substitute and so returns. It is precisely because X (e.g., the law) is a substitute for Y (e.g., bodily hurt) that when X is taken away (i.e., the law is broken) what returns is Y (bodily hurt). The relatively benign role of the written warning ultimately extends to the scriptures themselves: they are the framing artifact that subsumes as its content both the problematic deconstruction of creating in the scenes of wounding and the restructuring of creating in passages explicitly about material culture. Thus, even in the worst scenes of savagery, content is subordinate to form: the bodily substantiation of God's existence is itself recorded in a freestanding narrative artifact rather than recorded directly in the bodies of those who read, rather than hurt, in order to believe.

The material and materialized verbal artifacts of the Old Testament fall, then, into two groups: those made by man, the graven images, and those made by God, the passover images. The first confer on God a body and therefore materialize him; the second divest man of a body by producing a freestanding equivalent for that body. Thus the realm of created culture, the locus of God's body and man's body, threatens to shatter and transform the categorical separation of the two realms. If, as noted earlier, this shattering is (within the Judeo-Christian framework) only finally realized in the Christian narratives, if Jesus is the full materialization of what is only imminent in the prohibited and reluctantly granted Hebrew artifacts, it is also true that he is brought into being by them, that he must, at least in part, be recognized as being born forth out of the pressure that they exert, a pressure seldom long absent from the pages of the older writings. That is, if within the Hebrew imagination the artifact is in the very fact and form of itself a materialization of God and a dematerialization of man, then any artifact, say a story, any story that is told, regardless of its content, exists as a materialization of God and a dematerialization of man. But it is then perhaps inevitable that one day a particular story arises that becomes these things not only in the formal fact of itself but in its content as well, a story describing God's willing taking on of a body and, simultaneously, his willing disembodiment of men and women in earthly acts of healing, in the ultimate healing of the resurrected whole body, and in the elimination of the human body as a source of analogical verification. So that the form that was the benign substitute for its own content in the narratives of warning and hurt in the older writing now becomes itself the overt subject in the new. Thus the original exchange of categories sponsors another more sustained and profound version of itself in the Christian story, and Christ's body in turn by incorporating the transgression of these categories into the very act of belief makes the fact of continual ''making'' an inevitable and essential habit of mind.

It has been suggested here that the transition from the authoritarian emphasis of the Old Testament, where there is an inequality based on bodily representation, to the New, where that inequality is eliminated, has as its inevitable counterpart the greater sponsoring of material culture. It may well be that the anti-authoritarian impulse always has as its counterpart a correspondingly greater emphasis on materialism. Just as the Christian scripture is at once less authoritarian and more overtly material than the Hebrew, so too within Christianity, Protestantism has long been recognized as at once less authoritarian and more materially centered than Catholicism: that Luther's rebellion against and revision of the older version of Christianity was anti-hierarchical in its essential impulse is as widely accepted as is the identification of modern western materialism with what has come to be called "the Protestant work-ethic."[40] In turn, within Protestantism, various denominations can be differentiated from one another by the same counterparts: Quakerism, for example, has been elaborately and persuasively shown by sociologist Digby Baltzell to have been at once more egalitarian and more materially centered than Puritanism, and to have been the religious ideology that eventually came to hold sway in the obsessively democratic, obsessively materialistic, United States.[41] Thus certain characteristics of the revision of Judaism by Christianity may themselves have been repeated and extended in the revision of Catholicism by Protestantism, and in turn by the successive revisions within Protestantism that occurred through the emergence of various sects.

Whether material culture is itself a vehicle of democratization, whether an iconoclastic religious culture is inherently inconsistent with an egalitarian political philosophy, are questions too large to be entered into here. Further, to pursue the sequential modifications of both the hierarchical and the material through successive revisions would wrongly suggest that material making *always* preserves the benign ethical impulses of belief while simultaneously extending its benefits by eliminating the potentially brutal procedures for verification necessitated by unanchored belief. Although material "making" is a solution to some of the problems entailed in "believing," it may itself of course become the problem: that is, the deconstruction of categories that occurred at the site of the imagined object may occur at the site of the material object as well. Thus the very problems encountered by a religious culture come to be re-encountered by a secular culture.

The section that follows will focus on the character of materialism or material artifice as analyzed in the nineteenth-century philosophy of work. The discussion will have two parts. The first will show that Marx's fundamental conception of material creation echoes the conclusion about material creation found in the scriptures. The Judeo-Christian attribution to material culture of the deepest and most resonant of moral categories, our bodies and our God (our sentience and our revision of the facts of sentience through the projective work of creating) is consistent with and reaffirmed by Marx's own secular analysis of the material artifact: he too breaks open the sensuous object (now a table, now a wall of bricks, now a bolt of lace) and finds located in the interior structure of each our bodies and our gods. So humane is materialism in its premises that at no point does Marx ever imagine that the culture would be better served by a retraction of the impulse toward material making, even were such a retraction possible.

The second part of the discussion will move from Marx's celebratory affirmation of materialism to his criticism of one historical form of materialism that troubles him precisely because it distorts the basic premises of materialism itself. Although the overriding structure of making is benign, it has become the location of an injustice again based on problematic forms of substantiation: this injustice arises along a path of deconstruction similar to the one in the Judeo-Christian scriptures that was itself solved by the appearance of the material artifact. The existence of searing injustices in early nineteenth-century British industrialism has seldom, if ever, been disputed by anyone. Marx's overall analysis of the problem and conception of the solution are, in contrast, very disputable and will not be assessed here. What is from the point of view of the concerns raised in the present study of specific interest about his account of the collapse of making into unmaking is the fact that he identifies the problem as arising out of the *referential instability* of the primary sign of the weapon (or the sign of the artifact as it reverts into the sign of the weapon). The overall work of his philosophic analysis can (like the Judeo-Christian scriptures) be summarized as an attempt to restructure and stabilize this most potent of signs.

IV. The Construction and Deconstruction of Making within the Material Realm

Projection and Reciprocation: The Presence of the Human Body in the Artifact and the Remaking of the Human Body into an Artifact

It is important to become reacquainted with the interior structure of material objects because people in the West, though deeply committed to material objects in their actions and intuitions, often verbally disavow and discredit their own immersion in materialism, sometimes even scorning the tendency of less materially privileged cultures to aspire to the possession of these same objects: that blue jeans are cherished in the Soviet Union, that a picture from a Sears Roebuck catalogue should appear on the wall of a hut in Nairobi, that Sony recorders are prized in Iran, are events sometimes greeted by western populations with bewilderment, as though the universal aspiration toward such objects (both in countries where they are plentiful and countries where they are scarce) were a form of incomprehensible corruption or an act of senseless imitation rather than itself a confirmation and signal that something deep and transforming is intuitively felt to happen when one dwells in proximity to such objects.

The scriptural writings articulate for us, thousands of years later, the scale of our own spiritual investment in artifice: the deepest psychic categories, our bodies and our God, are at stake in the created realm of objects we inhabit. As was appropriate and necessary to a nineteenth-century reinvestigation of the structure of the human imagination, Marx eliminates from his description of material culture the outlines of God's presence, the entry of the Original Artifact into its successors, a work of entry accomplished and long since in place by the time he reopened the subject. But his assessment acknowledges the interior design whose shape is traced in the older account: he throughout his writings assumes that the made world is the human being's body and that, having projected that body into the made world, men and women are themselves disembodied, spiritualized. A made thing remade not to have a body, the person is himself an artifact. For Marx, material culture incorporates into itself the frailties of sentience, is the substitute recipient of the blows that would otherwise fall on that sentience, and thus continues in its colossal and collective form the "passover" activity of scriptural artifacts. Through this generous design the imagination performs her ongoing work of rescue, and because of that design Marx never disavows or discredits the western impulse toward material self-expression but is, instead, in deep sympathy with it. Each half of his assumption—first, the presence of the body in artifacts and second, the making of the human body into an artifact—will be looked at separately below before attending to their interaction.[42]

The human imagination has its collective expression in civilization: it is the thing created. But this created thing contains within itself the process of its own creation, the system of production and reproduction by which it comes into being, sustains and perpetuates itself. It is civilization conceived in this way—not as a stable and completed object to be externally assessed in its freestanding activity but as something that seems at once interior to that thing, the process residing within it that brings it about, and yet exterior to the thing, the vast artifact in which all other artifacts (pitchers, plates, cities, and systems, all objects collectively designated "civilization") are made and modified—that is Marx's subject. That this should be his subject is itself crucial, for it means that he is addressing at every moment that dimension of civilization in which the originally disparate conditions of sentience and the imagination are visibly conflated. On the one hand, the "system of production" is a materialization of the imagination's own activity of "making" (just as in the older writings the Primary Artifact, God, is itself the objectification of the human power of "creating" with all the ethical requirements and complications of that power brought fully into view). On the other hand, it is an artful extension of the metabolic and genetic secrets of the human body.

The economic system of production is for him a vast body any part of which has its physical equivalent. In the Preface to the First German Edition of the first volume of *Capital*, for example, he describes previous analyses of economic systems that lack any consideration of the nature of singular artifacts (a bed, a bushel of corn, a shirt) as attempts to understand the body without understanding first the cells of that body: the commodity is the "economic cell-form"; dismissed as minutiae, it is no more minutiae than is the "subject matter of microscopic anatomy."[43] When he moves from his discussion of single objects to their later and larger appearance as capital, he moves from the smallest structural unit of the cell to the body's generalized powers of self-magnification: "It is the old story: Abraham begat Isaac, Isaac begat Jacob, and so on. The original capital of £10,000 brings in a surplus-value of £2,000 which is capitalized. The new capital of £2,000 brings in a surplus-value of £400 and this too is capitalized, transformed."[44] Again in the *Grundrisse*, the seven notebooks of the late 1850s in which many of the working assumptions of *Capital* are formulated, he writes:

> The simultaneity of the process of capital in different phases of the process is possible only through its division and break-up into parts, each of which is capital, but capital in a different aspect. This change of form and matter is like that in the organic body. If one says e.g. the body reproduces itself in 24 hours, this does not mean it does it all at once, but rather the shedding in one form and renewal in the other is distributed, takes place simultaneously. Incidentally, in the body the skeleton is the fixed capital; it does not renew itself in the same period of time as flesh, blood. There are different degrees of speed of consumption (self-consumption) and hence of reproduction.[45]

The extent to which such passages, as well as his generalized use of the bodily language of "production," "consumption," "reproduction," and "circulation," are to be understood literally may seem unclear. Certainly the idiom is in part rhetorical: the body is sometimes invoked to make a vast and therefore invisible entity visible through its condensed approximation. Certainly, too, it is in part invoked to strengthen the recognition (a recognition Marx repeatedly called for by emphasizing the scientific character of his work) that the realm of artifice is as susceptible to rigorous analysis as is the realm of natural occurrence studied by the physical sciences.[46] Certainly, finally, the sense that the analogy is one of philosophic import and deep consequence is undercut by the breezy introduction of "It is the old story" or the quick parentheses of "Incidentally, in the body the skeleton is the fixed capital," stylistic acts of distancing that may alternatively be attributed to his assumption that his assertion is too self-evident to require labored argument, or to a general ambivalence arising from the fact that he is at once describing a universal artifact, a material system of production, that he perceives as profoundly beneficent, and a particularized version of that general artifact, the capitalist system of production, which he perceives as profoundly problematic. Yet the frequency of the allusions to the body, his structural dependence on them in arriving at his overall political critique, carries us far beyond such qualifications. In their cumulative weight, such passages announce his sober recognition that the large Artifact has about it the character of living matter.

If metaphor is perceived to reside here, it is crucial to specify exactly how it is occurring. To say that in the system of production, the privileges and problems of the human body are carried out into the made world is to say that the system is itself a materialized metaphor of, or substitution for, the body: the *artifact itself* rather than Marx's description of the artifact is metaphorical; the description merely records the presence of the metaphor.[47] Thus his allusions cannot be discounted on the grounds that they are "metaphorical," for that argument only recapitulates the very point made by the allusions themselves. Yet clearly the vast artifact does not have the strict equivalence in the human body that any single object has: it does not like a chair absorb an identifiable attribute of sentience into itself. A lone artifact like a chair must be held within a larger framework of artifice that brings it about and perpetuates it in the way that the total body carries all of its lone physical attributes such as "weight" within the larger framework of its metabolic and generative processes, and thus those metabolic and generative processes are appropriately recognized as being projected into the economic system of production. But the strict matching of commodity to cell or one form of capital to the skeleton and another to flesh and blood only has literal integrity in its specification of the relation between part and whole. One may say, then, that the vast artifact is a recreation of the body or instead, with slightly more precision, that it is a recreation of the way in which the single artifact recreates the body. If it is accurately identified as a metaphor for the body, it is with more accuracy identified as a metaphor of the single object—it performs metaphorically the activity of absorption that the single object much more literally performs.

Although, therefore, the body remains present in the largest phenomena Marx describes—phenomena larger even than the system of production, for his conviction that the cultural, political, and philosophic "superstructure" emerges out of, reiterates, and sustains the economic system (as the economic system emerges out of, reiterates, and sustains the commodity, and as the commodity emerges out of and reiterates in order to sustain the human body) is not a discrediting of the aesthetic and spiritual capacities of the human soul but simply a restatement of his belief that the strategies of artifice remain constant across such changes in scale, and in their ever greater extremity of sublimation reinforce, magnify, and protect the solutions to the problem of the body discovered on more elementary levels of artifice—yet it is the body's presence in those elementary sites that can most easily be described by Marx in a way wholly free of rhetorical and philosophic complication.

It is when in *Grundrisse* or *Capital* Marx has occasion to describe the commodity, the most elementary site within our own richly elaborated economic system, that the body's presence in the made object (e.g., a bolt of woven cloth) is soberly, often movingly, pointed to again and again. Marx's designation of the single artifact as a "body" is at some moments based on the concept of use value (the woven cloth refers to the human body because it has "use to" the living body, at once objectifying and eliminating the sentient problems of temperature instability and nakedness) and is at other moments based on its being the materialized objectification of bodily labor (the woven cloth is a material memorialization of the embodied work of spinning, for it endures long after the physical activity has itself ceased: "the worker has spun and the product is a spinning"[48]). Marx moves freely back and forth between the two without explanation or apology. Particular sentient actions (spinning) or particular sentient attributes (temperature instability) are projected there rather than the most generalized and fundamental characteristic of sentience, "aliveness." But the artifact, precisely because it arises out of and has no original existence other than in its contact with the human being, almost comes to borrow this characteristic as well. Thus the activity of "making" comes to be the activity of "animating the external world," either described as a willed projection of aliveness ("Yarn with which we neither weave nor knit is cotton wasted. Living labour must seize on these things, awaken them from the dead"[49]) or as a more passive occurrence arising from sheer proximity to real human tissue ("Now the raw material merely serves to absorb a definite quantity of labour. By being soaked in labour, the raw material is in fact changed into yarn"[50]).

Similarly, when Marx moves back behind our own economic constructions to speculate about earlier forms of economic activity, he describes sites of imaginative projection more elementary than the commodity—"the land" and "the tool"—in this same way. In such passages, the body's presence is not qualified by the glacine and cutting reflex of his style, nor is it even introduced with "like" or "as": it is instead registered in simple and moving acknowledgments. For a younger humanity, the land began to be altered at the moment it was mentally appropriated and became "the original instrument of labour, both laboratory and repository of raw materials." At the end of the fourth and throughout the fifth notebooks of *Grundrisse* the land is consistently described as a "prolongation" of the worker's body:

> the relationship of the worker to the objective conditions of his labour is one of ownership: this is the natural unity of labour with its material prerequisites.[51]

> the relationship of the individual to the *natural* conditions of labour and reproduction, the inorganic nature which he finds and makes his own, the objective body of his subjectivity . . .[52]

> The individual simply regards the objective conditions of labour as his own, as the inorganic nature of his subjectivity. . . . The chief objective condition of labour itself appears not as the *product* of labour, but occurs as *nature*. On the one hand we have the living individual; on the other the earth, as the objective condition of his

reproduction . . . .[His ownership of earth] which is *antecedent* to his activity and does not appear as its mere consequence, and is as much a precondition of his activity as his skin, his senses, for whole skin and sense organs are also developed, reproduced, etc., in the process of life, they are also presupposed by it.[53]

For, just as the working subject is a natural individual, a natural being, so the first objective condition of his labour appears as nature, earth, as an inorganic body. He himself is not only the organic body, but also inorganic nature as a subject.[54]

Thus originally *property* means no more than man's attitude to his natural conditions of production as belonging to him, as the *prerequisites of his own existence*; his attitude to them as *natural prerequisites* of himself, which constitute, as it were, a prolongation of his body.[55]

What turns the soil into a prolongation of the body of the individual is agriculture.[56]

In short, the instrument of labour is still so intimately merged with living labour . . . that it does not truly circulate.[57]

The recognition that the land and the tool are a literal prolongation of the working body may ultimately have its source in the immediately available confirmation of phenomenological experience: in his classic text on perception, for example, James Gibson calls attention to the at once startling and (once stated) wholly familiar fact that a person can literally "feel" at the end of a walking stick the grass and stones that are three feet away from his hand, just as a person holding the handle of a scissors actually feels the "cutting action" of the blades a few inches away.[58]

Both the brevity and the recurrence of Marx's announcement reveal his assumption that the sentient continuity between subject and object is too self-evident to require extended explanation. This assumed continuity (which comes to have a structural centrality in his whole political philosophy) has, however, many verbal and visual elaborations in the literature[59] and painting of the nineteenth century, which thus help to make compellingly available to the senses the truth of the observation Marx himself assumes. In paintings by Millet, for example, the laborer's physical presence stops not at what would be conventionally understood as the boundaries of his or her own body, but at the boundaries of the canvas. In the *Winnower*, the dark mergence of man and world creates a uniform blue and brown surface on which there float luminous white pools of grain, hands, ankles, back—the shared radiance of the materials of work and those parts of the human body most acutely engaged in the activity. In the *Peasant Women Carrying Firewood*, a rhythm is established by the two human figures, each merging with the other and with the wood they carry: the foremost carrier is still distinguishable from the wood she carries; the second is much less so, and behind her looms up the forest itself which merges with her load and becomes the snowfilled thing she (and by extension, her comrade) carries on her back. In his 1850–51 black crayon sketch of *Two Gleaners*, two or three women (their number is visually indeterminate, for again the edges of the individual body no longer have a discrete terminus) bend in unison over the ground and move forward together leaning toward the right-hand edge of the canvas; behind them the colossal mounds of hay (the collective outcome of their own and others' labor, as large as the vegetable fragments picked up in the gleaning act appear in isolation minimal) themselves lean slightly toward the left-hand edge of the canvas, and thus appear to be pulled along by the forward movement of the women, underscoring the monumental scale of the world-alteration actually entailed in their careful scanning of the ground beneath their faces. Again in *Man With a Hoe*, the huge dark figure pressing his heavy weight down across his tool into the equally dark soil that engulfs him has its extension in the rising combustion of hay, smoke, light, and air immediately in back of him: its upward thrust becomes the recorded outcome of his own downward thrust across the lever that reaches and lifts not a three inch square of soil, but the whole visible surface of earth. It is the recognition of this same reach of sentience, and the unity of sentience with the things it reaches, that moves in and out of Marx's own account of the original relation between men and women and the world they are about to remake.

The discrepancy between the tone of Marx's writing when he is acknowledging the body's presence in the elementary sites of artifice (raw materials, tool, material object) and the tone of his writing when he is acknowledging its presence in much more sophisticated sites (money, fixed capital, circulating capital, productive capital, interest-bearing capital) arises not only from the fact that its occurrence is much more literal in the first instance but also from the fact that its occurrence there is much less politically stipulated. That is, "the land" and "the tool" and "the material object" are the primary, inevitable, and universal locations of artifice in any actual or theoretical system of self-materialization, whereas larger, more extended phenomena such as capital are neither omnipresent nor inevitable elements in cultural materialism. The first three are identical with the site at which they occur, whereas capital is not identical with the site at which it occurs; a very different kind of extended version of artifice could reside here (Marx repeatedly calls attention to what he calls the "bourgeois error" of believing capital to be the fixed and necessary form of the more extended site of making). Consequently, only in the latter case do Marx's perceptions about the nature of the universal site itself become conflated with his perceptions about and attitude toward the particular tenant of the site; the attributes it has by being an artifact and those it has by being a particular version of the artifact of which he disapproves sometimes become confused. When, for example, he registers the equivalence between the self-magnifying powers of capital and the self-magnifying powers of the human body (in the tonally unstable language of "It is the old story: Abraham begat Isaac . . . "), that observation is of course occurring in the midst of an extensive analysis of capital that is laden with disapproval. That he disapproves of capital and that capital re-enacts the activity of the body are both true, but the first does not follow from the second because the second would be equally characteristic of a much more benign occupant of the site.[60]

The tone of Marx's descriptions of the elementary sites is much steadier precisely because his critique of the imagination and his political critique are here not running at cross purposes. This is not to say that such descriptions are outside his political argument: that the land, the tool, and the material object are all three artifacts into which the maker's body is extended and with which the maker's body exists in an original integrity is the perception that stands at the very center of his revelations about the operations of the imagination but simultaneously becomes the basis for his explanation of why the maker's later separation from these sites of self-extension leaves him in the midst of his activity of making with control over only one isolated element of that extended action, his own labor. It is the identification of the materials of earth as "a prolongation" of the worker's body that leads Marx to designate "private property" as a key problem for civilization: through private property, the maker is separated from the materials of earth, from the inorganic prolongation of his own activity, and therefore enters into the processes of artifice as one who cannot sell what he makes (coats, bricks) but can only sell his own now truncated activity of making. Even his choice to sell or not sell that lone activity is eliminated with the emergence of laws prohibiting strikes.[61] Thus the disturbingly graphic concept of the severing of the worker from his own extended body becomes central to *Capital*, though it usually occurs in the more abstract phrasing of "the separation of the worker from the means of production" and as a difference between the capacity to "sell the products of labour" and to sell "labour power."[62]

But while Marx's political critique has its foundation in his assessment of the nature of artifice, his assessment of artifice at these early sites can be understood separately from the critique; and thus it is here that the body's place is most widely recognized in the writings of others.[63] Even, however, at that site where it is most widely accepted—the individual material object (such as a smooth floor, a chair, and a table)—it is of course not universally accepted, and is sometimes criticized as a romantic interpretation of materialism. But such an objection is itself being uttered in a material context. The weight of the person thinking this thought is dispersed out onto three horizontal boards: one intercepts his body from below; another, two feet higher than the first intercepts and lifts his body from behind; and another, a foot higher than the second, intercepts his body from the front, bearing the weight of his arms and the book he holds. These in turn occur within a larger lattice of surfaces, an elaborate series of vertical planes that surround him in all directions, in some directions recurring in such dense intervals that five or six or perhaps twelve layers separate him from the outside world. They stabilize his surroundings, mediating the approach or departure of everything human and inhuman, though he does not think of their work except perhaps as it is echoed in the tangible, preconscious reassurance of the light pressure of the cloth that falls across his back. He has woven himself into a tapestry of artifice that regulates the appearance and disappearance of all ''natural'' presence including his own. If his own body should intrude itself into his thoughts by signaling him that it is thirsty, he leans slightly to one side, turns his wrist and water appears before him (and the bodily intrusion is gone). He leans slightly to the other side, twirls his fingers and his child's voice enters the room. He reads and listens confidently for the sound on the stairs of a brother who yesterday was thousands of miles away and, coming across a sentence claiming that the human body is lifted out into the material world, is at that moment so free of the pressures and limitations of embodiment that looking up, he surveys the objects around him and concludes that the perception recorded in that sentence arises out of the distant tenets of a romantic philosophy rather than out of any observation that can be experientially confirmed.

The presence of the body in the realm of artifice has as its counterpart the presence of artifice in the body, the recognition that in making the world, man remakes himself. This second thesis, which has for Marx its philosophic origins in Hegel's doctrine of the self-creating power of work, is so implicit in the first thesis and is itself, independent of the first thesis, so widely understood to be central to Marx's account of civilization that it need not be dwelt on at length here. It is enough to note briefly the range of contexts in which it appears and the depth of significance it gradually acquires.

At its most modest, man's recreation of himself occurs in his activity of providing himself with sustenance to renew his body each day. The activity of eating, regarded as an exclusively natural process when initiated and controlled by animal instinct, becomes in its entry into human consciousness the starting place of self-artifice, the first occasion of man's assumption of his responsibility for his own making and remaking:

> . . . originally the act of producing by the individual is confined to the reproduction of his own body through the appropriation of ready-made objects prepared by nature for consumption.[64]

> [In slavery or serfdom] . . . the original conditions of production appear as natural prerequisites, *natural conditions of existence of the producer*, just as his living body, however reproduced and developed by him, is not originally established by himself, but appears as his *prerequisite*; his own (physical) being is a natural prerequisite, not established by himself.[65]

Throughout Marx's descriptions of early economic culture, the ''working'' subject is the ''working (producing) subject (or subject reproducing himself)'';[66] and the phrase ''reproduction of the individual'' and the specifications of conditions ''for the labourer to reproduce himself''[67] recur throughout. Although in Marx's account of civilization man's remaking of his own body goes far beyond the activity necessary for cellular self-renewal, the original intimacy of "eating" and "self-artifice" is everywhere saluted in the proximity (and at moments near inextricability) of the central terms "consumption" and "production."[68]

When Marx does move beyond cellular self-renewal to more compelling forms of self-artifice (language, material objects, moral and political consciousness), it continues to be the actual living body itself that is altered.[69] Thus, however simple the early nutritional accomplishments of humanity, they are in their outcome a precocious anticipation of the final outcome of the most elaborate (ancient or modern) civilizations:

> Among the ancients we discover no single enquiry as to which form of landed property, etc., is the most productive, which creates maximum wealth. . . . The enquiry is always about what kind of property creates the best citizens. . . .
>
> Thus the ancient conception, in which man always appears (in however narrowly national, religious or political a definition) as the aim of production, seems very much more exalted than the modern world, in which production is the aim of man and wealth the aim of production. In fact, however, when the narrow bourgeois form has been peeled away, *what is wealth, if not the universality of needs, capacities, enjoyments, productive powers, etc., of individuals, produced in universal exchange?*[70]

To say that needs and capabilities are projected out into a place where they can be "universally exchanged" may seem at first only to reconfirm the earlier thesis that the body is projected out into civilization; but this projection in turn entails a fundamental change in the living source out of which the projection arose. If the conditions of sentience are objectified, made social, placed in universal exchange, one of the most essential facts about sentience has been eliminated— its deep privacy, its confinement of its own experience of itself to itself, its being felt only by the one whose feeling it is. In civilization, as in the early altars of a religious culture, the body is turned inside-out and made sharable.

That sentient beings move around in an external space where their sentience is objectified means their bodies themselves are changed. That this is an actual physical alteration can be better grasped after turning for a moment to more graphic instances of bodily recreation. Perhaps the single most striking formulation occurs in Frederick Engels's essay, "The Part Played by Labour in the Transition from Ape to Man." His provocative essay is now understood to be in some of its arguments much less contestable than it was earlier supposed to be, and in other of its arguments much more contestable. Engels's speculation that the crucial location of the transition from ape to man had been in the hand, the organ of making, rather than in the skull, the attendant of the organ of thinking, has after many years been confirmed by the discoveries of anthropologists. Steven Jay Gould calls attention to the fact that the scientific search for the "missing link" was for a long time subverted by the search for the wrong body part (skull rather than hand), a mistake itself arising from a faulty (and, according to Gould, ideologically stipulated) emphasis on man-as-intellect rather than on man-as-creator, man-as-maker, or man-as-worker.[71] Engels also introduces into the essay the idea that the hand is itself an artifact, gradually altered by its own activity of altering the external world. He writes, "Before the first flint was fashioned into a knife by human hands, a period of time must have elapsed in comparison with which the historical time known to us appears insignificant," and he then continues: "Thus the hand is not only the organ of labour, *it is also the product of labour*. Labour, adaptation to ever new operations, the inheritance of muscles, ligaments, and, over longer periods of time, bones that had undergone special development and the ever-renewed employment of this inherited finesse in new, more and more complicated operations, have given the human hand the high degree of perfection."[72]

Although Engels goes on to put forward similar claims about language and the steady refinement of the senses, the hand, the direct agent of making itself remade, remains his central focus. Thus the hand—whose internal delicacy is externalized not only in the complexities and intricacies of the paintings Engels cites but even in the frozen delicacy of finger movements recorded in an ordinary piece of lace—has been, through long engagement with the resistant surfaces of the world, itself woven into an intricate weave of tendons, ligaments, muscles, and bones. It is this interior disposition that is made visible and celebrated in the paintings or the lace: whatever the specific subject matter of a particular canvas (the sea breaking on the shore, flowers opening in a vase), part of its subject matter, part of what it makes available to the viewer, is the shape of the interior complexities and precisions of the sentient tissue that held the brush. What may threaten to undercut the power of Engels's observation is his particular phrasing that suggests an evolutionary mechanism for the passing on of such alterations to flesh and bone acquired during the individual's lifetime: the Lamarckian concept of the inheritance of acquired characteristics, though not disproven, has never been proven and has, in recent decades, been deeply discredited.[73]

If, however, modern science has not identified any mechanism that supports Engels's immediate claim, it has on the other hand increasingly revealed that the remaking of the human body is an ultimate aim of artifice, since the construction of artificial hearts, hips, wombs, eyes, grafts, immunization systems, each year extends the confidence with which we intervene in the human tissue itself. When medicine was only at the stage of predicting such implants, people often expressed aversion to the anticipated presence of a machine or artificial part, however delicate, within the body; but the actual accomplishment of such medical outcomes has instead been greeted with almost every human emotion—happiness, wonder, surprise, the poise of inevitability—except aversion. The presence of such man-made implants and mechanisms within the body does not compromise or "dehumanize" a creature who has always located his or her humanity in self-artifice. If we do not yet have the descriptive mechanism that can account for the way in which human beings unconsciously assumed responsibility for their own bodily evolution, evolution has at least brought us to a place where that unconscious goal, buried and inarticulate at our beginnings, has finally surfaced and becomes indisputable.

If one compares the living human body with the altered surrogate of the body residing in the material artifact, one can say that the second almost always has this advantage over the first: human beings can direct the changes the second undergoes much more easily than they can the first. Two versions of the body stand side by side and the one is in its susceptibility to control an improved version of the other. The chair (which assists the work of the skeleton and compensates for its inadequacies) can over centuries be continually reconceived, redesigned, improved, and repaired (in both its form and its materials) much more easily than the skeleton itself can be internally reconceived, despite the fact that the continual modifications of the chair ultimately climax in, and thus may be seen as rehearsals for, civilization's direct intervention into and modification of the skeleton itself. Even before the climactic moment of medical miracle, the redesigning of the chair allows the benefits of "repairing the live body" to accrue to the body without jeopardizing the body by making it subject to the not yet perfected powers of invention. So, too, to return to the body location of greatest interest to Marx and Engels,[74] the hand may itself be altered, redesigned, repaired through, for example, an asbestos glove (allowing the hand to act on materials as though it were indifferent to temperatures of 500°), a baseball mitt (allowing the hand to receive continual concussion as though immune to concussion), a scythe (magnifying the scale and cutting action of the cupped hand many times over), a pencil (endowing the hand with a voice that has more permanence than the speaking voice, and relieving communication of the requirement that speaker and listener be physically present in the same space), and so on, through hundreds of other objects. The natural hand (burnable, breakable, small, and silent) now becomes the artifact-hand (unburnable, unbreakable, large, and endlessly vocal).

Like the making of a chair, the making of asbestos gloves, baseball mitt, scythe, and pencils (as well as countless other extensions of the hand—violins, shovels, scepters, surgeon's gloves, tennis rackets, fans, and tweezers) may in part be understood as "practice for" the redesign of the tissue of the hand itself: the surgical remaking of an arthritic hand into a nonarthritic hand, the rejoining of severed fingers to the palm, the invention of an artificial limb that has the same sensory powers and intricacies of movement as the missing hand it replaces. In turn, the act of medical repair may itself sponsor more universally beneficent inventions: the sensory capacities of the artificial limbs now being worked on will probably give rise to artifacts of general usefulness as inevitably as Alexander

Graham Bell's efforts to assist the hard of hearing resulted in the widely useful telephone, which changed the range of "normal hearing" from a distance of several hundred feet to thousands of miles (or alternatively, revised the definition of hard of hearing from "unable to hear a voice originating anywhere in the room" to "unable to hear a voice originating from any continent on earth").

Even if, however, the collective acts of verbal and material making had never culminated in a civilization's medical ability to resculpt directly the interior of the human body, it would still be the case that the freestanding objects remake the live body itself. Such objects, by eliminating the limitations of sentience or, as it can with equal accuracy be phrased, magnifying its powers (the ear trumpet, hearing aid, sign language, telephone, songs, poetry, telegraph, victrola, stereo system, radio, tape recorder, sonar, acoustically precise symphony hall and so forth all extend the range and acuity of the ear), make sentience itself an artifact. In turn, while the specific attributes of sentience that are remade will vary from object to object (as one object may extend the range of audition; another, the range of temperature that may be withstood; and another, the kind of ground that may be walked over without jeopardy), the one bodily attribute that will in each and every instance have been altered by the object's existence is privacy.

What differentiates men and women from other creatures is neither the natural acuity of our sentience nor the natural frailty of the organic tissue in which it resides but instead the fact that ours is, to a vastly greater degree than that of any other animal, objectified in language and material objects and is thus fundamentally transformed to be communicable and endlessly sharable.[75] The socialization of sentience—which is itself as profound a change as if one were to open the body physically and redirect the path of neuronal flow, rearrange the small bones into a new pattern, remodel the ear drum—is one of Marx's major emphases. Sense organs, skin, and body tissue have themselves been recreated to experience themselves in terms of their own objectification. It is this now essentially altered biological being that, in going on to remake himself or herself in other ways, enters into that act of remaking as one whose sentience is socialized, fundamentally restructured to be relieved of its privacy:

> [The objects of civilization and culture] are thereby posited as property, as the organic social body within which the individuals reproduce themselves as individuals, but as social individuals. The conditions which allow them to exist in this way in the reproduction of their life, in their productive life's process, have been posited only by the historic economic process itself.[76]

To say that this subsequent process of self-artifice entails the continual recreation not just of the body but of the spirit and mind is tautological; for the very notion of spirit and mind—the very fact that a person looking straight forward at her physical image in the mirror or looking down at her own embodied circumference "sees" that she is "not just" and "much more than" a body—is itself at its origins a profound registration of the fact that physical sentience has, after first projecting itself outward, then absorbed back into its own interior content the externalized objectifications of itself. That is, human beings project their bodily powers and frailties into external objects such as telephones, chairs, gods, poems, medicine, institutions, and political forms, and then those objects in turn become the object of perceptions that are taken back into the interior of human consciousness where they now reside as part of the mind or soul, and this revised conception of oneself—as a creature relatively untroubled by the problem of weight (chair), as one able to hear voices coming from the other side of a continent (telephone), as one who has direct access to an unlimited principle of creating (prayer)—is now actually "felt" to be located inside the boundaries of one's own skin where one is in immediate contact with an elaborate constellation of interior cultural fragments that seem to have displaced the dense molecules of physical matter. Behind the surface of the face in the mirror is blood and bone and tissue but also friends, cities, grandmothers, novels, gods, numbers, and jokes; and it is likely to be the second group (the socialization of sentience) rather than the first (the privacy of sentience) that she at that moment "senses" as the washcloth in the mirror moves back and forth over the illuminated surface of the skin.

The notion that everyone is alike by having a body and that what differentiates one person from another is the soul or intellect or personality can mislead one into thinking that the body is "shared" and the other part is "private" when exactly the opposite is the case. The mute facts of sentience (deprived of cultural externalization) are wholly self-isolating. Only in the culture of language, ideas, and objects does sharing originate. For Marx, the more extended and sublimated sites of making should extend this attribute of sharability: the interaction made possible by a freestanding object is amplified as that object now becomes a "commodity" interacting with other objects and so increasing the number of persons who are in contact with one another; the socialization of sentience should continue to be amplified as one moves to more extended economic (money, capital) and political artifacts. Marx's whole work is devoted to his belief that this does not happen. If extended economic sites continued this work of amplification, they would collectively constitute a "socialist economy," one compatible with and magnifying the "socialization of sentience" occurring at the earlier sites. Instead, the "capitalist economy" reverts to an emphasis on "privacy," contracting rather than extending the number of human makers who will be disembodied by their own acts of making, and thereby subverting the essential impulse and intention of what imagining was at its origins.

The Deconstruction of Making: The Severing of Projection and Reciprocation

For Marx, material making is a recreation of the body and the body is itself recreated in that activity. He thus identifies an interior structure that, though very different in idiom from that presented in the Judeo-Christian scriptures, is close in idea. In each, the material artifact is a surrogate or substitute for the human body, and the human body in turn becomes an artifact; in each, the object is a displacement of sentient pain by a materialized clarification of creation; in each, the object is the locus of a reciprocal action. It is not, however, Marx's appreciation of this structure that motivates the writings of his works but his apprehension of a failure within contemporary nineteenth-century economic culture to perpetuate this structure. What is striking about his analysis of the "unmaking of the structure of material making" is that it takes a form similar to the "deconstruction of creation" that occurred in the troubled relation between the Israelites and their Original Artifact in the problematic scenes of wounding. Although the western impulse toward materialism diminishes the culture's reliance on the phenomenon of "analogical verification," it does not by itself guarantee the disappearance of that phenomenon. The very conflation between wounding and creating that earlier had as its solution the expanded realm of material artifice *now itself recurs within the material realm.*

For Marx, the nineteenth-century industrial world departs in its large economic structures from the model of making summarized above. The words "model" and "departs from" should be briefly clarified before looking at the departure itself. The "model" is not one abstractly conceived: it is not a platonic form or "ideal" standard invoked to show the failure of the "real" world to approximate a perfect paradigm nowhere existing on earth. Though Marx's work has "utopian" elements (particularly in his imagined solutions to contemporary social and political distress), the model of making itself has a structure of activity immediately apprehensible in the concrete actions and sensuous outcome of the craftsman's and farmer's and laborer's own work, whether contemporary or ancient. A model may, as Nelson Goodman points out, be external to the thing that is compared to it (an abstract idea, a mathematical design), or it may instead be internal to the thing that can then be expected to have its characteristics: a swatch of cloth, for example, may be taken as a model of the attributes that the whole bolt of cloth can then be anticipated to have on a larger scale.[77] The model explicitly or (far more often) implicitly present on every page of Marx's work is much closer to the second than to the first kind of model.

The difficulty in articulating the meaning of the word "model" is perhaps matched by the difficulty in articulating the meaning of the phrase "departure from the model." Marx's own resonant term "contradiction," though primarily applied to the specific internal inconsistencies within capitalism, can be invoked here: the large productive system contradicts the nature of production at its original locus. But like the word "departure," the word "contradiction" in this context does not make obvious the ethical pressure of the problem: intellectual arguments if "inconsistent" may cease to be intellectual arguments, but human behavior, if inconsistent, does not cease to be human behavior; and economic systems are closer to being extended and materialized forms of behavior than to being intellectual arguments. Thus to identify them as "contradictory" or "inconsistent" does not announce the alarming character of the dislocation that Marx actually attempts to convey. Similarly, to describe the departure from the model as its "falsification" would be more appropriate if the model were bodying forth the nature of "truth" rather than the nature of "fictions and made things."

The departure requires the language appropriate to the model itself. If a particular political philosopher believed that people's essential humanness resided in their relation to God, and if he found that the larger social structures were characterized by the absence of the attributes of that relation, that "godlessness" might be seen by him as jeopardizing the entire existence of human beings. Or if a political philosopher identified men's and women's humanity as residing centrally in their "rationality," a social structure that departed from the principle of rationality would be perceived by him as a very serious threat on the grounds of its "irrationality." Because Marx understands men's and women's fundamental human identity to reside in their existence as "creators," "imaginers," and "makers," the social system that departs from this ground is uncreative, anti-imaginative, destructive, a deconstruction of making. This deconstruction has as its most immediate evidence and outcome the widespread physical suffering of the industrial population, and Marx reacts to the deconstruction with anger, with fury, with embarrassment, and above all, with the metaphysical incredulity of a good craftsman looking at a bad piece of work.

If the monumentally complex substance of *Capital* were to be described in a single sentence, it could be described as an exhausting analysis of the steps and stages by which the obligatory referentiality of fictions ceases to be obligatory: it is an elaborate retracing of the path along which the reciprocity of artifice has lost its way back to its human source. To recreate the path of loss is to attempt to recreate the original obligation. Like the old Celt who tried to shove to the side a misdirected tide, or like the ancient Egyptian who each year submerged in the Nile a slender papyrus contract reawakening the river to the hour and direction of its tidal responsibilities, Marx stands in the presence of the colossal artifact, shaking his fist in fury and lecturing it hour upon hour on its forgotten largesse. His project, perhaps more prosaic than theirs, has the special genius of a legitimacy theirs wholly lacked; for if like the silt-laden spill of a holy river and the fish-laden surf of a mythic sea, the industrial economic system is for all who dwell in its presence the primal ground of physical and spiritual sustenance, it is, unlike the river and sea, man-made. The inventions of Cuchulainn (his weapons, his dreams) and of the Egyptian priest (his papyrus, his ink, his words, his idea of contract) only mediated between the human being and his primal ground, a primal ground whose existence preceded the existence of man or man's ink or his dreams. Now, however, in the industrial world, the primal ground is one that has itself been invented. The dream of an obligatory reciprocity is not this time dreamed up after the fact. The *contractual premise* accompanied the artifact's birth, occasioned that birth, and is itself recorded in the hidden intricacies of its own interior.

The large all-embracing artifact, the capitalist economic system, is itself generated out of smaller artifacts that continually disappear and reappear in new forms: out of the bodies of women and men, material objects emerge; out of material objects, commodities emerge; out of commodities, money emerges; out of money, capital emerges. In the first phase, the original work of creation entailed a double consequence, the projection of the body into the material object and the reprojection of the object's power of disembodiment back onto the about-to-be-remade human body. (The first would have no purpose if it were not accompanied by the second: there would be no point in a person projecting the nature of "seeing" into the lenses of eyeglasses, if that person or another could not in turn put on the eyeglasses and be physically remade into one who sees better.) When, however, the material object then goes on to generate new versions of itself, one of these two consequences remains stable throughout its successive forms and the other becomes unstable. In its final as in its first form, the artifact is a projection of the human body; but in its final form, unlike its first, it does not refer back to the human body because in each subsequent phase it has taken as the thing to which it refers only that form of the artifact immediately preceding its own appearance. Thus, while in the original phase the double consequence looks like this:

person $\longleftrightarrow$ object (projection of human body)
(sentience
remade)

in subsequent phases, it looks like this:

persons $\longrightarrow$ objects $\longleftrightarrow$ commodity

persons $\longrightarrow$ objects $\longrightarrow$ commodities $\longleftrightarrow$ money

persons $\longrightarrow$ objects $\longrightarrow$ commodities $\longrightarrow$ money $\longleftrightarrow$ capital

The overall work of its successive forms is to steadily extend the first consequence (capital is, like the solitary pair of eyeglasses or any other made object, the projected form of bodily labor and needs) and to steadily contract the second: each new phase enables the line of reciprocity to pull back further and further from its human source until the growing space between the artifact and its creator is at last too great to be spanned either in fact or in an act of perception. In the end, that creator has not only ceased to be the recipient of his creation's beneficent disembodying powers but, in some ways more radically, has even ceased to be recognizable to himself and to others as the "creator." It ceases to be self-evident that the tens of thousands of workers who have made the commodities have also collectively made the excessive residue of value that appears in later forms as money and capital.

To enter into the interior of capital is thus to enter into a deeply flawed artifact in which the two originally inseparable consequences of creation have been interrupted by the artifact's own capacity to become internally referential.[78] The overall artifact (the capitalist economic system) is, of course, only temporarily self-referential. Its apparent autonomy signals not its abiding freedom to be self-contained (for it has no such freedom: women and men have control of the object world and an object that refuses to surrender its referential power will be destroyed or discarded) but, rather, its eventual freedom to take as its external referent someone other than the human beings out of whom or on behalf of whom it was originally projected.[79] Because the path of reciprocity has contracted back from its human source, in the end when it will once more be made externally referential, made to refer to something beyond its own boundaries, its referential powers can be pointed in a wholly different direction, at those either absent from or only marginally participating in its actual creation. Thus, if the overall economic system preserved the original integrity of making, it would look like this:

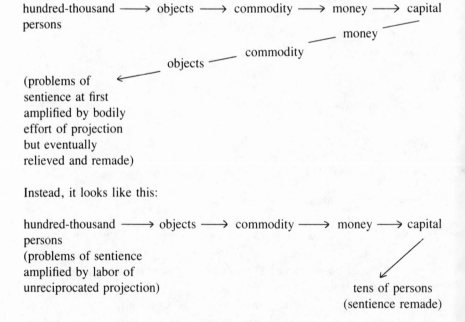

For Marx, the contemporary system of production deconstructs the imagination's own structure of activity by bringing into relation two groups of people (one very large, the other very small) who in their lifelong monetary as well as cultural and philosophic negotiations will forever confront one another across the central fact that one of them is deeply embodied and the other is disembodied.

It is not the existence of a conflict that is central to the present discussion (the conflict between laborer and capitalist is too familiar to warrant recitation) but the particular formulation of the relation as one between two disparate levels of embodiment. This formulation surfaces in many ways in Marx's work, most explicitly in the numerous single sentence summaries of that relation. He writes in the 1844 manuscripts, for example, "In general, we should note that where worker and capitalist both suffer, the worker suffers in his very existence while the capitalist suffers in the profit on his dead mammon."[80] Despite the contemptuous phrasing of "dead mammon," wealth itself is not the source of Marx's concern (for wealth is, potentially, "the universality of needs ... produced in universal exchange"), nor is it that wealth has become the substitute recipient of the blows that would otherwise fall on sentience. It is rather that the surrogate has been created through a process that requires the intensification of the sentience of one people, and then actually performs its function as surrogate for a second, much smaller number. Thus the double consequence of creation, in which simultaneously the body is projected out into artifice and the human body itself becomes an artifact, exists no longer as an inextricable occurrence, for now the first consequence describes what occurs to one group of people and the other describes what occurs to a second group.

This dislocation in the structure of creation entails a discrepancy in the mode of suffering to which each is vulnerable, and this discrepancy in turn entails discrepancies in all other human activities and modes of consciousness. The situation can be pictured by returning to the language of intentionality introduced earlier. All intentional states—physical, emotional, mental—take intentional objects: the more completely the object expresses and fulfills (objectifies) the state, the more it permits a self-transformation out of that embodied state; conversely, the more the state is deprived of an adequate object, the more it approaches the condition of physical pain. Thus to say that one's wealth is precarious or subject to diminution ("the capitalist suffers in his dead mammon") is to say that the objects of sentience are subject to diminution, that the rich complexity of consciousness expressing itself across the density of artifacts (books, pianos, schools, vehicles, clothing, theatres, heated homes) is subject to diminution, and that as various dimensions of consciousness lose their objecthood, these states (as well as the persons they "belong" to) are moving in on themselves in the direction of utter objectlessness, which in its final and severe form manifests itself as physical pain. But it is also to say—as is implicit in the benign identification of wealth with "excess," which is in its deepest implication an "excess" of self-extension—that when one's existence becomes entangled with the protective generosity of abundant artifice, many layers of objectification separate one from the state of objectlessness, many tiers of projected embodiedness, many successive surfaces of, in effect, "artificial body tissue" must be lost or altered before the alteration will record itself in the objects directly outside the body (like shelter, furniture,[81] and food), which in turn must be passed through before the force of alteration records itself directly in the body (the disease and accident and exhaustion to which those who are shelterless, furnitureless, foodless, are immediately subject).

Not just suffering but all forms of consciousness are involved in the difference between belonging to the people who are disembodied and belonging to those who are radically embodied, for the very end point of the one's precariousness (after many tiers of objecthood are crossed) is the starting and stable point of the other's existence: the second endures in near objectlessness; all his psychic states are without, or nearly without, objectification; hence in all his or her life activities, he or she stands in the vicinity of physical pain. To describe a difference in the mode of suffering to which one is vulnerable (the capitalist suffers in his money; the worker suffers in his very existence) is to describe a difference in the density of objecthood, the thickness of the margin of artifice, that separates oneself from oneself. Further, this already "given" difference in the degree of self-extension is self-amplifying: it conditions the degree to which new forms of self-extension can be initiated, acts of initiation that are variously expressed by the words "aspiration," "desire," "will," "risk-taking," "creation," and "self-recreation." The lending out of the self to develop new objects requires temporarily some inattentiveness to the already created aspects of self until the new form of self is in turn objectified or materialized, existing in a form where it has a stable or enduring existence without requiring the immediacy of human action or daily reinvention. If by this inattentiveness one puts at risk one of thirty or three hundred layers of surrogate embodiedness, this is not the same as putting at risk one frail layer of intervention that is the only layer (e.g., enough money to buy bread for the next two days).

Thus the difference in the density of self-extension carries with it a difference in the capacity for new and willed forms of self-extension whether occurring in acquisition, education, invention, or other forms of spiritual adventure. For this reason the explicit and recurring single-sentence summaries of the relation between the two groups of persons in Marx are sometimes expressed as a difference in the mode of suffering (the workers suffer in x way; the capitalists in y way) and are at other times expressed as a difference in various forms of projection like "desiring," "aspiring," or "risk-taking" (the worker's aspiration takes the form x; the capitalist's aspiration takes the form y). He writes, for example, "An increase in wages arouses in the worker the same desire to get rich as in the capitalist, but he can only satisfy this desire by sacrificing his mind and body,"[82] or again—

All economists, when they come to discuss the prevailing relation of capital and wage labour, of profit and wages, and when they demonstrate to the worker that he has no legitimate claim to share in the risks of gain, when they wish to pacify him generally about his subordinate role *vis-à-vis* the capitalist, lay stress on pointing out to him that, in contrast to the capitalist, he possesses a certain fixity of income more or less independent of the great adventures of capital. Just as Don Quixote consoles Sancho Panza with the thought that, although of course he takes all the beatings, at least he is not required to be brave.[83]

Marx's frequent and sometimes tonally bitter reminders almost always take some version of the form, "These people have *x* and these people have *y*." This form, though intended to signal a fundamental dislocation in the nature of making, sometimes seems in accounts of the Marxist argument, or again in everyday descriptions of social inequalities wholly separate from Marx's patronage, to have been lifted away from its explanatory framework and to have become vulnerable to being misunderstood or misheard as merely a petulant pointing at a difference that, though perhaps recognizable as "unfair" or "unjust," is too easily restatable and dismissible as "too bad," a shame. But for Marx it entails a serious dislocation in the species itself, for it announces that the original relation between sentience and self-extension (between hurting and imagining, between body and voice, or body and artifice) has been split apart and the two locations of self have begun to work against one another. The problem of economic distribution (whether, as in nineteenth-century Britain, taking the form of drastic inequalities experienced by the population within a single country, or as is more often the case in the twentieth century, taking the form of drastic inequalities between the different populations of separate countries[84]) is the problem of distributing the power of artifacts to remake sentience. To summarize this, as we habitually do in everyday conversation, as the problem of "the haves and have-nots" is inadequate to express its concussiveness unless it is understood that what is had and had not is the human body.[85]

This dislocation in the structure of artifice has in Marx's writings a much more sustained registration than the periodically recurring single-sentence announcements. By *Capital 1*, it records itself not in explicit summaries but in the frame of the entire work: it determines the most elementary and pervasive characteristic of that text. Although Marx's monolithic intelligence gives a singular tone and sweep to the entire work, the argument moves slowly back and forth between two very different kinds of analytic texture: for long stretches, it pursues exhaustively the direction and rapidity of metamorphosis in abstract processes and structures; for other long stretches, it becomes a methodical recitation of the concrete and banal facts of day-by-day existence (not, for example, the general appearance of a room but its exact height, length, and width, and the precise number of women who are working within). This stylistic duality, omnipresent throughout all of *Capital 1*, allows the volume as a whole to make visible the problem of doubleness that is much more fleetingly portrayed in the single sentence.

In the single-sentence summaries looked at a moment ago, two kinds of persons appear, even though the sentence asserts that one of the two is not physically present in person: one suffers, desires, and risks in his body and the other suffers, desires, and risks in "his" artifice. Both persons are for a moment present in the sentence before one of the two is subtracted out and replaced by that which (in the actual social reality to which the sentence refers) replaces his embodied participation in the shared cultural enterprise. But in the larger structure of *Capital*, the capitalist is from the outset simply absent. Disembodied, his suffering, risks, and desires only enter insofar as they are implicit in the swellings and contractions of commodities, in the rates and masses and circulations and stoppings of alternating forms of value, in the accumulation and diffusions, the constancies and vagaries of money and machines and surpluses—occurrences that in their compelling assumption of a direction seem almost yearning, occurrences that in the triumph of their continual re-arrival at new appearances seem full of self-delight, occurrences that in their sequences of transformations seem to contain the attributes of the mysterious gropings and growth of sentient tissue forever assuming new forms of life. Capital. It is colossal. It is magnificent. And it is the capitalist's body. It is his body not because it has come into being through the solitary projection of his own bodily labor, but rather because it bestows its reciprocating power on him, relieving his sentience, acting as his surrogate. He "owns" it—which is to say he exists in such a relation to it that it substitutes for himself in his interactions with the wider world of persons (as it also substitutes for him in Marx's account of that world).

When very occasionally in the midst of these massive, sensuous descriptions of animated capital, the subject of the capitalists themselves arises (when the account approaches the neighborhood of some *psychological* category such as "competition," there is a pressure to speak in terms of some actual human presence, a pressure that in such passages rhythmically builds and is then dispersed), Marx mentions them only long enough to remind us that they are absent, that they are being omitted from the account only because they are absent from the thing to which the account refers. He writes, for example, "While it is not our intention here to consider the way in which the immanent laws of capitalist production manifest themselves in the external movement of the individual capitals, assert themselves as the coercive laws of competition, and therefore enter into the consciousness of the individual capitalist as the motives which drive him forward, this much is clear: a scientific analysis of competition is possible only if we can grasp the inner nature of capital."[86] Our subject, he says, is not competition, but if it were competition, it could only be understood in terms of the actions happening within the artifact rather than in terms of the psychological attributes of the people themselves. The reflex of this passage—in which a subject is arrived at that seems to require it to be described in terms of some human agent or source, a requirement that is then explained to be inappropriate because the source resides in the process itself—is characteristic of most instances in which the capitalist is explicitly mentioned.

There are four categories of occasion on which the capitalist is briefly introduced: the psychological, the functional, the transformational, and the spiritual. Just as the realm of motivation (competition) is embodied in the large artifact, so too are most other *psychological* attributes: arrogance, he writes, is not the capitalist's but is "the arrogance of capital" (298, 926); respectability is not the capitalist's for "only as a personification of capital is the capitalist respectable" (739); any personal desires for pleasure are only "personally" his at the moment when he ceases to be a capitalist (254, 739, 740); cruelty toward, or physical degradation of, other men and women does not depend on the will or intention of the individual capitalist (739, 740). Second, the *functions* of work sometimes called his instead belong to capital: the "work of directing, superintending, and adjusting" (449) as well as the "command" (448) and "leadership of industry" (450) are attributes of the artifact; "the functions fulfilled by the capitalist are no more than the functions of capital" (989).[87] Third, what are misidentified as his *transformations*—from "money-owner" to "capitalist" (426), for example, or from "buyer" to "seller" (301, 302)—are transformations that originate within the process of circulation rather than within his own person, mind, or psyche. His changes thus follow from changes within the artifact rather than originating with him and being projected into an objectified transformation in capital that he authors. He acquires such attributes through his relation to capital; and capital has acquired them independent of and prior to that relation. (It has acquired them through its relation with its collective makers, whose projected characteristics and powers it now has.) Finally, the largest, perhaps *spiritual*, expressions of personhood, what are called his "soul," his "consciousness and will," his "historical existence" are instead capital's, and only come to belong to the capitalist insofar as he is, in the phrase that occurs almost every time that the word "capitalist" occurs, the "personification" or "incarnation" of capital (254, 342, 423, 739, 740, 1053).

Marx's point is substantive rather than polemical. He does not mean that a person who is a capitalist does not compete, has no pride, arrogance, or respectability, has no pleasure, no cruelty, experiences no changes, and has no consciousness, will, or historical existence. Only in his capacity as capitalist does he lack these things. Such attributes may manifest themselves in his personal life but do not enter into the economic system, which already has these attributes independent of him. So, too, he of course personally has a body but it is absent from the production process, and when he dies a "private" death it will not be an outcome of his participation in industry.[88] In fact, it is when a person has a relation to the system of production that allows him to survive without risking his own embodied psyche, will, and consciousness in that survival—has a relation to the system of production that frees his psyche, will, and consciousness for arenas other than that system of production—that he *is* a capitalist. That liberating relation, that attribute of nonparticipation, is summarized by the word "capitalist." When used as a rubric to identify a person, the term has almost no meaning other than "exempted person." Because the person has been exempted or absented from the process, his name follows from the material artifact ("capital") rather than (as in the case of the "worker," "maker," "laborer") designating an action prior to the existence of the artifact. Long stretches of Marx's analysis (sometimes tens, sometimes hundreds, of pages) of the interior attributes of capital intervene between the several-sentence-long introductions and dismissal of the capitalist. Even to have mentioned him at this length misrepresents *Capital*, for it endows him with a minimal density of presence that is simply and starkly absent from the endless pages of that large volume.

In contrast, the worker's motives, desires, pride, cruelties, consciousness, will, historical existence are absorbed into the system of production. The concreteness of his presence there is registered by the immediacy of his presence in *Capital*—not only in those attributes just enumerated but, most important, in that part of himself whose counterpart in the capitalist is never mentioned even long enough to be rejected, the physical body.[89] So, too, the most intimate extension of personhood, the family, continually enters bodily into the text; for like all other attributes of the worker, his or her children and spouse are present in the factory, field, or mine, just as the capitalist's family is wholly absent, exempt, from those arenas.

The activity the worker invests into the made world is always specific: the pages of *Capital*, like the nineteenth-century world *Capital* describes, contain not just workers and laborers but makers, miners, growers, and gatherers; and not just these but more specifically, nail-makers, needle-makers, brick-makers, brick-layers, coal miners, straw-plaiters, lace-makers, linen-makers, match-makers, silk-weavers, shirt makers, dress makers, iron workers, glass-bottle makers, steel-pen makers, wheat growers, corn millers, watch makers, wall builders, wallpaper makers.[90] The daily projection of his or her body into the artifact has its record in the artifacts themselves, in the recreation or rearrangement of the material world: clay rearranged into bricks (its spreading pointlessness transformed into small, hand-held fragments of materialized geometry); dispersed bricks rearranged into a wall (functionless geometry transformed into an object that protects); thread rearranged into lace (its nearly invisible linear persistence converted into the complex form of a visible presence); coal rearranged, relocated, to the visible surface of the earth (its deep refusal to surrender its capture of an ancient sun now converted into an immediately accessible source of incandescence); and so on through hundreds of other objects, the spill of tiny wheels rearranged into a watch, the soft and dangerous miscellany of rags rearranged into piles and eventually into paper.

But this projection of the worker's body into artifice does not carry with it its implicit counterpart of recreating his body into an artifact. If, as a consequence of his eight or ten or fourteen hours of making, he was empowered to "make" a room with heat, light, and furniture to accommodate and diminish that body, if he were able to provide enough food so that his children were exempt from their own small but exhausting extension of themselves into industrial artifice (often ten hours long) and were thus free to learn reading and writing, then the double consequence of making would be intact. With the money he earns, he reproduces himself in none of these ways. He only reproduces in himself the capacity for bodily work so that he can sell it again the next day (343). His self-recreation cannot exceed the activity of cellular self-renewal because, though he will continually do more work than is required by "the narrow circle of his needs" (425), he will always be receiving less remuneration than is required by that narrow circle of needs. If the double consequence of making were intact, then his own alterations in the material world would be accompanied by alterations in himself in the direction of disembodiment; instead, his unreciprocated labor of projection heightens and intensifies the problems of sentience.

Thus, whatever the particular made object Marx happens to be describing (coal, lace, bricks, watches), it tends to be accompanied in the text by an equally particular form of bodily extremity. For example, by the side of a pile of miscellaneous rags (in the industry called "rag sorting") stands a human being undergoing a constellation of bodily alterations summarized by the word "smallpox" (593); by the side of a pile of made objects called matches is the intensified embodiedness summarized by the word "tetanus" (356); the body working in an airless room of shirts and sewing machines asserts its magnified claim by momentarily eclipsing consciousness in repeated waves (602); the small child working on the plaited straw has each day a new array of small cuts around her mouth and on her hands (598). The specific artifact and the specific form of heightened embodiedness occur together. But to see the occupational disease, or the particular form of infection residing in the material out of which the artifact is made, as itself the problem is to underestimate the reality of what Marx brings forward in such juxtapositions; for the disease is in each case only an extension and radical summary of the margin of fatigue and malnutrition which, modest in its increment on any given day, is over sixty or six hundred days growing in its scale and rate of increase until, with the entry of the specific disease, the rate and progress of bodily alteration suddenly becomes alarmingly visible, as though the overall progress has just been refilmed in time-lapse photography. In the young industrial world, the hours of the working day are computed against the shortening of physical life by decades (343). Whatever its nameable form (smallpox, tetanus, fainting, cuts) and however much it manifests itself in "wasting-away," occupational disease has as its cause the extremity of projected and unreciprocated labor and takes the form of the "magnified body"—the body enlarging its claims, engulfing all other aspects of consciousness and finally eliminating them. The counterpoint of object and body throughout *Capital* (whether taking a specific form such as matches and tetanus, rags and smallpox, or instead the general form, commodity and embodied worker, money and embodied worker, capital and embodied worker) makes visible the fact that the creator, though exhaustingly engaged in acts of artifice, is exiled from the realm of self-artifice. Failing to accomplish any self-recreation except in its most minimal form, cellular self-renewal, the worker gradually fails to perform even that most modest act.

The pressures of the body are conveyed in Marx's writings not by sensory or sensuous description but by numerical recitations. The arithmetic factualness of his tables, lists, and summaries of government reports is the factualness of sentience:

(1) Houses

Vulcan Street, No. 122	1 room	16 persons
Lumley Street, No. 13	1 "	11 "
Bower Street, No. 41	1 "	11 "
Portland Street, No. 112	1 "	10 "
Hardy Street, No. 17	1 "	10 "
North Street, No. 18	1 "	16 "
North Street, No. 17	1 "	3 "
Wymer Street, No. 19	1 "	8 adults
Jowett Street, No. 56	1 "	12 persons
George Street, No. 150	1 "	3 families
. . .		

(2) Cellars

Regent Street	1 cellar	8 persons
Acre Street	1 "	7 persons
33 Roberts Court	1 "	7 "
Back Pratt Street, used as a brazier's shop	1 "	7 "
27 Ebenezer Street	1 "	6 "

(817)

Kitchen:	9 ft 5	by 8 ft 11	by 6 ft 6.
Scullery:	8 ft 6	by 4 ft 6	by 6 ft 6.
Bedroom:	8 ft 5	by 5 ft 10	by 6 ft 3.
Bedroom:	8 ft 3	by 8 ft 4	by 6 ft 3.

(847)

Although each of these two tables (excerpted from Public Health Reports of the mid-1860s) is prefaced by the explicit naming of the problem of embodiedness, "fever" in the first instance and "diphtheria" in the second, such lists even when unaccompanied by prefatory medical labels themselves make direct statements about the problem of embodiedness. The radius of the body at rest is a specifiable size; its multiplication by the number of persons present is a specifiable size; and the radius of the room containing those persons is a specifiable size. The radius of the resting body's requirements (or more concretely, for example, the radius of the lungs, the radius of the air required by those lungs) is a specifiable figure. The radius of the body in motion (the body as it moves in and out of various postures, the modest motion appropriate to indoor living) is also a specifiable figure, as is again its multiplication by the number of bodies present, and as is again the scale of the framing shelter that must somehow accommodate the earlier figures. ''Vulcan Street, No.122, 1 room, 16 persons'' or ''Bedroom: 8 ft. 5 by 5 ft. 10 by 6 ft. 3'' are neutral, concrete, arithmetic acknowledgments of the body's presence; and the iteration into a list quietly asserts and re-enacts the pressure of the acknowledgment.

When, therefore, Marx stops to specify that the brickmaker is twenty-four years old and makes each day two thousand bricks and that she is assisted by two children who in repeated trips throughout the course of the day carry ten tons of clay up the 30 feet of the wet sides of the clay pit and then for a distance of 210 feet (593), their bodily participation in the production of the walls of civilization (before they return home to walls that will not perform their function of moderating and regulating human presence because there are already sixteen people lying on the floor) is recorded numerically precisely because the magnification of sentience takes the form of a specific weight sustained over a specific number of steps repeated over a specifiable number of hours. Just as it is difficult to imagine how the description of the worker (as it occurs in Marx or in the parliamentary reports on which he consistently draws) could take any more appropriate form, so too the intention of reform legislation to diminish that magnified sentience is necessarily formulated in concrete, neutral, designatable, quantifiable terms. Such legislation specifies the age of children prohibited from work; it specifies the maximum length of the working day; it specifies the minimum number of minutes required for eating and resting; it specifies the required minimum of light and air; it does so because the body on behalf of which such specifications are made has a calculable, specifiable, nonnegotiable set of minimum requirements for survival. So, too, the violations of such legislation, the border crossings into the concrete texture of the physically living body, must again be recorded in the same terms. Marx cites a sequence of reports made by government factory inspectors between 1856 and 1859:

'The fraudulent mill-owner begins work a quarter of an hour (sometimes more, sometimes less) before 6 a.m., and leaves off a quarter of an hour (sometimes more, sometimes less) after 6 p.m. He takes 5 minutes from the beginning and from the end of the half hour nominally allowed for breakfast, and 10 minutes at the beginning and end of the hour nominally allowed for dinner. He works for a quarter of an hour (sometimes more, sometimes less) after 2 p.m. on Saturday. Thus his gain is:

Before 6 a.m.	15 minutes
After 6 p.m.	15 minutes
At breakfast time	10 minutes
At dinner time	20 minutes
	60 minutes

Total for five days	300 minutes
On Saturday before 6 a.m.	15 minutes
At breakfast time	10 minutes
After 2 p.m.	15 minutes
	40 minutes
Weekly total	340 minutes

Or 5 hours and 40 minutes weekly, which, multiplied by 50 working weeks in the year (allowing two for holidays and occasional stoppages) is equal to 27 working days.'

'Five minutes a day's increased work, multiplied by weeks, are equal to two and a half days of produce in the year.' 'An additional hour a day gained by small installments before 6 a.m., after 6 p.m., and at the beginning and end of the times nominally fixed for meals, is nearly equivalent to working 13 months in the year.' (350)

Regardless of whether the work is physical labor or craft or school teaching (e.g., 644), the worker's participation in the system of production is bodily, and alterations in his or her situation will entail either the diminution of that bodily presence (reform legislation) or its magnification (the pre-legal working place, or post-legal working place that violates legislation). Counting, as was noticed at an earlier moment in this chapter, has a fixed place in the landscape of emergency.

The double consequence of artifice—to project sentience out onto the made world and in turn to make sentience itself into a complex living artifact—is thus fractured, neatly fractured, into two separable consequences, one of which (projection) belongs to one group of people, and the other of which (reciprocation) belongs to another group of people, and this shattering of the original integrity of projection-reciprocation into a double location has its most sustained registration in the *texture of analysis that alternates* between an almost *sensuous rendering of the inner design and movements of capital* (the large Artifact) on the one hand and an almost *arithmetic recording of amplified human embodiedness* on the other. Though the interior value of capital is projected there through the collective labor of the workers, it now (by becoming internally self-referential, and when once more externally referential, referring to a much smaller group of people whom it now disembodies and exempts from the process of production) stands apart from and against its own inventors. Thus the original relation between an object and its maker (object + problems of sentience diminished) is deconstructed into its opposite (object + problems of sentience amplified). As in the series of noneconomic contexts examined in early parts of this book, a problem of justice is signaled by the juxtaposed extremes of body and construct.

The alternation in the analytic texture sometimes appears within small compositional units, passages that occur over a sequence of pages; and as such passages recur throughout *Capital*, so does the alternation. The juxtaposition of the "labor process" and the "valorization process," however, also generates the design of the larger structural units and, finally, the design of the volume as a whole. *Capital 1* consists of eight Parts, each containing between three and eight chapters, each chapter in turn subdivided into titled sections. The disposition of sections within chapters, chapters within Parts, and most crucially, Parts within the whole volume, ensures that there is constantly held visible the dislocation within the realm of making that entails the extremes of the increasingly intensified body and an increasingly magnified artifact. Chapter 24, for example, "The Transformation of Surplus Value into Capital," contains both a section describing that transformation as it occurs within the boundaries of its own reappearance (that is, within the ever-changing artifact itself) and a section describing that transformation as it arises out of the magnification of the increasingly collective labor, the intensification of unremunerated work. In turn, the way the double subject determines the larger sequencing of chapters is exemplified by the progression through Chapters 9, 10, and 11: Marx splits apart his analysis of the growth of "Absolute Surplus Value" into "Chapter 9: The Rate of [Absolute] Surplus Value" and "Chapter 11: The Rate and Mass of [Absolute] Surplus Value" so that he can interrupt them and place at their center "Chapter 10: The Working Day" with its meticulous recording of the minutes and hours of the 8 or 10 or 12 or 14 or 16 or 18 hour period of various specified forms of exertion in rainy fields or in workrooms thick with abscesses, thick with pulmonary infections, thick with phosphorous, thick with dwarfed and misshapen human forms who have left their beds at 7 a.m. or 4 a.m. or 3 a.m. that morning, or was it two mornings ago. The same logic surfaces in the layout of the three chapters that together make up Part 5: the middle chapter chronicling the relation between surplus value and the concrete working day interrupts, or is framed on either side by, two chapters on the almost magical flows, rates, and interactions of two forms of surplus value.

There is nothing arbitrary about the specific aspect of the worker's activity and the specific form of the artifact that are, at any given moment, laid side by side. Marx is not alternating his attention between now any one aspect of capital and now any one aspect of labor. The placement of "The Working Day" in the sequencing of the ninth, tenth, and eleventh chapters illustrates his own work of construction and his insistence that the structure and texture of the volume themselves help to carry the economic and philosophic argument. The originality of his analysis of surplus value rises from his distinction between "absolute surplus value" and "relative surplus value," the first of which depends on the length of the working day and the second on the intensification of either the activity or the means of production. Consequently, it is inevitable that his discussion of absolute surplus value will include not just references to the workers' general existence but references to the specific aspect of their existence, their working day. But while the introduction, even the repeated introduction, of the *descriptive term* "working day" is therefore at this point in the analysis inevitable (there can be no analysis without it), what is not at all inevitable is his conscious and careful inclusion here of an eighty-page long chapter on the *concrete reality* of the working day, and the consequences for the human form moving through such a day. Similarly, when toward the end of *Capital 1*, he shifts his focus to the freestanding artifact (not now the amorphous Surplus Value of Part 5 but the Accumulated Capital of Parts 7 and 8), the negative equivalent is itself lifted into the center of the analysis by now becoming not a section or a chapter but a major Part, "Part 6: Wages"—or the workers' negative counterpart to accumulation, "Part 6: Inadequate Wages," "Part 6: Unremunerated Work." Capital is the materialized reciprocity of the collective work of making, and one form of capital is wages; only this part of the created object world is openly referential to the original artificer; the smallness or modesty of capital in this form of appearance allows the largesse of its other appearances, for which the original artificer has ceased to be the referent. Marx only added "Part 6: Wages" in the final version of *Capital 1*, eliminating what was originally intended for that space,[91] a revision that calls attention to the self-consciousness of the sustained juxtaposition.

It is possible to imagine *Capital 1* divided into two large parts, the first half devoted to the worker's embodied existence and the second to the inner working of the freestanding artifact, its human referent repeatedly mentioned but not itself concretely accompanying the analysis. Marx at one point contemplated a separate volume on wage labor but rejected it in favor of having the embodied labor process and the disembodied valorization process run side by side throughout a single volume.[92] The distortion of the cultural project of creating results from the instability of the artifact's referential obligation. Marx's own exhausting labor is the work of restoring the original referent, not just pointing to the human authors again and again, but carrying their portraits forward into the analysis, so that the sentient origins of the made world stay visible and accompany the progressively spiritualized or sublimated reappearances of that object world. The overall movement within *Capital 1* as well as within *Capital 2* and *Capital 3* progresses from the earliest and most concrete form of the capitalist product (commodity) through its subsequent, increasingly dematerialized and amplified forms:

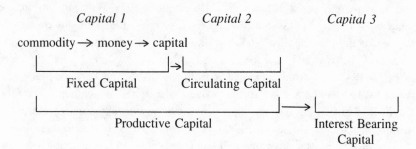

But running side by side with the successive reappearances of capital in the carefully finished structuring of *Capital 1* is the concrete and sustained appearance of the worker, thereby holding visible the fact that the worker's presence is equally emphatic at all levels of transformation. Only scale varies: the larger and more dematerialized the artifact, the greater and more collective the number of persons who have projected themselves into its making. As in the much older writings looked at in the first half of this chapter, embodied humanity and their artifact, or Body and Voice, exist as two distinguishable horizontal ribbons of occurrence, the intensified bodily reality of the lower band bestowing its reality (or as it is now called in the revised, economic idiom, its "value") onto the upper band, creating and substantiating the artifact, bringing it into existence and renewing its existence each day. Rather than being recognizable as themselves the creators, they are (as in the older text) instead perceived as the offspring of the invented object: the worker experiences himself or herself as a "commodity" produced by the capitalist system.[93]

One additional manifestation of the way the work of restoring the human referent is built into the structure of *Capital 1* is the nature of its opening and its closing. In the notebooks of *Grundrisse,* the commodity, though a continual subject, never itself becomes a structural division. During the writing of *Capital,* Marx increasingly thought of it as warranting a self-contained unit of analysis, considered including it as an introduction to the volume, then finally made it the opening Part, and called attention in the preface to the importance of its placement by explaining that all economic analysis depends on recognizing the commodity as the fundamental building block, the cell form, of the whole. This isolation of the commodity as the opening focus is also critical to Marx's own labor of restoring the referent, for though the worker is equally implicated at every level of the artifact, his work of projection is most easily recognizable, and thus most easily restorable, here. During one period while the volume was still in rough manuscript form, Marx intended to end *Capital 1* by returning to the commodity. His opening sentence to that proposed final section reads: "As the elementary form of bourgeois wealth, the *commodity* was our point of departure, the prerequisite for the emergence of *capital*. On the other hand, *commodities* appear now as the *product of capital*."[94] The return to the subject of the commodity at the end of the volume would have meant that at the point of exiting from the text, as at the point of entering, the reader would have had in front of his eyes that form of the artifact in which the worker's responsibility for creation was most sensuously obvious (and the failure to credit that responsibility was thus most immediately recognizable as a failure).

The way he finally structured the conclusion, however, may accomplish this task just as effectively. Although a material commodity is an early site of imaginative projection, there are (as the Fourth and Fifth Notebooks of the *Grundrisse* stress), both historically and conceptually, two earlier sites of imaginative self-extension: the land (or raw materials) and the tool. Were *Capital 1* to open with the path of the worker's own act of making (rather than with *that part* of the path that first takes a distinctly capitalist form), the starting sequence of the analysis would not have been

$$[\text{Worker}] \longrightarrow \text{Commodity} \longrightarrow \text{Money} \longrightarrow \text{Surplus Value}$$

but instead this—

$$[\text{Worker}] \longrightarrow \text{Tool} \longrightarrow \text{Land (raw} \longrightarrow \text{Material Object} \longrightarrow \text{Money} \longrightarrow \text{Surplus}$$
$$\text{materials)} \qquad \text{(commodity)} \qquad\qquad \text{Value}$$

The worker projects his embodied labor out across the tool[95] onto the raw material that, so acted upon, becomes a material object, which will (within a capitalist economy) then itself become a commodity, and then money, and so forth. The sequence from the human creator through the three elementary sites (tool-land-material object) is, however, itself memorialized in the structure of *Capital,* but occurs there in a reversed order. A large portion of the center of the volume, Part 4, describes the nature of tools (hand tools, machine tools, machines, and the factory, itself a large tool). In turn, a large portion of the volume's close, Part 8, describes the problem of the land. The appearance of each is occasioned by the part it takes in the creation and perpetuation of capital, but this arrangement simultaneously permits us to move back through progressively earlier sites (even though represented in their modern form; not, for example, a rake but a sewing machine). That is, while the analysis begins with the Material Object (Commodity) and moves forward through later and later transformations, it simultaneously in its Opening, Middle, and Final Parts moves backward from the Material Object to the earlier Tool and then to the still earlier Land:

$$\begin{array}{ccc} 1 & 4 & 8 \end{array}$$

$$\text{Commodity} \longrightarrow \text{Money} \longrightarrow \text{Surplus Value} \longrightarrow \text{Accumulated Capital}$$

$$\text{Material Object} \longrightarrow \text{Tool} \longrightarrow \text{Land}$$

Not only, then, in its sequences of passages, sections, and chapters, but even in its largest outlines, the work holds body and artifact steadily in relation to one another. As the artifact becomes increasingly sublimated, separated from its source, it is accompanied by the increasingly elementary ground of its own occurrence. The recoverable referentiality of artifice, sustained throughout, is made especially pronounced at the moments when the reader's entry into or exit out of the text occurs.

If the dislocation in the structure of creation were, as in the Judeo-Christian scriptures, described in narrative rather than embedded in two disparate cognitive styles (the sensous rendering of the interior of capital and the arithmetic rendering of the embodied worker), if it were rendered imagistically rather than analytically, the action of the story told might be summarized in this way. In the midst of a vast industrial plain stood an artifact, a commodity, a pile of luminous coal so glittering with reflected sunlight that it seemed to belong to the world of heat, yet so deep and dark in its purple and blue that its blackness seemed not just its color but the very thing that it once must have been, something far removed from the sunlit surface of the plain. Two men crossed the plain, approached the commodity, and stood on either side of it. The one extended his arm and touched the artifact and, as he did so, his body grew larger and more vivid until all attention to his personhood or personality or spirit was made impossible by the compelling vibrancy of his knees, back, hands, neck, belly, lungs: even the interior of his body stood revealed in small cuts and larger wounds. Simultaneously, the other extended his arm and touched the artifact and as he did so, his body began to evaporate, grow airy: he was spiritualized, and disappeared. A name was given to each of the two: in his bodily magnification, the first was called by the name "worker"; in his bodily evaporation, the second was called by the name "capitalist." The two belonged to two tribes who, though they inhabited the same plain (where one produced coal and the other was warmed by it), never confronted one another face to face, for though the location of the first was apparent to the second, the location of the second was unknown to the first: having no body, he was grounded in no specifiable location; his face was hidden, invisible; or rather, he did not have a face if by face is meant bodily tissue subject to frostbite. Scattered across this part of the plain were many such artifacts (some were piles of coal, others bolts of cloth, others books, others beds) framed by many such pairs of men. Elsewhere, not far away, on a different part of the same plain stood a pile of money. Like the coal pile, the money pile was an artifact, though far less sensuous in its appearance than the first (for on this plain of the relation between persons and artifacts, it is not just men and women but artifacts themselves that display varying degrees of de-sensualization and dematerialization). It too was framed by men, five on one side and one on the other. As at the earlier site, all extended their arms and touched the artifact and, as they did so, the five on one side became larger and more intense in their physical presence, while the one on the other side was sustained in his disappearance. Still further along on this same vast plain was a small pile of paper-halfway-transformed-into-an-idea. Although this third artifact was less compelling in its sensuous appearance than the second, as the second was less sensuous than the first, its powers to regulate the appearance and disappearance of persons was correspondingly greater than the second, as the powers of the second were greater than those of the first. Framing this pile of paper were fifty men on one side and one on the other, and once more the arms of all were extended: the fifty persons became fifty enlarged bodies and the one disappeared from the plain and could not be found, though his voice was present, even omnipresent, in the social rules, legislation, and philosophic assumptions that swirled across the plain like an angry wind that was felt on the embodied faces of those who remained.

If rather than being merely summarized, the hypothetical narrative were told with splendor and conviction, it would bear a compelling resemblance to the Old Testament scenes of wounding. As there man and God confront one another across the intermediate fact that one has a body and the other does not, so in this later culture person and person confront one another across the intermediate fact that one is embodied and the other is disembodied, an intermediate fact that makes mediation for a time impossible. As in the early narrative scenes of hurt where human makers, rather than being disembodied by their Artifact, are now required to undergo more severe bodily distress in order to substantiate and sustain the original Artifact, so now in the later story men and women stand in the presence of the economic system collectively made to relieve them of the problems of sentience and must instead undergo increasingly severe bodily alterations to sustain and perpetuate its existence. Only the vocabulary changes: in the first case, the bodily attribute that must be perpetually re-projected onto the Artifact is its "realness"; in the second case, the bodily attribute that must be perpetually re-projected into the Artifact is its "valuableness" or "value." In both stories, the large Artifact (God in one, the collective economic system in the other) continues to be a projection of human capacities but has ceased to perform the counterpart of projection, reciprocation: God does not answer prayers (the peoples' voices are heard instead as godless murmuring and complaint); the economic system does not warm, clothe, and feed (the peoples' voices are again heard as murmuring and complaint rather than as an announcement from the creators themselves that a problem in the interior structure of artifice has arisen). The work of creation, which always has at its center the work of rescue, has broken down. As the solution to the deconstruction of creation in the scenes of wounding is material artifice—the materialization, or the embodying, or the humanization of the principle of creating so that the double consequence of projection and reciprocation will be once more intact—so now within the material realm itself the two have for a time again become severed into two stranded categories. Together, the two texts suggest that civilization's task of clarifying the structure of making is ongoing, for the very solution at one moment may become the site of difficulty at another: thus the material solution to the problem of belief itself becomes the problem in nineteenth-century industrialism, and so that site comes to require a new form of repair, as in the multiple strategies of distribution that have, since that time, begun to arise from both capitalist and socialist sources.

The similarities between the two (in so many other ways, radically disparate) writings have been attended to in this chapter not so much because either illuminates the other as because they together help to expose the nature of creating. Whether the Judeo-Christian scriptures are a precocious anticipation of Marx's account of the young industrial world, or Marx's account is a recapitulation of the philosophy of materialism embedded in the Judeo-Christian framework of which he (like it or not) is a child, are questions beyond the reach of this study. The intent here is not to credit Judeo-Christianity in the eyes of those who trust Marx's historical description, nor to credit Marx in the eyes of Judeo-Christians. They are presented as companion texts less because of their consistency with one another than because of what they together begin to make visible about the interior consistency in the structure of making and again in the structure of unmaking: in conjunction with the very different contexts looked at in earlier chapters, the parallels between them suggest that when the action of creating is occurring, it occurs (regardless of context) along a single path; and again, when its action fails or breaks down, that failure also occurs along a single path. It may be, as suggested earlier, that no other two texts in western civilization contain such sustained and passionate meditations on the nature of the human imagination. Their shared conviction that the ''problem of suffering'' takes place and must be understood within the more expansive frame of the ''problem of creating'' may at the very least be taken as an invitation to attend, with more commitment, to the subject of making, a subject whose philosophic and ethical import we do not yet fully understand.

5

THE INTERIOR STRUCTURE OF
THE ARTIFACT

W HEN one suddenly finds oneself in the midst of a complicated political situation, it is hard not only to assess the "rightness" and "wrongness" of what is taking place but even to perform the much more elementary task of identifying, descriptively, what it is that is taking place. The fact that torture, whose activity has a structure accurately summarized by the word *"stupidity,"*[1] should ever even for a moment successfully present itself to the outside world as an activity of *"intelligence*-gathering" is not an aimless piece of irony but an indication of the angle of error (in this case, 180°) that may separate a description of an event from the event itself.

The instability of our powers of perception and description may be even greater in situations that are not so simply, starkly, radically immoral as this one. That war, relentlessly centered in the reciprocal activity of *injuring* and only distinguishable from other means of arriving at a winner and loser by the specific nature of injury itself, should so often be described as though *injuring were absent* from or, at most, secondary to its structure, again indicates the ease with which our descriptive powers break down in the presence of a concussive occurrence, and may lead one to worry how we can set about to answer ethically complex questions about war when even the phenomenology of the event so successfully eludes us. The two historical moments contemplated in the previous chapter, though introduced primarily for their revelations about making, themselves include instances of the same perceptual problem. In the Old Testament scenes of hurt, what should be recognizable as simple and unequivocal acts of divine *immorality* (the willful and repeated infliction of human hurt) are instead perceived as revelations of his *superior morality:* the problem is presented not as the artifact's unreality, unbelievability, but as the people's disobedience; the pain-filled solution is presented not as analogical verification but as punishment. So, too, in the young industrial world described by Marx, the exclusion of the women and men who are the *creators* of made objects from the benefits of those objects is perceived as resulting from their *inferior creativity* (spiritedness, interest in education, capacity to create good lives, capacity for risk-taking and adventure). In each of the four instances, central rather than peripheral attributes are eclipsed and displaced not simply by alternative attributes but by attributes that are their very antithesis. The recurrence of such inverted descriptions suggests the existence of a general phenomenon that goes beyond these four instances: as physical pain destroys the mental content and language of the person in pain, so it also tends to appropriate and destroy the conceptualization abilities and language of persons who only observe the pain.

Political power—as is widely recognized and as has been periodically noticed throughout this book—entails the power of self-description. The mistaken descriptions cited above are in each instance articulated either by or on behalf of those who are directly inflicting, or actively permitting the infliction of, bodily hurt. But the failure to recognize what is occurring inside a concussive situation cannot be simply explained in terms of who controls the sources of description, for an observer may stand *safely* outside the space controlled and described by the torturer, by the proponents of a particular war, by the priests of an angry God, or by a temporally distant ruling class. Our susceptibility to the prevailing description must in part be attributed to the instability of perception itself: the dissolution of one's own powers of description contributes to the seductiveness of any existing description.

In turn, the instability of our descriptive powers results from the absence of appropriate interpretive categories that might act as "perceptual stays" in moments of emergency: we enter such events uncompanioned by any pre-existing habits of mind that would make it possible to go on "seeing" what is taking place before our eyes. The possible character of those needed-but-missing interpretive categories is suggested by the preceding chapters, for each of the human events examined there was found to be inextricably merged with questions of making and unmaking: torture and war are not simply occurrences which incidentally deconstruct the made world but occurrences which deconstruct the structure of making itself; conversely, western religion and materialism suggest that the ongoing work of civilization is not simply making *x* or *y* but "making making" itself, "remaking making," rescuing, repairing, and restoring it to its proper path each time it threatens to collapse into, or become conflated with, its opposite. These same interpretive categories would, if themselves unfolded and developed, also make it possible to enter and understand other concussive events, whether arising on the unreachable ground of a distant past or on the more important (because reachable and repairable) ground of an approaching future.

It is part of the work of this book to suggest that achieving an understanding of political justice may require that we first arrive at an understanding of making and unmaking. As in an earlier century the most searing questions of right and wrong were perceived to be bound up with questions of "truth," so in the coming time these same, still-searing questions of right and wrong must be reperceived as centrally bound up with questions about "fictions." Knowledge about the character of creating and created objects is at present in a state of conceptual infancy. Its illumination will require a richness of work far beyond the frame of any single study: like the activity of "making," the activity of "understanding making" will be a collective rather than a solitary labor.

Although an array of attributes belonging to "making" have emerged in the present discussion, they can be summarized in three overarching statements. First, the phenomenon of creating resides in and arises out of the framing intentional relation between physical pain on the one hand and imagined objects on the other, a framing relation that as it enters the visible world from the privacy of the human interior becomes work and its worked object (Chapter 3). Second, the now freestanding made object is a projection of the live body that itself reciprocates the live body: regardless of the peculiarities of the object's size, shape, or color, and regardless of the ground on which it is broken open (the sands of the Old Testament, the plains of nineteenth-century industrialism, or the vibrant and shifting ground on which we now stand), it will be found to contain within its interior a material record of the nature of human sentience out of which it in turn derives its power to act on sentience and recreate it (Chapter 4). Third, as is implicit in the overlay of the first two statements, the created object itself takes two different forms, the imagined object and the materialized object: that is, "making" entails the two conceptually distinct stages of "making-up" and "making-real." In the first of these, the imagination's work is self-announcing while in the second she completes her work by disguising her own activity. This may also be phrased in the following way: the imagination first "makes a fictional object" and then "makes a fictional object into a nonfictional object"; or, the imagination first remakes objectlessness (pure sentience) into an object, and then remakes the fictional object into a real one, one containing its own freestanding source of substantiation. Thus the benign pretense that "nothing" is "something" becomes the even more benign pretense that that "something" is not a pretense but has all the sturdiness and vibrancy of presence of the natural world (which it is at that moment in the midst of displacing). Recognizing the two as two conceptually (and often chronologically) distinct stages is especially important because—as has become evident in the preceding pages—the deconstruction of creating and aping of its activity may occur either at the "making" stage (where the decomposition and displacement of pain by made objects becomes instead the decomposition and displacement of objects by made pain) or at the "making-real" stage (where benevolent procedures of verification and reality-conferring are displaced by the procedure of borrowing the "realness" of the live human body).

Even if these three overarching statements are an accurate description of the structure of creating, they make visible that structure only in its skeletal outline. That structure has tens—in all probability hundreds—of smaller attributes that themselves require clarification. The closing pages of this book will briefly consider two of them. It is important to keep the scale of what will be discussed in perspective. If, for example, the skeletal structure summarized above were visually depicted as a large and miraculous suspension bridge, then all of what now follows would be the equivalent of describing, for example, the character of the metal in a few of its pins or the pressure in its weave of cables in one small section of its gigantic tracery. Once inside any one attribute (which may itself turn out to consist of many still smaller attributes), one may of course become lost in its own intricacies and complexities; but what we are lost in is a few square inches of something far more magnificent in scale.

The two attributes that will in this chapter be attended to are the following. Returning to the idea that a made object is a projection of the human body, Section I will summarize briefly the multiple ways in which this has been formulated in earlier pages, and will then explore and assess the legitimacy of the most radical formulation, demonstrating that artifacts are (in spite of their inertness) perhaps most accurately perceived as "a making sentient of the external world." While there is no part of making that is empty of ethical content, this particular attribute carries within it a very special kind of moral pressure. Section II will focus on the relation between the made object as a site of projection and the same object as a site of reciprocation. Although projection and reciprocation are (except in deconstructed making) *inseparable* counterparts, they are *not equal* counterparts: the work of reciprocation is ordinarily vastly in excess of the work of projection. The kinds of problems that arise when the two are wrongly assumed to be equal will be briefly introduced to underscore the importance of taking as a normative (or model) object one whose capacity to disembody the human being greatly exceeds the degree of intensified embodiedness required to bring the object into being. Each of the two sections will, then, attend to some aspect of the nature of projection and reciprocation (the second of the three-part skeletal structure of making summarized above). It is thus the incompleteness of the emerging model of making that is being acknowledged in the brief chapter which brings the present study to its end.

I. Artifacts: the Making Sentient of the External World

The recognition that a made object is a projection of the human body has been formulated throughout these pages in three different ways. The first of the three makes the relation between sentience and its objectifications compellingly visible by describing the phenomenon of projection in terms of specifiable body parts.

When, for example, the woven gauze of a bandage is placed over an open wound, it is immediately apparent that its delicate fibers mime and substitute for the missing *skin,* just as in less drastic circumstances the same weave of threads (called now "clothing" rather than "bandage," though their kinship is verbally registered in the words "dress" and "dressing") will continue to duplicate and magnify the protective work of the skin, extending even its secondary and tertiary attributes so that, for example, any newly arrived observer would not say people come in hues of yellow, pink, brown, and black but that they come in forest green and white, kelly green and gold, yellow and brown stripes, pink and black squares, varying shades of magenta, mauve, red, orange, blue —in other words, they are creatures whose color tends on the average to change every twenty-four hours, and thus creatures that must be described as independent of any fixed surface color.[2] As the skin has many equivalents in the external made world, so does every other body part. Eyeglasses, microscopes, telescopes, and cameras are, as Freud notes in passing in *Civilization and Its Discontents,* projected materializations of the *lens* of the human eye.[3] Freud's own work is primarily devoted not to skin or lens but to a third body part that he has taught us habitually to recognize in successive circles of sublimation: people living in a post-Freudian era effortlessly and unembarrassedly identify the *phallus* in dream objects, domestic objects, and civil objects. It is apparently "out there" in dream sticks, dream vultures, materialized pipes, hats, drills, swords, skyscrapers, obelisks, and rockets, where it is companioned by equally pervasive materializations of its female counterpart, the *womb,* which reappears in multiple forms of sheaths, shields, dwelling-places, and shelters. Similarly, the human *heart,* generously lending its name[4] to anything that in its location or significance is perceived to be central ("the heart of the matter," "the heart of the poem," "the heart of the problem," "the heart of the experiment," "the heart of the city," "the heart of the nation"), has also lent its structure of action to a mechanism which was invented in ancient Egypt and Rome to bring the benefits of water to a waterless terrain, was then reinvented and developed in the sixteenth century to clear underground coal and metal mines of dangerous waters, and went on to have hundred of other modifications and uses. In turn, the pump, as Jonathan Miller observes, provided a freestanding technological model which allowed William Harvey to identify correctly the "pumping" action of the bodily organ whose existence preceded by many centuries that of its artificial counterpart.[5] Again, the *electrical nervous system* of the live body has, according to Jeremy Bernstein, its materialized objectification in the computer. While the translation of skin into clothing, phallus into obelisk, or heart into pump may in each instance have originally arisen out of unconscious acts of projection, the translation of nervous system into computer was highly self-conscious. In his celebrated work on the computer, John von Neumann consciously drew on the work of two physiologists who had analyzed the structure of the body's neuronal system.[6] The bodily sources of culture are, then, multiple: skin, lens, phallus, womb, heart, and nervous system form a very partial list, to which there can be added many others, such as hand, ear, lungs, stomach, skeleton, teeth, leg muscles, and hinge joints, some of whose cultural objectifications have been encountered in earlier chapters.

The second way of formulating the phenomenon of projection is to identify in the made object bodily capacities and needs rather than the concrete shape or mechanism of a specifiable body part. The first formulation has the advantage of making the relation more graphic and thus more immediately apprehensible, yet it has three obvious disadvantages. First, certain complex characteristics of the embodied human being have no (or as yet, no known) physical location or mechanism. The printing press, the institutionalized convention of written history, photographs, libraries, films, tape recordings, and Xerox machines are all materializations of the elusive embodied capacity for *memory,* rather than materializations of, for example, one cubic inch of brain matter located above the left ear. They together make a relatively ahistorical creature into an historical one, one whose memory extends far back beyond the opening of its own individual lived experience, one who anticipates being itself remembered far beyond the close of its own individual lived experience, and one who accomplishes all this without each day devoting its awakened brain to rehearsals and recitations of all information it needs to keep available to itself. Similarly, we routinely speak of certain artifacts as "expressing the human *spirit,*" a statement that would be impossible to formulate in terms of bodily location. Second and conversely, many inventions exist that have no specifiable precedent in the body: perhaps the wheel astonishes us in part because we do not "recognize" it—that is, because we intuitively sense that it has no prior existence within the boundaries of our own sentient matter (the ball and socket joint and the rotary mechanism of some insect wings notwithstanding). Although machine tools have been widely described as taking over the work of the muscles, any one-to-one equation is often impossible. The work of the steam engine in magnifying the bodily capacity for *movement* does not require a mechanistic equivalent in the body; it is perhaps enough simply to know that, for example, at the moment the steam engine first burst forth into John Fitch's imagination, he was, by his own account, limping.[7] Third and most important, even if every made object did have a bodily counterpart (an improbable proposition given the fragile dimensions of the human body and the robust dimensions of culture), it would even then be more accurate to formulate the projection in terms of "attributes" rather than "parts," since creating is undertaken to assist, amplify, or alter the felt-experience of sentience rather than merely to populate the external world with shapes and mechanisms already dwelling within us.

Even when a given artifact bears an obvious kinship to a bodily part, it will usually be more productive to articulate that kinship in terms of sentient attributes. So, for example, all of the artifacts invoked a moment ago as mimetic of "parts" can now be reinvoked as mimetic of "attributes." What most amazes Jeremy Bernstein as he meditates on the computer is not that it is an external material-ization of our electrical and neuronal pathways but that it is an external mater-ialization of our interior capacity for *self-replication* and *self-modification*. Only a small fraction of Freud's work can be summarized in terms of the projected shapes of phallus and womb, whereas almost all of it can be summarized in terms of the projection of sentient *desire:* it is the presence of complex structures of desire that he has taught us to recognize in dreams, in externalized patterns of family and civic behavior, in the art works of Sophocles and da Vinci, in the materialized and verbalized products of civilization. Similarly, Marx's writing— in which the shape of hand and back have, if only implicitly, something of the same primacy that phallus and womb have in the writings of Freud—must be centrally described in terms of the bodily capacity for *labor:* he teaches us to recognize human labor in successive circles of self-extension, from its obvious presence in single, individually crafted objects, to its less obvious, because more collective, presence in money, and so on out through increasingly sublimated economic and ideological structures. Because Freud and Marx are generally recognized as the two cultural philosophers of greatest importance to the modern world, it is appropriate to notice that the work of each has been primarily devoted to making an embodied attribute (desire, labor) the recoverable referent of the freestanding structures of civilization that are their materialized counterparts.

One final example of the difference between formulating the phenomenon of projection in terms of concrete body parts or instead in terms of more elusive interior attributes is the phenomenon of projection itself. That is, while the human being is a seeing, moving, breathing, hearing, hungering, desiring, working, self-replicating, remembering, blood-pumping creature (who projects all these attributes outward), he is therefore also a projecting creature. This has, here and there in earlier pages, been expressed in terms of discrete bodily location: the human being has an outside surface and an inside surface, and creating may be expressed as a reversing of these two bodily linings. There exist both verbal artifacts (e.g., the scriptures) and material artifacts (e.g., altar) that objectify the act of believing, imagining, or creating as a sometimes graphically repre-sented turning of the body inside-out. But what is expressed in terms of body part is, as those cited contexts themselves make clear, more accurately formulated as the endowing of interior sensory events with a metaphysical referent. The interchange of inside and outside surfaces requires *not* the literal reversal of bodily linings but the making of what is originally interior and private into something exterior and sharable, and, conversely, the reabsorption of what is now exterior and sharable into the intimate recesses of individual consciousness.

When the pure fact of "projection" is articulated in terms of bodily location (inside and outside surfaces), it takes a much more extreme form than when the projection of any specific attribute (e.g., vision) is articulated in terms of bodily location: as startling as it is to think of the lens of the eye being lifted away from the body and carried out into the external world, it is much more alarming to contemplate, however briefly, the turning of the body inside-out. This greater extremity of imagistic representation occurs because the overall framing fact of projection (which moves between the extreme boundary conditions of physical pain and created objects) *is* more radical than the projection of any specific intermediate attribute. That is, a particular dimension of sentience will, by being projected, undergo an alteration in degree: the power of vision is amplified when supplemented by microscope and telescope, as the problem of hunger is diminished and regulated through the strategies of artifice. But the inclusive phenomenon of projection entails not simply an alteration in degree but a much more extraordinary form of revision in which the original given is utterly eliminated and replaced by something wholly other than itself. What is wholly absent in the interior (the missing objects in the pure sentient condition of utter object-lessness) is made present (through objectification), as conversely, what is wholly present in that interior state (pain) is (when projection is successful) now made absent. Thus, the reversal of inside and outside surfaces ultimately suggests that by transporting the external object world into the sentient interior, that interior gains some small share of the blissful immunity of inert inanimate objecthood; and conversely, by transporting pain out onto the external world, that external environment is deprived of its immunity to, unmindfulness of, and indifference toward the problems of sentience.

This last statement carries us forward to the third and, in the end, most accurate way of formulating the phenomenon of projection; for it calls attention to the fact that it is part of the work of creating *to deprive the external world of the privilege of being inanimate*—of, in other words, its privilege of being irresponsible to its sentient inhabitants on the basis that it is itself nonsentient. To say that the "inanimateness" of the external world is diminished, is *almost* to say (but is *not* to say) that the external world is made animate. The rest of this section will try to define that "almost" with more precision.

As one moves through the three ways of formulating the phenomenon of projection, the "body" becomes progressively more interior in its conceptualization. To conceive of the body as parts, shapes, and mechanisms is to conceive of it from the outside: though the body contains pump and lens, "pumpness" and "lensness" are not part of the felt-experience of being a sentient being. To instead conceive of the body in terms of capacities and needs (not now "lens" but "seeing," not now "pump" but "having a beating heart" or, more specifically, "desiring" or "fearing") is to move further in toward the interior of felt-experience. To, finally, conceive of the body as "aliveness" or "awareness of aliveness" is to reside at last within the felt-experience of sentience; and it is this most interior phenomenon that will now be considered. "Aliveness" or "awareness of aliveness," it will be argued, is in some very qualified sense projected out onto the object world.

When, as in old mythologies or religions, nonsentient objects such as rocks or rivers or statues or images of gods are themselves spoken about as though they were sentient (or alternatively, themselves endowed with the power of sentient speech) this is called "animism." Again, when poets or painters perform the same act of animation, it is called "pathetic fallacy." But as will very gradually become apparent here, to dismiss this phenomenon as mistake or fallacy is very possibly to miss the important revelation about creation exposed there. The habit of poets and ancient dreamers to project their own aliveness onto nonalive things itself suggests that *it is* the basic work of creation to bring about this very projection of aliveness; in other words, while the poet pretends or wishes that the inert external world had his or her own capacity for sentient awareness, civilization works to make this so. What in the poet is recognizable as a fiction is in civilization unrecognizable because it has come true.

It should be registered from the outset that this habit of mind is restricted to neither poetic nor mythological forms of perception. Perhaps no one who attends closely to artifacts is wholly free of the suspicion that they are, though not animate, not quite inanimate. Marx, for example, who periodically in *Capital* rails brilliantly against "animism" and "fetishism," is himself constantly tempted in his analysis of economic objects to describe their attributes in the language of "aliveness."[8] In fact, as every reader of that volume will have noticed, the pages crediting the "alive-like" character of commodities, money, and capital so vastly outnumber the pages on which this characteristic is successfully bracketed off as "fetishism" or "reism" that we can only think what Marx periodically tells us to think by ignoring what he elsewhere and everywhere shows us to be the case. This is not to say that Marx is himself a fetishist or reist. It is rather to say that Marx and the reists are differentiated not by the former's insistence that objects are inanimate and the latter's insistence that they are animate, but by the radically different implications the two discover in object-animism: the reist takes that apparent-aliveness as a basis for revering the object world; Marx takes that apparent-aliveness as a basis for revering the actual-aliveness of the human source of that projected attribute. Given the ease with which these two positions might become confused in a reader's mind, Marx had every reason to avoid the "aliveness" idiom altogether in his own account of artifacts. That he did not do so suggests that he could not do so, that the idiom is, for reasons that will eventually become clear, unavoidable.

One additional instance of overtly fetishistic animism will be cited here to underscore from the outset that this habit of perception is neither exclusively ancient (the event took place in 1976) nor exclusively poetic (the event is emphatically anti-poetic), and thus cannot be attributed to acute sensitivity nor, as it is sometimes phrased, to romantic sentimentality. The Brookings Institute study, *Force Without War*, describes an 18 August 1976 incident in which "two American officers supervising the pruning of a tree in the Korean demilitarized zone were attacked by North Korean soldiers and killed." This event gave rise to a series of actions, culminating in the following: "Finally a few days after the initial incident, a large force of American and South Korean soldiers entered the demilitarized zone and cut down the offending tree while armed helicopters circled overhead and B-52 bombers flew near the border."[9] The elaborately dramatized assumption of object-responsibility might be formulated as a legal statute: any tree that protests being pruned by taking (or by permitting in its vicinity the taking of) human life will be subject to the penalty of death by more-radical-pruning. Were it not framed on one side by the deaths of two men and on the other side by the absence of armed conflict which it perhaps helped to prevent, the incident could be simply enjoyed for the spectacular scale—a large force of soldiers, helicopters, B-52s—on which its atavistic premises are unembarrassedly acted out. It is introduced here not to credit the animistic impulse (for it is more likely to expose that impulse as foolish), but simply to suggest the multiplicity of paths by which animism is arrived at: one may get there by way of the darkness of superstition, the exquisite insight of poetry, the rigors of economic analysis, or the strategic resourcefulness of military frustration.

The eventual goal here is to identify exactly what within our willful recreation of the external world repeatedly beguiles us into crediting it with awareness and hence with responsibility for its actions. The answer to this question will be coaxed into clarity by first turning back to the Old Testament where the inner logic of the animistic impulse is unfolded before our eyes in stark outlines, and then turning forward to contemporary legal formulations of object-responsibility where the same inner logic is articulated in a more familiar idiom.

It was earlier observed that material objects in the Old Testament fall into two categories, graven images and passover artifacts, the first of which confer on God a body and the second of which relieve man of his body. The entry of God's body and man's body into the material artifact can be stated, as it has been in the opening half of this sentence, as though it were a symmetrical occurrence when of course it is not. The coming into being of the passover objects is eagerly accepted by humanity, while the making of graven objects is never overtly accepted, often condemned and destroyed, by God. This asymmetry of response occurs because the artifacts made by God relieve man of the necessity of being wounded, whereas those made by man wound God: with them, man is literally[10] relieved of his pain; by them, God is literally put in pain.

This juxtaposition only appears to entail human harshness if one forgets that it is not human tissue that is put in pain, that these are not two autonomous actors conferring on the nature of created objects, but that God himself is the original created object, now itself being altered. The putting of the Original

Artifact in pain acts out the essential premise of the entire undertaking of the imagination: it is the benign, almost certainly heroic, and in any case absolute intention of all human making to distribute the facts of sentience outward onto the created realm of artifice, and it is only by doing so that men and women are themselves relieved of the privacy and problems of that sentience. This intention was surely present in the initial apprehension, invention, of "a God" capable of rescuing them from their own sentience (capable, that is, of existing as a metaphysical referent to which their individual sentient experiences could be referred and thus reread in terms of a collective or communally shared objectification), and then again present in the introduction of graven images which bring forth an intensification of that projected sentience in order to bring about in him a more compassionate accommodation of their own sentience.

The story that is told provides the language in which the story of the making of all the artifacts of civilization can be retold. A chair, as though it were itself put in pain, as though it knew from the inside the problem of body weight, will only then accommodate and eliminate the problem. A woven blanket or solid wall internalize within their design the recognition of the instability of body temperature and the precariousness of nakedness, and only by absorbing the knowledge of these conditions into themselves (by, as it were, being themselves subject to these forms of distress), absorb them out of the human body. A city, as though it incorporates into its unbroken surfaces of sand and stone a sentient uneasiness in the presence of organic growth and decay (the tyranny of green things that has more than once led people to the desert whose mineral expanse is now mimed in every modern urban oasis) will only then divest human beings of that uneasiness, divest them to such an extent that they may even come to celebrate and champion that green world, reintroducing it into their midst in the delicate spray of an asparagus fern or in a breathtaking framed photograph of the Andes. A clock or watch, as though it were itself sentient, as though it knew from the inside the tendency of individual sentient creatures to become engulfed in their own private bodily rhythms, and simultaneously knew of their acute and frustrated desire to be on a shared rhythm with other sentient creatures, will only then empower them to coordinate their activities, to meet for a meal, to meet to be schooled, to meet to be healed, after which the clock can be turned to the wall and the watch can be taken off, for these objects also incorporate into their (set-asidable) designs an awareness of sentient distress at having to live exclusively on shared time.

The naturally existing external world—whose staggering powers and beauty need not be rehearsed here—is wholly ignorant of the "hurtability" of human beings. Immune, inanimate, inhuman, it indifferently manifests itself in the thunderbolt and hailstorm, rabid bat, smallpox microbe, and ice crystal. The human imagination reconceives the external world, divesting it of its immunity and irresponsibility not by literally putting it in pain or making it animate but by, quite *literally, "making it" as knowledgable about human pain as if it were itself animate and in pain.* When the roar of the flood waters comes, water and rocks and trees are mutely indifferent, but when the mythmaker recounts the story of the flood, the tree is invested with the capacity of compassionate speech: "I too feel the waters rising, and see that you will drown; take hold of this branch." His fiction of object-responsiveness anticipates the actuality of object-responsibility, for though the tree does not speak, when it is itself remade into raft or boat (as when the indifferent rocks are rearranged into a dam), the world outside the body is made as compassionately effective as if every line and nuance of its materialized design were speaking those words. We come to expect this of the world. Thus, the tree in Korea was inappropriately unsusceptible to "pruning," to being domesticated, civilized, remade. Had it been a proper tree, it would have heard the North Korean planes approaching, seen the men beneath its branches, and sent up some form of protective shield. At the very least, it would have given a signal ("They are coming: leave, run, hurry") as civilized trees, with their radar branches, routinely do. This expectation is as old as the human imagination. The "tree of knowledge," the "tree of life," *is* the "tree of artifice." The tree in Eve's garden never said to her, "I see how frightened, overwhelmed, you are by believing yourself to be nakedly exposed to One who has no body, and advise you to cover yourself as you are when you stand hidden here within my branches." But when she remade the tree into an apron of leaves, she restructured the grove into a structure of materialized compassion.

Thus, the literalness of the claim that creation entails the projection of the "awareness of aliveness" becomes immediately intelligible. A material or verbal artifact is not an alive, sentient, percipient creature, and thus can neither itself experience discomfort nor recognize discomfort in others. But though it cannot be sentiently aware of pain, it is in the essential fact of itself the objectification of *that awareness*; itself incapable of the act of perceiving, its design, its structure, *is* the *structure of a perception*. So, for example, the chair encountered so often in the previous chapter, can—if projection is being formulated in terms of body part—be recognized as mimetic of the spine; it can instead—if projection is being formulated in terms of physical attributes—be recognized as mimetic of body weight; it can finally and most accurately, however, be recognized as mimetic of sentient awareness, as will be elaborated below.

If one imagines one human being seeing another human being in pain, one human being perceiving in another discomfort and in the same moment wishing the other to be relieved of the discomfort, something in that fraction of a second is occurring inside the first person's brain involving the complex action of many neurons that is, importantly, not just a perception of an actuality (the second person's pain) but an alteration of that actuality (for embedded in the perception is the sorrow that it is so, the wish that it were otherwise). Though this interior event must be expressed as a conjunctive duality, "seeing the pain and wishing it gone," it is a single percipient event in which the reality of pain and the unreality of imagining are already conflated. Neither can occur without the other: if the person does not perceive the distress, neither will he wish it gone; conversely, if he does not wish it gone, he cannot have perceived the pain itself (he may at that moment be experiencing something else, such as his own physical advantage, or his own resistance to having to attend to another person, but he cannot be perceiving the pain, for pain is in its essential nature "aversiveness," and thus even within technical medical definitions is recognized as something which cannot be felt without being wished unfelt[11]). If this complex, mysterious, invisible percipient event, happening somewhere between the eyes and the brain and engaging the entire psyche, could be made visible, could be lifted out of the body and endowed with an external shape, that shape would be the shape of a chair (or, depending on the circumstance, a lightbulb, a coat, an ingestible form of willow bark). The shape of the chair is not the shape of the skeleton, the shape of body weight, nor even the shape of pain-perceived, but the shape of perceived-pain-wished-gone.

The chair is therefore the materialized structure of a perception; it is sentient awareness materialized into a freestanding design. If one pictures the person in *the action* of making a chair—standing in one place, moving away, coming back, lifting then letting fall his arm, kneeling then standing, kneeling, half-kneeling, stooping, looking, extending his arm, pulling it back—and if one pictures all these actions as occurring without a tool or block of wood before him, that is, if one pictures only the man and his embodied actions, what one at that moment has before one is *not* the *act of perception* (his seeing of another's discomfort and wishing it gone) but *the structure of the act of perception visibly enacted*. What was originally an invisible aspect of consciousness (compassion) has now been translated into the realm of visible but disappearing action. The interior moment of perceiving has been translated into a willed series of successive actions, as if it were a dance, a dance entitled "body weight begone." Perhaps as he dances, his continual bodily readjustments relieve him of his awareness of his own weight; or perhaps as he dances before his pregnant wife, he (by his expression of concern) half relieves her own problem of body weight, assuring her that she is not alone, engulfed, in her adversity. In any event, the dance is not the original percipient event but that percipient event endowed with a communicable form.

If, now, the tool is placed back in his hand and the wood placed beneath that tool, a second translation occurs, for the action, direction, and pressure of his dance move down across the tool and are recorded in the surface of the wood. The two levels of projection are transformations: first from an invisible aspect of consciousness to a visible but disappearing action; second, from a disappearing action to an enduring material form. Thus in work, a perception is danced; in the chair, a danced-perception is sculpted.

Each stage of transformation sustains and amplifies the artifice that was present at the beginning. Even in the interior of consciousness, pain is "remade" by being wished away; in the external action, the private wish is made sharable; finally in the artifact, the shared wish comes true. With each successive recreation, compassion is itself recreated to be more powerful: in the end, it has made real what it at first only passively wanted to be so. For if the chair is a "successful" object, it will relieve her of the distress of her weight far better than did the dance (or alternatively, far better than a verbalized expression of sympathy). Even if, however, it relieves her distress only to the same degree as the expressive dance, it has two striking advantages over its antecedent action. First, the chair itself memorializes the dance, endures through time: to produce the same outcome, the dance would have to be repeated each day, thus requiring that the man enter and sustain the aversive intensity of labor (his sharing of the pain) without cessation, and thereby only redistributing, rather than diminishing, the pain itself. This does not mean that "active sentient compassion" (live human caring) and "compassion made effective" (the freestanding artifact) are at odds with one another, that we are in any sense asked to choose between friendly human presences or instead the companionship of objects. The existence of the second merely extends the range of subjects that can be entered into by the first: when both persons are free of the problem of her weight, they share endless other concerns, work to eliminate other pains, so that increasingly the pleasure of world-building rather than pain is the occasion of their union.

The second advantage of chair over sympathetic expression is that once it is in existence, the diminution of the woman's problem no longer depends on the goodwill of whatever other human being co-inhabits her world. She may have the good fortune to have a compassionate mate; she may instead have an indifferent one; it is also not impossible that she may have one who wishes her ill. The general distribution of material objects to a population means that a certain *minimum* level of objectified human compassion is built into the revised structure of the external world, and does not depend on the day-by-day generosity of other inhabitants which itself cannot be legislated. This is why, as the films of Ingmar Bergman so frequently suggest,[12] the first act of tyrants and other egoists is often to replace a materially bountiful world (with its implicit, if anonymous, human wish for the individual's basic comfort) with a starkly empty one in which each nuance of comfort depends on the vagaries of the egoist's own disposition. This is also why a woman imprisoned under a hostile regime in Chile once clung passionately to a white linen handkerchief slipped to her from another country, for she recognized within the object the collective human salute that is implicit in the very manufacture of such objects;[13] just as this same salute has been recognized by many prisoners of torture who mention (often with an intensity of gratitude that may at first sound puzzling) the solitary blanket or freshly whitewashed walls one day introduced into their midst by the quiet machinations of the International Red Cross.[14] It is almost universally the case in everyday life that the most cherished object is one that has been hand-made by a friend: there is no mystery about this, for the object's material attributes themselves record and memorialize the intensely personal, extraordinary because exclusive, interior feelings of the maker for just this person—This is for you. But anonymous, mass-produced objects contain a collective and equally extraordinary message: Whoever you are, and whether or not I personally like or even know you, in at least this small way, be well. Thus, within the realm of objects, objects-made-for-anyone bear the same relation to objects-made-for-someone that, within the human realm, caritas bears to eros. Whether they reach someone in the extreme conditions of imprisonment or in the benign and ordinary conditions of everyday life, the handkerchief, blanket, and bucket of white paint contain within them the wish for well-being: "Don't cry; be warm; watch now, in a few minutes even these constricting walls will look more spacious."

Although the Old Testament account of the artifact as the meeting place of man's body and God's body may to a secular mind sound alien, one basis for the formulation is that the artifact is a conflated projection of the fact of physical pain (our bodies) and a counterfactual wish (our gods), that itself makes the realness of pain unreal by making the unrealness of the wish real (embodied). A lightbulb transforms the human being from a creature who would spend approximately a third of each day groping in the dark, to one who sees simply by wanting to see: its impossibly fragile, milky-white globe curved protectively around an even more fragile, upright-then-folding filament of wire is the materialization of neither retina, nor pupil, nor day-seeing, nor night-seeing; it is the materialization of a counterfactual perception about the dependence of human sight on the rhythm of the earth's rotation; no wonder it is in its form so beautiful. There would be no need to introduce this example into the expansive company of all the preceding examples except that in this one instance we overtly reveal our recognition that the artifact is a materialization of perception by the widely shared convention of inserting it back inside a drawing of the human head where it stands for the moment when a problem is reconceived in terms of its solution. Itself a materialized projection of an *instance* of that form of perception, it is now, iconographically, pushed back into its original location, where it comes to stand for the *generic* event of problem-solving. A much less widely shared manifestation of this same phenomenon is the tendency of certain artists to reinsert an artifact into a portrait of the human interior at a moment when they are attempting to express some difficult-to-express event in the history of the live human body: so in the pages of Miguel Asturias, a man dies when the "penny-whistle of his heart" gives way (he could not have said "pump," for had it been a pump, it would not so easily have given way),[15] as in the pages of Charles Dickens, the body in its final minutes is made to contain within its interior a wagon (or in another instance, a clothes press), whose labored movements now objectify the labored and exhausting efforts of the dying body to breathe, to work, to pump, to stay alive.[16] These objects are mimetic not simply of body parts but of percipient awareness: Asturias and Dickens transport them back into the body at a moment when they are attempting to elicit the reader's compassion because the objects are themselves already compassion-bearing.

These examples are unusual because in them our recognition that external objects are mimetic of percipience is overtly announced. More often, the recognition is expressed by indirection and inversion: that is, rather than overtly celebrating such objects when they successfully perform this work of mimesis, we disapprove of, discredit, and even "punish" them if they fail to perform that mimesis. This habit of *taking object-awareness as the norm and object-unawareness as an aberrant and unacceptable occurrence* reveals the depth of our expectations more eloquently than would any overt celebration. Though this expectation has many manifestations, it is nowhere so clear as in the law.

The "statutes of homicide" in Plato's *Laws* begin by requiring that a person convicted of murder be put to death: his body will be taken outside the city and stoned, and then carried to the frontier (873).[17] This same course of action is then extended to animals which, upon conviction of having killed a person, "shall be put to death and cast out beyond the frontier" (873e). The same action is then extended to inanimate objects:

> If an inanimate thing cause the loss of a human life—an exception being made for lightning or other such visitation of God—any object which causes death by its falling upon a man or his falling against it shall be sat upon in judgment by the nearest neighbor, at the invitation of the next of kin, who shall hereby acquit himself and the whole family of their obligation—on conviction the guilty object to be cast beyond the frontier, as was directed in the case of a beast [as well as of a person]. (873e, 874a)

Two observations are immediately relevant to the present discussion. First, when Plato exempts certain objects from this statute ("an exception being made for lightning and other such visitation of God"), he might have said, "an exception being made for those aspects of the naturally given world that are beyond the reach of civilization." That is, "lightning and other such visitation of God" are privileged not because they are unpunishable (though they are, of course, unpunishable: the lightning cannot be carried beyond the edge of the city) but because of the prior fact that, unlike most aspects of the external world, they are unsusceptible to being reconceived and remade by the human imagination, and thus, unlike most aspects of the external world, cannot be held responsible for their ignorance about and thus harm done to human tissue: that they are unpunishable (i.e., unregenerate) is itself only one form of the larger fact of their being unreconstructable (i.e., unregenerate in the wider sense). Second, it may seem that the sequence in the statutes from persons, to beasts, to objects would permit one to dismiss the inclusion of objects by some version of the following argument: Plato's expectations about the responsibilities of living presences (human animals and other animals) at the last minute spill over into the realm of the nonliving. It is therefore worth noticing that Plato might have, with equal intelligence, presented the sequence in the reverse order. Thus the statutes might have read as follows:

> In civilization, the inanimate external world is reconceived and invested with the responsibility of existing as though it had sentient awareness. Any object, therefore, which exposes its absence of sentient awareness (announces its inanimate objecthood or its object stupidity) by lethally hurting a human being will not be permitted to continue dwelling within civilization and will be carried to the frontier. Further, should a sentient animal lapse into this same object stupidity and kill someone, it will, upon conviction, be deprived of the sentience it has already been guilty of lacking and will be removed from civilization. Finally, as unlikely as it is that any human being should ever lapse into this object stupidity, should this happen, the person will, upon conviction, be similarly deprived and removed.

It should be recalled here, as at a very early point in this book, that the word "stupidity" is not being used as a term of rhetorical contempt for those who willfully hurt others but as a descriptive term for the "nonsentience" or "the lack of sentient awareness," or most precisely, the "inability to sense the sentience of other persons" that is incontestably present in the act of hurting another person. Maximum expectations (e.g., aliveness) begin with persons and may be extended to objects, but minimum expectations begin with objects and may be extended to persons. The sequence of the statutes can be inverted because, in some very real way, the logic underlying civilization's prohibition of homicide proceeds from objects to persons: if this most *minimal* expectation (not to kill a person) can be required of even the only animate-like, inanimate world, how much more reasonable is it to require this minimal expectation of things that are actually animate (beasts) and finally of persons themselves. In other words, if civilization can ask an object not to act like an object, surely it can ask a person not to act like an object.

Oliver Wendell Holmes in his opening lecture on liability in *The Common Law* attends to the presumption of object-responsibility in American and English legal procedures, as well as in their German and Roman antecedents. He finds that "if a man fell from a tree, the tree was deodand. If he drowned in a well, the well was to be filled up,"[18] and notices that the animistic impulse tends to be especially pronounced if the object has the attribute of "motion." While motion is present in a moving cart, a falling house, and endless other objects, it is especially characteristic of a ship that comes to be regarded as "the most living of inanimate things," regarded that way to such an extent that, according to Holmes, it is impossible to decipher the complexities and apparent contradictions of maritime law unless one recognizes as the key to the code, the presumption of aliveness: "It is only by supposing the ship to have been treated as if endowed with personality, that the arbitrary seeming peculiarities of the maritime law can be made intelligible, and on that supposition they at once become consistent and logical."[19]

Holmes's overall purpose in this essay is to demonstrate that the concept of liability as it occurs in both criminal law and the law of torts originates in a moral impulse and invokes a moral standard, even though in its modern transformations the explicitly moral language comes to be rephrased in a more "external or objective" idiom.[20] While, for example, in modern damage suits the presumption of object-responsibility is presented in terms of owner- or manufacturer-responsibility, it was originally the object itself that was blamed. Nor can this object-blaming be understood as a short cut to, or substitute for, owner-blaming, for Holmes cites court decisions in which this interpretation is explicitly rejected: Chief Justice Marshall, for example, writes, "This is not a proceeding against the owner; it is a proceeding against the vessel for an offense committed by the vessel."[21] It may be as accurate to think of modern owner-blaming as a way of reaching the object, as to think of older object-blaming as a way of reaching the owner. Again, while in a modern proceeding the goal of the suit is compensation, Holmes draws on many historical instances to persuade us that the original goal was not compensation but revenge, revenge against whatever fragment of the external world inflicted death or pain.

Although the identification of the psychological and moral phenomenon of revenge successfully works to clarify Holmes's point, it may at first work to obscure ours, and should therefore be attended to for a moment longer. Damage suits usually arise when someone has been killed, paralyzed, or caused prolonged pain, yet the revenge impulse is visible even when one has been only very modestly hurt and is more familiar to us in this form: it is present in the "hatred for anything giving us pain, which wreaks itself on the manifest cause, and which leads even civilized man to kick a door when it pinches his finger."[22] The problem with the revenge vocabulary is that it may mislead us into thinking that it is only at the instant of being hurt that the person projects sentient awareness onto the object, that the act of animism arises within, and is carried outward by, the retaliatory act of revenge: the man is pinched and in the next split-second he projects aliveness onto the door and assumes it will suffer as much by being kicked as he just suffered by being pinched. But it seems instead the case that the act of revenge is itself premised on the prior assumption of animism and must be seen within a much wider frame. Our behavior toward objects at the exceptional moment when they hurt us must be seen within the context of our normal relations with them. The ongoing, day-to-day norm is that an object is mimetic of sentient awareness: the chair routinely relieves the problem of weight. Should the object prove insufficiently mimetic of awareness, insufficiently capable of accommodating the problem of weight (i.e., if the chair is uncomfortable—an animistic phrase we use to mean if "the person is uncomfortable in the chair"), the object will be discarded or set aside. Only now do we reach the third and most atypical occurrence in which the object neither eliminates the problem of sentience, nor even simply passively fails to eliminate the problem of sentience, but instead actually amplifies the problem of sentience by inflicting hurt: the legs of the chair suddenly break beneath the weight of the person and he is hurt. The very reason the chair's object-stupidity strikes all who witness its collapse as a surprise, an outrage, is that it has normally been wholly innocent of such object-stupidity. In fact, it is crucial to notice that if the person now picks up a fragment of that object and hurls it against the wall (as though it could be made to feel the hurt it just inflicted), the person is actually continuing to act out of the context of the normal situation (in which the chair indeed has the mimetic attributes of sentient awareness) rather than out of the immediate moment (in which the chair has just exposed its object-obliviousness).

Thus the moment of revenge merely occasions the dramatization of the ongoing assumption of animism rather than occasioning the animism itself. The retaliatory drama that takes place between Holmes's man-pinching door and door-kicking man must be seen within the wider frame of the fact that nine times out of ten (or, if the man is skilled at opening doors, nine hundred and ninety-nine times out of a thousand), the door has acted as though it were percipiently aware, and has done so because its design is a material registration of the awareness that human beings both need the protection of solid walls and need to walk through solid walls at will. The door not only seems capable of transforming itself back and forth between the two states of wallness and nonwallness but, more remarkably, seems capable of understanding which of the two states the man wants it to be at any given moment—it recognizes what he wants not by requiring from him elaborate paragraphs of self-revelation but only a minimal signal, the turning of his wrist. If the door exists in a realm where people can be anticipated to be incapable of performing this signal (such as when they are carrying groceries), the door may be free of even this small form of communication: it may "sense" that the person wants it to disappear merely by "noticing" that the person is walking in its direction.

The fact that object-awareness is the acceptable, expectable, and uncelebrated condition of civilization, while object-unawareness is the unusual and unacceptable condition is stressed here because it is possible to forget that when one encounters an object in a legal proceeding, one will be encountering it only in its aberrant condition. The brass-knobbed door whose magically correct sense of timing seems "sensitive" to human sentience will never turn up in court; the door that merely fails to be fully sensitive to sentience (e.g., blows open whenever it rains) will never turn up in court but will instead be endured, repaired, or replaced; it is only the door that by pinching produced blood-poisoning, or the door that let a three-year-old walk into a dangerous boiler room, that may end up there, and if it ends up there, the jury may decide that the door *should have known better* or, alternatively, that the manufacturer should have made the object to be an object that *knew better*. Although our cultural expectations about artifacts are visible in the wording of legal statutes (e.g., Plato), and more clearly visible in philosophic meditations on the moral psychology underlying such statutes (e.g., Holmes), they are most clearly exposed in the structure of the product liability trial itself. Such trials are characterized by three major attributes, the third of which is most important to the present discussion.[23]

First, though such a trial is often astonishing in the range of subjects it comes to include, that diverse subject matter will be organized around the skeletal structure of a discrete action, the path of an accident, a sequence of occurrences that (for the plaintiff) carried the whole world from being normal to being abnormal. Here is such a sequence: a man is standing and an eleven-year-old girl is sitting in the kitchen of their home which is attached to a second room that serves as a grocery store; the man moves away from the sink, walks to the stove, tries to light it, fails, tries again to light it, and there is an explosion (or, in the Sicilian-English of the man that was to echo throughout *Foresta* v. *Philadelphia Gas Works*,[24] "explosione"). One may accurately say that this sequence of actions has the same structural centrality to the trial that a plot (which Aristotle identified as the soul of drama) has to a play, except that it differs from a plot in the following important ways.[25]

While the duration of a play (e.g., three hours) is much briefer than the duration of the action it represents (the plot may extend over several years and is ordinarily not shorter than twenty-four hours), the duration of a trial (lasting between several days and several weeks) is much longer than the action it represents: the real-life duration of the action cited above was approximately forty-five seconds, and the generic "product liability" story is often one encompassing an action of between fifteen and ninety seconds. Although the relatively short play contains a relatively long story, within the play itself the two are made commensurate: the plot of Oedipus Rex begins and ends exactly when the play *Oedipus Rex* begins and ends. Similarly, the relatively long trial and its relatively short story will be made commensurate: the "path of the accident" will be enduringly present from the trial's opening through its closing days. But rather than (as in the play) being told once, it will be told ten, or forty, or two hundred times, sometimes in its forty-five second entirety, and other times in one of its ten-second or five-second subunits. The lawyer for the plaintiffs will introduce it in some form in the opening argument. One of the plaintiffs may be asked, by either or both lawyers, to tell the story in its entirety. The defense attorney may later ask, "The first time you tried to light the stove, Mr. Foresta, did you notice . . . ," and now for the next forty minutes the jury will be suspended in contemplation of the interior intricacies of that ten-second interval. Emergency hospital workers, though not present in the home, will find themselves retelling the story, for among the many things they will be asked is the question, "What were you told when the man and the child were first brought to the hospital?" Nor, once we enter the days of medical testimony, is it only the final unit of the story, "explosion," which is held before the jury's eyes, for the injured bodies will themselves bear the record of earlier intervals. Although, for example, three-quarters of the surface of the girl's body has been burned, they will hear that there is a discrete narrow white band of healthy tissue across part of her torso where the metal back of the chair intervened between stove and child: the jury is transported back to and recalls the beginning of the story; with new clarity and concreteness they understand the opening sentence, "a man is standing and an eleven-year-old girl is sitting . . ." So, too, during the many days when the gas itself, the stove, and the plumbing fixtures are described and assessed (the suit has been brought against PGW, but PGW has brought suit against Roper and Mars who are for a time co-defendants), the story will re-emerge many times. When a fire expert, for example, is called upon to deduce the cause of the explosion, he must, in order to make the deduction, verbally reconstruct the story and its timing in its entirety once more. The story must be told, and retold, and retold, because only by entering it countless times and from countless directions does the jury learn what it must learn: was someone hurt (but this is not all), was there a defective product (but this is not all), most important for the legal question that must be answered, given the first two, is it the case that the second is the proximate cause of the first; did the two meet on "the path of the accident," did the two meet at "the crossroad of the catastrophe"?

This is, of course, the most crucial difference between the "unifying plot action" of a play and the "unifying plot action" of the trial. The action of the first is complete and cannot be altered; its audience must passively bear it. The action of the trial is incomplete and can be mimetically altered; its audience, the jury, is empowered to in some sense reverse it, and it is *only* because this possibility exists that the story is being retold. That is, the audience of *Oedipus Rex* or *Hamlet* can only mentally reverse it: they will be engaged in the counterfactual wish, let Oedipus not move down that road, let him not marry the Queen, let Polonius this time not be behind the curtain, let Hamlet at least not act moronic to Ophelia.[26] But the trial audience, the jury, is there to "make-real" what the audience of a play can ordinarily only "make-up."

The overall skeletal action may be summarized in this way. In the generic plot of a liability trial, the world has slipped from the ordinary to the extraordinary by the short path of a passive and unfathomable slippage that is resurrected into recoverable intelligibility by being subdivided into a sequence of discrete actions (standing, sitting, lighting, failing, not smelling gas, lighting again, exploding, and so forth). Implicit in this mimesis of restorability is the belief that catastrophes are themselves (not simply narratively but actually) reconstructable, the belief that the world can exist, usually does exist, should in this instance have existed, and may in this instance be "remakable" to exist, without such slippage. This belief in the counterfactual is on one level shared by everyone present in the courtroom, all of whom, by their participation in a civilization that conducts such trials, credit the possibility that this *may*, in this particular case, be the appropriate legal outcome. But the varying populations within the courtroom are also differentiated by their varying relations to the counterfactual. That is, everyone (whether present for the defense or the plaintiff) will—like the audience in a play—have the *passive wish* that what is so were otherwise: no one hearing the story twenty-one times will, as they sense it about to resurface in its twenty-second iteration, be empty of the thought, "let it not be so," "let her this time not have been so burned." But it is the very particular burden of the plaintiff's counsel to raise that collective passive wish into an objectifiable form, or object form, by showing that it is in the nature of responsible human making or manufacture that this need not have been so: while the wish, "let her not have been hurt," may be translated into, and may float aimlessly among, many other equally passive wishes (let her not be in the kitchen, let them not have moved to the United States, let Mr. Foresta not have a daughter, let the man not move toward the stove), the plaintiff's lawyer must show that the only site of an actual reversal is the artifact, and the only sensible wish, "let the gas not be defective." Finally, it is the particular burden of the jury (one not shared by anyone else in the courtroom) to determine whether it is legally appropriate *to further* objectify the counterfactual by "bringing in" (that is, bringing into the about-to-be completed action of the trial) a verdict for the plaintiff, *to further* make-real or materialize the counterfactual by endowing it with the material form of monetary compensation.

That "the making real of the counterfactual" is centrally at issue in the legal contest and differentiates the defense and plaintiff positions becomes most overt in the closing arguments. The lawyer for the defense will often in such a case attempt to persuade the jury that they are powerless in this regard[27] by saying some version of the following statement: "A terrible accident has happened; we all wish it weren't so; but there is nothing anyone can do that will change the fact that it happened." The lawyer for the plaintiff, in contrast, will often take great care to remind the jury that they indeed have at this moment a very special power ("I try to give jurors a feeling of royalty," explains one of New York City's leading plaintiff lawyers[28]), that some of the remaining body damage can be reversed and undone by medical care, that the problems of medical costs can be reversed by being paid for, that the problems of being out of work can be reversed or diminished by being paid for, that even the objectlessness of acute suffering can in some sense be mimetically reversed by a more bountiful object world,[29] that, in effect, the first two hundred recitations of the story they have heard can be displaced by a two-hundred-and-first recitation in which the story of the failure of artifice can be displaced by a story about the medically and psychologically curative strategies of artifice. If the jury brings in a verdict for the plaintiff, and if they then bring in a compensatory award,[30] they by their first action announce that the story they have heard is the story of object-irresponsibility ("product liability"), and they by their second action act to convert the story about object-irresponsibilty into a story about object-responsibility.

Even in what Holmes identifies as the earlier strategy of "revenge" (against the offending object) rather than "compensation" (for the offending object), there is a mimesis of counterfactual reversal. To kick the door a split second after it has inflicted pain is to immediately change the location of hurt from its human victim to its cause, and thus is to (however ineffectively) mimetically undo or reverse the path of the prior action. Compensation, though again only a mimetic rather than an actual undoing, comes closer to actualizing it, for it quite literally allows the external environment of the hurt person to be reconstructed into one where objects relieve rather than amplify the problems of sentience. This outcome is clearer if one moves from the first attribute of such a trial, its unifying action or story, to its second and third major attributes, the human and inhuman characters in the story.

As noted earlier,the range and complexity of information brought before the jury is often very great, and only the comparative simplicity of the gradually clarified story line works to control and contain that information. During the course of this trial, for example, the jury will be educated about many different institutions: they will mentally enter into the interior of two different Philadelphia hospitals, one Philadelphia courtroom, and three different businesses. They will learn about the complex construction of one particular stove, of one brand of stove, of American stoves in general, and of Sicilian stoves, as they will learn also about the complex construction of a bed that can suspend the body in the air while only touching a small portion of its surface. They will learn about the difficulties nurses have working on a hospital floor where there is every day a child crying; they will learn about first-generation immigrant employment; they will learn why Lady Justice is blindfolded; they will learn about the path of gas through underground pipes and into homes; and they will learn about the opportunism of microorganisms toward a body that is missing its protective skin (let Janice not have been hurt). They will come to understand the difference between the "beyond a reasonable doubt" of a criminal suit and the "tipping of the scales" of a civil suit; they will come to understand that there is a legal distinction between "a service" and "a product," as they will also come to understand whether gas in particular is a service or a product. They will hear precise descriptions of intricate feats of surgery that have already occurred, and of many more that are going to occur; they will hear descriptions of the medical difference between being severely burned on parts of your body (the man) and being severely burned over a great deal of your body (the girl); and they will hear descriptions of the special medical difficulty of repairing tissue that is covering bones that are still growing (let her not have been hurt). They will learn about clear arguments, about unclear arguments, about court interruptions, and about the way the presence of co-defendants changes the question from "is some object responsible" to "which of the three objects is responsible." They will learn that gas is itself odorless and that gas is routinely odorized to signal a person that it is leaking; they will learn that small gas leaks are benign and omnipresent; they will learn that gas cannot be *so* odorized as to call attention to itself when it is leaking a little but must be odorized *enough* so that it will call attention to itself if it is leaking a lot (let the gas not be leaking a lot; let the gas that is leaking a lot be odorized enough to be noticeable). They will learn a great deal about the role of judge, and about this particular judge; they will see that it is the way of a judge to allow different versions of different facts, as they will also see that out of hundreds of facts there comes to be one privileged fact about which this judge will tolerate no second version.[31] They will learn that once human flesh has healed, a black rubberized suit that is worn over the head and body by day and by night can eventually reduce two-inch thick scar tissue to half-inch thick scar tissue (let Janice not have been hurt), as they will also learn what Philadelphia school children say about a scarred body and a black rubber suit. They will learn about a mechanical device by which the gas company monitors the amount of "odor" in gas (let the gas have been odorized), as they will also learn that because of the imprecision of such instruments, the gas company has employees that personally conduct what is called a "sniff test" (let the gas have been odorized); they will learn that a "sniff test" happens twelve times a day every day throughout the year in random homes in every neighborhood of the city, as they will also learn that no "sniff test" had been regularly conducted anywhere in this particular neighborhood for nine days preceding the explosion (let there have been a sniff test, let them not have been hurt).

The jury will learn these things, and many more things, and in much greater detail than can be recited here; but as this list makes clear, there are, in the midst of so much complexity, only two real subjects, the nature of the human body and the nature of artifice, the ease with which "hurting" occurs and the responsibility with which "making" must therefore occur.

As the conflict of a play requires a protagonist and an antagonist, so the contest of the trial requires a plaintiff and a defendant. But though in a product liability suit the structural positions of plaintiff and defendant can be named by many different names (the names of the attorneys, the names of the persons and the companies they respectively represent), the two that will at every moment stand side-by-side are the human body and the artifact: the man, the girl, and the gas, as elsewhere a woman named Sophie and her industrial cleaning cart, as elsewhere a boy who climbs chestnut trees and an electric cable, as elsewhere a man who installs compressors and the compressor, as elsewhere other individuals and other objects, washing machine, forklift, soft-drink bottle, grain hopper, or headrest. As in other contexts of deconstructed making encountered at earlier moments in this book, the juxtaposed extremes of body and construct signal a radical dislocation in the structure of creation. Two characteristics of this distress-filled drama of object responsibility are apparent from the miscellaneous catalogue of courtroom subjects enumerated above.

First, though it was earlier stressed that in a court case we are encountering cultural expectations about objects in their unordinary because unfulfilled form, it is also true that even within the court case "civilization" has entered in its benign and ordinary form. The plaintiff's attorney need not remind the jurors what an artifact ordinarily is, for what it ordinarily "is" is everywhere before their eyes. The allegedly defective gas is not a solitary "product" in an empty room: it is surrounded by verbal and material artifacts, stoves, schools, neighborhoods, legal arguments, special beds, philosophic categories like "stoicism," psychological rubrics to explain how young adolescents perceive the world, the image of Lady Justice, rubber body suits, company structures, underground pipelines, skilled acts of surgery, the chairs on which they are sitting, the institutionalized role of judge. That made things *ordinarily* exist *on behalf of* sentient persons need not be overtly called attention to, for the courtroom itself is benignly cluttered with "evidence" of the ordinary. Further, though there is a dispute taking place, the dispute is not about whether made things ought to accommodate sentience: the defense attorneys do not argue that made things ought not to do so, nor that they ought not to be *expected* to do so: they assume that objects should (at least up to a certain point[32]) do so, and argue that this particular object did fulfill its responsibilities, though they will allow that there may have been some other object in the kitchen that did not do so (and, importantly, *should* have done so). So, too, the defendant gas company itself makes manifest the shared norm of civilization at large, for in its worries about too much and too little odor, its odometers, its titrologs, and its usual procedure for the twelve-times-a-day-every-day-every-neighborhood "sniff test," it demonstrates its own assumptions about the level of responsible awareness that must be built into the design of a particular product. The sometimes bitter antagonism, contradictions, interruptions, arguments and counterarguments are, then, all taking place within the frame of shared object expectation.

Second, what becomes clear in the legal contemplation of objects is not simply that they must internalize within their design an active "awareness" of human beings but that this "awareness" is not limited to, or coextensive with, their use. Up to this point, the fact that objects are mimetic of sentient awareness has been articulated only in terms of the specific sentient problem the object exists to eliminate: the chair must "know" about the problem of weight; the lightbulb must "know" about the problem of seeing at night. But it here becomes noticeable that artifacts must know a great deal more about their human makers than the particular needs they accommodate: while the gas must know (that is, must be itself a material registration of the awareness) about the problem of being cold and the problem of allowing raw objects to enter the body, it must know other things as well; it must "know" that it is in its original state unsmellable by human beings; it must "know" that it is, when unsmellable, dangerous to those human beings. The miming of sentient awareness is made more pronounced by the fact that it is often the knowledge of some sensory attribute of persons that the object is legally expected to have, and by means of which it is expected "to communicate" with, or announce its presence, to persons. As in this trial the "smellability" of gas became a central issue, so in other cases it is sometimes the seeability or the audibility of the object that becomes crucial, even though the object was not primarily invented to assist vision or audition.[33] The object may even be required to communicate verbally with persons by bearing a written label. A stepladder, for example, not only "knows" (incorporates into its design the knowledge that) human beings are shorter than they often need to be, but also "knows" that human beings tend to overstep themselves when lost in trying to be taller than they are: the top step may bear the words, "Do not step onto this step" (i.e., "I know that you will fall, even if you do not know that at this moment"). An object must be *self*-aware: its design must not only anticipate how it will be used (and even, how it might be oddly used) but how it will be installed and eventually removed. The Beloit manufacturers of compressors will genuinely merit our deepest respect when they demonstrate the many blueprints of the object's interior that show the way the makers made it to be both useful and safely useable; but when the plaintiff's lawyer asks to see the blueprint of the precautionary design for the weld and brace that would bear the weight of the person installing the compressor, and there is none, we may guess that these earnest craftsmen will eventually lose this case and that, among other things, they will have to return to and supplement their already great labor of design research.[34]

The frequency of suits in the United States has led some observers to identify us as the most litigious of societies, and it is with good reason that this widespread habit of "legal action" is so often lamented. Though there are many cases in which someone has been terribly hurt, there are others in which the plaintiff has not been hurt, or has been hurt but not by the defendant product, and sometimes the person bringing suit has such a history of suits that the trial comes to seem a deeply unpleasant way of attempting to raise money. American jurors have, however, repeatedly shown themselves to be skilled at distinguishing among these different kinds of suits: the jury system is itself an artifact that exists to allow society to have the benefits of appropriate legal action while protecting it from conscious or unconscious misuse of such an action. The cultural habit of suing, though perhaps partly anchored in the contemporary psychology of "blaming,"[35] can also be understood in other ways. When large awards are brought in, they almost always go to individuals who have been severely hurt; further, in the majority of suits, the defendant is a large entity such as a company (and it is not the defendant manufacturing company but its usually even larger insurance company that pays the award).[36] For this reason, the United States courts themselves have observed that such suits must be understood not only in the idiom of *legal action* but in the idiom of *economic redistribution*: the extreme costs to individuals of living in a complex industrial society are redistributed to that sector of society that can absorb those costs (or, as it might be reformulated here, the extreme vulnerability of sentience is projected out onto the object world of the company structure where losses and gains will be registered in profit fluctuation rather than alterations in embodied consciousness).[37] If the present litigious era is someday in the far future looked back upon, the account given of our legal reflex will probably not be entirely negative, for it will no doubt be remarked that the century of unprecedented making and manufacture was also a century of unprecedented speculation about the ethical responsibilities inherent in the act of manufacture, the act of making, the act of creating. For this reason, the product liability trial (which will one day be turned to for its cultural revelations the way we now gaze back on the Greek stage) has been presented here less as a *legal action* or a form of *economic redistribution* than as a form of *cultural self-dramatization*: the courtroom is a communal arena in which civilization's ongoing expectations about objects are overtly (and sometimes noisily) announced; the trial does not occasion the expectation but merely occasions the objectification of the expectation; and though it may be itself a concussive and exceptional occurrence, it only makes audible what is actually a very quiet, very widely shared, very deep, and in its own way quite magnificent intuition about the nature of creation.

Thus, after looking at objects in a legal context, one can return to objects in nonlegal contexts and see them more clearly. Everyday artifacts (which may never have been the subject of litigation nor even consumer pressure) are themselves usually characterized by forms of materialized awareness that go far beyond their most immediate use: the door to the boiler room that includes in its design a childproof latch is not only able to ''understand'' and accommodate the timing of the person's erratic wish that it be now-a-wall-now-not-a-wall, but is also able to ''differentiate'' small persons from persons in general, and ''knows'' that the former is a special subcategory of the latter whose wishes should not be accommodated. Sometimes in a technological and automated society, the mimesis of sentient awareness may become so elaborate that the object may become frightening: the computer has startled and disturbed one generation of adults, though the offspring of those adults seem to perceive the computer as perfectly consistent with, rather than disruptive of, the ordinary external object world. Computers differ only in the elaborateness rather than in the fact of mimesis, and they are not singular even in their elaborateness. Novels, for example, produce this same inanimate fiction of speaking, feeling, thinking, and are perhaps less startling and suspect only because we have lived with them a much longer time. What is (to return to an earlier subject) peculiar about the charge of "pathetic fallacy" is that it is only invoked by a literary commentator if the artist has made a tree speak, but is not invoked in the more extreme and (for an artist) habitual act of making a nonexistent presence (Catherine, Tess, Anna) speak, and speak with such complexity and palpable sentience. What is of still more importance to notice, however, is that the apparent knowingness of the computer (which is the projected knowingness of both its hardware and software designers) and the apparent knowingness of Tess (which is the projected knowingness of her maker) are themselves only radical versions of the apparent knowingness that surrounds us everywhere in the recreated external world (and that is again the projected knowingness of its collective makers). What is it that this aspirin bottle—with its long history in the bark of the willow tree and the bowl of the Indian peacepipe—"knows" about the human world? It knows about the chemical and neuronal structure of small aches and pains, and about the human desire to be free of those aches and pains. It knows the size of the hand that will reach out to relieve those aches and pains. It knows that it is itself dangerous to those human beings if taken in large doses. It knows that these human beings know how to read and communicates with them on the subject of amounts through language. It also knows that some human beings do not yet know how to read or read only a different language. It deals with this problem by further knowing how human beings intuitively and habitually take caps off bottles, and by being itself counterintuitive in its own cap. Thus only someone who knows how to read (or who knows someone else who knows how to read) can take off the cap and successfully reach the aspirin which, because the person not only knows how to read but has been made to stop and be reminded to read, will be taken in the right dosage. It contains within its design a test for helping to ensure responsible usage that has all the elegance of a simple three-step mathematical proof. Civilization restructures the naturally existing external environment to be laden with humane awareness, and when a given object is empty of such awareness, we routinely request that the garbage collector (himself a direct emissary of the platonic realm of ideal civilization) carry it away to the frontier, beyond the gates of the beloved city.

The aspirin bottle with its counterintuitive lid has been chosen as a final representative artifact in order to recall and underscore the fact that it is the work of the object realm to diminish the aversiveness of sentience, not to diminish sentience itself. The mental, verbal, and material objects of civilization collectively work to vastly extend the powers of sentience, not only by magnifying the range and acuity of the senses but by endowing consciousness with a complexity and large-mindedness that would be impossible if persons were forever engulfed in problematic contingencies of the body.

It is not objects but human beings who require champions, but the realm of objects has been briefly celebrated here because they are themselves, however modestly, the champions of human beings. The interior structure of the artifact is being attended to in this discussion for two reasons. First, human indifference to other persons is often explained and implicitly excused by pointing out that those who are indifferent are absorbed by their material wealth. But it is a deconstruction of the very nature of material wealth to permit, let alone excuse, this inattention. We sustain this deconstruction by simultaneously surrounding ourselves with material objects in everyday life while philosophically divesting ourselves of them, verbally dismissing and discrediting the importance of the material realm. This act of philosophic divestiture does not work to diminish or even regulate our own desire for objects but only works to permit us to be free of worrying about the objectlessness of other persons. If we cling to objects, we should trust our own clinging impulse; and once we trust that impulse we will acknowledge that such objects are precious; and once we confess that they are precious we will begin to articulate *why* they are precious; and once we articulate why they are precious, it will be self-evident why our desire for them must be regulated and why their benefits must be equitably distributed throughout the world. It is by crediting them that we will reach the insight that we only pretend to reach when we discredit them.

Second, it is assumed here that the project of understanding the nature of human responsibility will be assisted by coming to understand the human imagination. But the action of the imagination is mysterious, invisible, and only disclosed in the material and verbal residues she leaves behind. The interior structure of the object has been attended to because it contains the material record of the interior of this invisible action. Thus it is the work of the imagination (rather than the object) to make the inanimate world animate-like, to make the world outside the body as responsible as if it were not oblivious to sentience. This is only one attribute in a composite portrait that contains many, many attributes, and which can only be uncovered in piecemeal fashion. This solitary attribute carries with it two others. The imagination is not, as has often been wrongly suggested, amoral: though she is certainly indifferent to many subjects that have in one era or another been designated "moral," the realm of her labor is centrally bound up with the elementary moral distinction between hurting and not hurting; she is simply, centrally, and indefatigably at work on behalf of sentience, eliminating its aversiveness and extending its acuity in forms as abundant, extravagantly variable, and startlingly unexpected as her ethical strictness is monotonous and narrowly consistent. The work of the imagination also overlaps with another interior human event that is usually articulated in a separate vocabulary, for it has become evident that at least at a certain moment in her life cycle, she is mixed up with (is in fact almost indistinguishable from) the phenomenon of compassion, and only differs from compassion in that in her maturer form she grows tired of the passivity of wishful thinking. These attributes are surrounded by many others, a small number of which will emerge in the following section, which returns to the interior of the object world.

II. The Artifact as Lever: Reciprocation Exceeds Projection

A perception about human sentience is, through labor, projected into the free-standing artifact (chair, coat, poem, telescope, medical vaccine), and in turn the artifact refers back to human sentience, either directly extending its powers and acuity (poem, telescope) or indirectly extending its powers and acuity by eliminating its aversiveness (chair, vaccine). The first has no meaning without the second: the human act of projection assumes the artifact's consequent act of reciprocation. In the attempt to understand making, attention cannot stop at the object (the coat, the poem), for *the object is only a fulcrum or lever across which the force of creation moves back onto the human site* and remakes the makers. The woman making the coat, for example, has no interest in making a coat per se but in making someone warm: her skilled attention to threads, materials, seams, linings are all objectifications of the fact that she is at work to remake human tissue to be free of the problem of being cold. She could do this by putting her arms around the shivering person (or by hugging her own body if it is her own warmth on behalf of which she works), but she instead more successfully accomplishes her goal by indirection—by making the freestanding object which then remakes the human site that is her actual object. So, too, the poet projects the private acuities of sentience into the sharable, because objectified, poem, which exists not for its own sake but to be read: its power now moves back from the object realm to the human realm where sentience itself is remade. We every day speak of reading the works of Sappho, Shakespeare, Keats, Brontë, Tolstoy, Yeats, as though by doing so we gain some of the "sensitivity" and "perceptual acuity" projected there; people even announce that they are reading Keats, for example, as though this makes them Keats-like, which is in some sense accurate. Like the coatmaker, the poet is working not to make the artifact (which is just the midpoint in the total action), but to remake human sentience; by means of the poem, he or she enters into and in some way alters the alive percipience of other persons.

Projection and reciprocation are (except in deconstructed making) so entailed in one another that one can rarely be speaking of one without simultaneously speaking of the other. In this section, however, they will be spoken of separately so that their interaction can be more clearly apprehended. The action of the one may have attributes not shared by the other. Furthermore, when a given artifact is undergoing successive revisions, it may be something about the nature of projection that is being revised or instead something about reciprocation. For example, if the woman finds a better needle or better window light, this may ease the action of projection without altering (either amplifying or diminishing) the action of reciprocation (the second coat, though more easily made, may be only equally good at warming as the first). Alternatively, a new synthesis of natural and artificial fibers may amplify the action of reciprocation (bringing greater warmth to the wearer) without altering in any way (either making more difficult or more easy) the nature of the projective action of coatmaking. It will also often happen that what affects one will affect the other: if her new needle makes the projective act easier, she may decide to make coats for two neighbors; thus the object's power to remake sentience (or, more accurately, the woman's power to reach beyond the object and act indirectly on human sentience) has been amplified even though each of the two coats in isolation brings the same warmth as the earlier coat made with the original needle. The sites of projection and reciprocation are, then, conceptually distinguishable even though the actions themselves are inseparable. This is true whether the made object is a coat, a god, a poem, a marriage oath, a vaccine, or a crop of corn.

Although the made objects that will be introduced as illustrations in this section are solitary and concrete artifacts, and therefore only very fragmentary pieces of the civilization at large (a single coat, a single poem), what is exposed about the two counterpart actions *may* be equally descriptive of a large assembly of objects such as exists in a library, a philosophic tradition, or a marketplace, or even of an immense and collective artifact such as a nation-state. It may therefore be helpful to suggest, very briefly, the way the sites of projection and reciprocation inhere in a large artifact like the nation-state before turning back to the realm of diminutive objects.

During the period of the Carter administration, there was in the United States a great deal of attention to the nature of human rights in other countries. One result of this attention was that people came to understand that a particular right present in one country might be absent in a second country, but that that second country might itself have successfully established a different right, not yet emphasized in the first country. In other words, more appropriate than the question, Does this particular country have this particular right?, is the question, What is the overall pattern of rights within the given country and how much is presently being done to supplement those existing rights with additional rights formerly absent?[38] This second way of perceiving the question is especially relevant in understanding relations between the United States and the Soviet Union, since comparative legal and political analysts have shown that the United States Constitution emphasizes certain "procedural rights" (right to vote, freedom of speech, free press, right of assembly), while the Soviet Constitution emphasizes "substantive rights" (the right to eat, the right to a job, the right to medical care, the right of education); further, such analyses have shown that the recent extension of rights in the United States has come to include greater attention to substantive rights (the "right to work," for example, is a relatively young concept in this country), while there is evidence that Russia has begun to attempt to include a greater attention to procedural rights. This distinction between *procedural* (U.S.A.) and *substantive* (U.S.S.R.) rights has sometimes been formulated as a difference between "civil and political rights" on the one hand and "social and economic rights" on the other; it has also often been formulated as a difference between "individual rights" and "collective rights."[39]

Within the context of the present discussion, however, it is appropriate to notice that this difference can also be formulated in terms of the distinction between the sites of projection and reciprocation. The individual, procedural, civil rights of the United States all attempt to protect the site of projection: the individuals' autonomy over their own participation in the collective artifact (the state) or, more precisely, the individuals' power to determine what kind of political artifact they will collectively project, is ensured by the rights of assembly, voting, free press, and so forth. Through these procedures, the nature of the artifact (the nation) is itself held continually open to revision and modification. The artifact's action of reciprocation (its ability to feed them, clothe them, cure them) is, of course, greatly influenced by the projective act itself, greatly determined by the particular polis that each generation creates. But there has not until recently existed a separate set of rights explicitly directed toward the action of reciprocation: the reciprocating powers of the made-world have been allowed to "fall where they may"; only relatively new social legislation has worked to guarantee that the disembodying powers of the artifact be equitably distributed, that, for example, a minimum level of medical care not be available to one sector of the population and unavailable to another. Conversely, the collective, substantive, social and economic rights of the Soviet Union have been explicitly directed at the site of reciprocation, for they work to ensure that whether the beneficent disembodying powers of the artifact are very great or very small, they are in any event equally distributed across the population. *What the state is* (projection) is, comparatively speaking, out of the hands of the citizens, but *whatever the state is*, its benefits (reciprocation) are in the hands of all its citizens. Thus the constitution of one country stresses rights that are conceptually and chronologically prior to the made object (the polis) while the constitution of the other country stresses rights that are posterior to the made object (polis).[40] The often repeated observation that Marx conceived of his strategies for socialism as taking place in Britain rather than Russia is important for just this reason: he emphasized the rights of reciprocation because he was imagining their being introduced into a country where the rights of projection were already in place, or on their way to being in place. This brief example has been contemplated here only to illustrate the way in which the attribute of the diminutive artifact may be equally characteristic of a vast one: like a coat, a nation-state is an intermediate object between the inseparable (however conceptually distinguishable) actions of projection and reciprocation.

The artifact as a material and visible locus between two actions—or, more precisely, the artifact as the materialized site at which the human action of creating now moves back on the human creators themselves—can be more clearly apprehended if the object is seen as one in a series of three: weapon, tool, artifact. The modification of weapon into tool was at earlier points (Chapters 3 and 4) presented as the alteration of a two-ended object into an object with only one end: what in the weapon are the double sites of pain and power become in the tool a single site where sentience and authority (the changed vocabulary reflects the unification, for pain made active and self-objectifying is more accurately called "sentience" and power made sentiently aware and therefore responsible is more accurately called "authority") occur together at one end and together act on a nonsentient surface.[41] Although this change is a very great one, and although it is certainly one that "literally" occurs, to summarize that change in terms of the new object now "having only one end" is itself a metaphorical description, since the object (hammer, hoe) is of course still a two-ended object. But the near-literalness of the summarizing description becomes more evident in the subsequent transition from a tool to an artifact, for the artifact (which is itself like the transitional few inches between the handle and head of the hammer or between the handle and blade of the hoe) has within its material form no ends (e.g., chair) or at most only a residual record of ends (e.g., lamp), and only has for its end points the *single site* of the human beings out of whom it came into being and back toward whom it now moves. It is as though hammer and hoe have been bent in the middle, and now any action introduced at one end arcs back on to the very site out of which the action arose.

Thus the artifact is in this section called a "lever" or "fulcrum" in order to underscore that it is itself only a midpoint in a total action: the act of human creating includes both the creating of the object and the object's recreating of the human being, and it is only because of the second that the first is undertaken: that "recreating" action is accomplished by the human makers and must be included in any account of the phenomenon of making. What the human maker projects into the made object may change from object to object (as a counterfactual perception about seeing is projected into the telescope while a counterfactual perception about skin is projected into the bandage), but what he or she will have always projected there is the power of creating itself: the object (coat, telescope, bandage) is invested with the power of creating and exists only to complete this task of recreating us (making us warm, extending vision, replacing absent skin with a present skin). It is precisely because objects routinely act to recreate us that the confusion (encountered in Chapter 4) arises in which the object is seen as a *freestanding* creator. Though, for example, human beings are themselves the creators of the Artifact (God), God now comes to be perceived as the creator of human beings; and, of course, the Object *is* their creator, for by making this Artifact they have recreated themselves, altered themselves profoundly and drastically. There would be no point in inventing a god if it did not in turn reinvent its makers: all that is untrue is that the power of recreation originates in the object, a mistake that occurs by attending only to the second half of the total arc of action. Again, Marx has described the way in which in British capitalism, the women and men who make commodities, money, and capital come to be wrongly perceived as themselves commodities made by the capitalist system. But this designation of people as commodities is a pernicious misinterpretation of a phenomenon that is on another level accurate: as Marx himself sees, made objects exist to remake human beings to be warm, healthy, rested, acutely conscious, large-minded (but not to remake them into commodities, which they only appear to do if the economic system is seen as a freestanding object that is itself uncreated, and this misinterpretation may be used to excuse the fact that the object world does not feed, clothe, and warm them). The conception that artifacts create people is right. The conception that that creative power originates in the artifact is wrong. Only the second half of the total arc of action is being seen.

This phenomenon is complicated by the fact that in many situations it is advantageous to eclipse from attention the first half of the arcing action. For example, a god can much better work to recreate its people if its ability to recreate them is not recognized as only an extension of their own projective actions—if, that is, the god is not recognized as a fiction or made thing. In fact, in general it is the case that when an artifact goes from what has been called here the "made-up" state to the "made-real" stage, one way in which this is done is by eclipsing or erasing the first half of the arcing action. This is why artifacts that are purposely allowed to remain in the made-up stage—artifacts that are not only permitted but intended to be recognizable as fictitious—have pronounced signatures attached to them, signatures assuring that the first half of the arcing action will be remembered, whereas artifacts that are not intended to be self-announcingly fictitious usually have no such signature attached to them. When "Ode to Autumn" acts on us, we know that it is actually John Keats who acts on us, whereas when with the same coming of autumn our coats begin to act on us, we do not overtly recognize that it is actually Mildred Keats (or any other specified coatmaker) who has reached out through the caritas of anonymous labor to make us warm.

Although the issue of signatures is a very complex one, in general one can say that at the moment an artifact is performing its reciprocating action, we are aware of the chronologically prior act of projection to varying degrees, and this "varying degree" depends on what the object was invented to do. If the invented object can only perform its task by seeming to have an ontological status, or degree of reality, *greater* than human beings themselves, then it will be important that the earlier arc of action be not only *unrecognizable* but even *unrecoverable*. Clearly, a god falls into this category, as would a divine right monarchy whose state laws must be accepted as though a naturally (or supernaturally) existing given of the world. Such objects will therefore be devoid not only of any personal signatures but also of the general human signature that tells us they are man-made. The object must have no seams or cutting marks that record and announce its human origins. The second and by far the largest category of objects contains artifacts (both material and verbal) that work by seeming real, or by interacting with persons without any question of their realness or unrealness: they (clothes, language) do not have a reality greater than or even nearly equal to human beings, but they are not so unreal as to be immediately recognizable as "made"; they are not, as it were, framed by their fictionality. While (like the objects in the first category) they are not on a day-to-day basis *recognizable* as invented, their inventedness is not (unlike the objects in the first category) *unrecoverable*. That is, as one maneuvers each day through the realm of tablecloths, dishes, potted plants, ideological structures, automobiles, newspapers, ideas about families, streetlights, language, city parks, one does not at each moment actively perceive the objects as humanly made; but if one for any reason stops and thinks about their origins, one can with varying degrees of ease recover the fact that they all have human makers, and this recognition will not jeopardize their usefulness. Though these objects (like those in the first category) usually have no *personal* signatures affixed to them, they will (unlike those in the first category) have a *general* human signature. Though the name of Mildred Keats will not be in the coat, there may be a label on the seam or the interior of the collar that says ILGWU, and even when there is no label, the seams and collar themselves will, on inspection, announce that the coat has a human maker. Thus, even had we never discovered the group signatures on the casing-blocks of the Meidum pyramid—"Stepped Pyramid Gang," "Boat Gang," "Vigorous Gang," "Sceptre Gang," "Enduring Gang," "North Gang," "South Gang"[42]—the seams and materials of the pyramid themselves announce the agony of human labor entailed in their construction. The individual person who is one of the life-risking builders of the Golden Gate Bridge will, as he crosses that object fifty years later, think to himself, "I've got my fingerprints all over that iron";[43] the rest of us, periodically struck with the recognition that this dazzling object is "made," will see the fingerprints too, though we will not know to whom they belong. The signature will be general, not specific.

Though different kinds of occasions will prompt the recoverable recognition that such objects are man-made, one of the most common is the moment when the object needs repair, revision, or reinforcement—a moment when its ongoing reality has slipped a little, and thus its fictionality or madeness comes into view. Thus one ordinarily thinks only of the warming (reciprocating) action of the coat, and the prior action of coatmaking comes before the mind only in the season when the seams are torn and the buttons need replacing. We ordinarily use language without contemplating its "madeness" (and such contemplation would intrude on our ability to get through a day of utterances), but when one has an infant in whom the labor of "making" language is beginning, or a friend who has lost language facility because of a small stroke and who must relearn, reform, this capacity, its "madeness" will be strikingly apparent. The citizens in a democracy, like the citizens in a divine right monarchy, will on a day-to-day basis interact with its political and legal structures as though they are a freestanding natural given of the world, but when a problem arises, or when the election booths suddenly reappear on the horizon of daily life, we will be reminded that we have our fingerprints all over this country, that its construction and revisions have a human origin, entail human responsibilities, bear a general human signature.

All of the objects cited here as representative of the second category are ones that habitually exist in a state of "realness" rather than "madeness," but the moment of needed repair calls attention to the fact that they are "made-real," and may also even remind us that before they were "made-real" they were "made-up." That is, each of the objects cited above was not only invested with a freestanding material or verbal form, but was before that mentally invented (if given a material or verbal form at this stage, it will be less substantial, more schematic, than in its made-real stage). Thus the coat not only has a maker but before that, a designer; the Golden Gate Bridge not only had builders but architects; the democracy, too, had designers and architects. What is crucial to notice is that while the "makers" are only recoverably visible in a generalized human signature (ILGWU, Enduring Gang, the voters), the designers are usually known by an individual signature. Though Mildred Keats's name is absent from the coat, in a new season when the design for the coat is first introduced, the designer's name will be announced at fashion shows, mentioned by buyers, and that name may even appear in the coat itself. Though it would be difficult to track down the names of the bridgebuilders, the architects' names will be a matter of public record and more easily accessible. Though the builders of a country are collective and anonymous, the names of the designers of the Declaration of Independence and the Constitution are well known: we even refer to those designers as "signers." This consistent difference between individual signatures at the made-up stage and general human signatures at the made-real stage is sometimes attributed to the fact that the first act is more difficult, entails more unusual talents, or is (rightly or wrongly) for some mysterious reason more greatly honored. But it seems more probable (especially when this distinction within the second category is itself seen within the larger pattern of the three categories) that this simply results from its structural position: at the stage where something is made-up, we allow the presence of an individual signature that reminds us it is "made" (the coat pattern, blueprint, and constitution have no chance of being taken for a coat, a bridge, or a country, so nothing is jeopardized by the signature that confesses its madeness); at the stage, however, where these objects must function as "real" or self-substantiating, they perform this work much more successfully if they are not at every moment confessing their origins as human projections, and thus will have either no signature or an only recoverable, generalized human signature.

As this second and largest category of *real* objects is framed on one side by a very, very small category of objects that are *super-real* (that is, artifacts that only function by seeming to have greater reality and authority than persons), so it is framed on the other side by another very, very small group of objects that are overtly *unreal*, the category of art. While the "madeness" of objects in the first category is both *unrecognized* and *unrecoverable*, and the "madeness" of the objects in the second category is *unrecognized* but, on reflection, *recoverable*, the "madeness" of the objects in the third category is not simply *recoverable* and *recognized* but *self-announcing*. Poems, films, paintings, sonatas are all framed by their fictionality: their made-upness surrounds them and remains available to us on an ongoing basis; though there may be moments when we forget their inventedness, this moment will be as atypical of our interaction with this object as, conversely, remembering the inventedness of the coat is atypical of our interaction with that object. Consequently, while the objects in the first category *have neither a personal nor general human signature*, and the objects in the second *have a general but not* (except in the brief making-up stage) *a personal signature*, the objects in the third *have personal signatures*. In fact, so inseparable from the artifact is the affixed signature that the object will often be named by the signature: pointing to two objects in the room, a person will say,"This is a Millet, and this is a Caro," just as when the person places the needle at the edge of the record he is likely to look at expectant eyes and say, "Mozart."

These three categories are introduced only to underscore the fact that at the moment when an artifact is recreating us, or reciprocating us, or being useful— that is, at the moment when an artifact is performing the second half of the arcing action—whether or not the first half of that action is visible will depend on whether that visibility will interfere with its reciprocating task. That visibility will jeopardize the work of the objects in the first category, will not jeopardize but will interfere with the work of the objects in the second category, and will neither jeopardize nor interfere but will instead assist the work of the objects in the third category (since those objects exist both to celebrate and help us to understand the nature of creating).

When, then, one is standing in the midst of the second half of the arcing action, the visibility of the first half will vary. When, however, one is attending to the first half of that action—that is, when one is attempting to understand the nature of creating—it will always be misleading to look at it in isolation; it can only be understood when seen in conjunction with its second half. It is, again, for this reason that the object is referred to here as a "lever": for regardless of which of the three categories it belongs to, and regardless of whether it is a god, coat, poem, nation-state, bridge, or vaccine, the concrete object will always be only the *midpoint* in the total action. But the descriptive word "lever" is also used here in order to call attention to another major attribute of the overall phenomenon: the midpoint, the discrete object, is also the site of a magnification. As will become evident in the following discussion, the action of reciprocation is ordinarily vastly in excess of the action of projection.

The large discrepancy between the degree of alteration that occurs at the active end of a weapon or a tool, and what occurs at the passive end of weapon or tool was noticed in an earlier chapter. So the change in the position of a finger at one end of a gun may bring about a change from life to death in the body at the other end. The degree of embodied alteration that occurs when a person shoots an arrow is vastly exceeded by the degree of alteration that the arrow in turn brings about: the wound in the tissue of the animal shot is not only a qualitatively more significant alteration than the arm, hand, and back contractions that brought it about but also has a much greater temporal duration. The pull and force of the arms on a rake have an outcome at its other end whose reach and duration exceed the reach and duration of the embodied raking motion: it is not just the vegetation beneath the radius of the arms that has been acted upon, but an area with a larger radius, and the effects of that action will continue to be visible hours and even days after the arms themselves have stopped their action. If one woman presses down on a piece of crumpled linen and a second woman with the same degree of force presses down across an iron onto a piece of linen, the first piece will be virtually unaltered while the second will be transformed into a smooth surface because the intervening tool has magnified her action. Although large changes occur in the transformation of a *weapon* into a *tool*, the phenomenon of magnification is one element that remains constant. Although there are again large changes in the transformation of the *tool* into a freestanding *artifact*, this phenomenon of magnification once more remains constant, for the degree to which the object *disembodies* or recreates the human makers will ordinarily greatly exceed the degree of heightened aversive *embodiment* required by the projective act of creating the object.

If one returns to the woman in the midst of her action of coatmaking, it is clear that her translation of a counterfactual wish ("perceiving her own susceptibility to cold and wishing it gone") into the projective act of labor requires the embodied aversiveness of controlled discomfort. Arms, mind, back, eyes, fingers, will all be concentrated on bringing about a certain outcome: the sustained mental and physical attention to seams, shapes, materials, is itself an interiorized objectification of the original counterfactual wish (which she may not even self-consciously think of at this point—her mind is filled not with thoughts of wind and snow and shivering but with getting this unwieldy edge of material to align itself with this other edge). But while her making of the coat (the first half of the total arc of action) requires a deepened embodiment, the coat's remaking of her (the second half of the total arc of action) will bring about her disembodiment, divesting her body of its vulnerability to external temperatures and therefore also freeing her mind of its absorption with this problem. In the total arc of action, then, she is first more intensively *embodied* (projection) and then *disembodied* (reciprocation); but clearly the level of the second is much greater than that of the first. If the second were the exact equivalent of the first—if the second relieved her of discomfort precisely to the same degree to which she had earlier willfully subjected herself to discomfort—it would have been senseless to make the coat: she might as well have remained wholly passive before her environment. Instead, the work of the second is vastly in excess of the first. The embodied discomfort, exhaustion, and concentration of projection has eliminated not merely discomfort but the possibility of dying when freezing temperatures arrive. One may argue that without the coat, she will not necessarily die, that she can, for example, stay near a fire (itself a made object). But this only leads to a similar conclusion: with the embodied discomfort of coatmaking she has eliminated the enslavement entailed in the necessity of never moving beyond the five-foot radius of the fire.

Even if, therefore, one is juxtaposing one hour of projection against one hour of reciprocation, the second is, in the nature of the alteration brought about, much greater than the first. But the introduction of the temporal element calls attention to a second form of excess: for several weeks of the discomfort of projection, she is reciprocated by fifteen months (i.e., three winters) of freedom from susceptibility to the cold. If the two were temporal equivalents, she might perhaps do just as well to perform the embodied motion of coatmaking (what was in the preceding section called the dance of labor) without making the object, for she could by this method of intense movement stay warm. But these patterned calisthenics would in actuality have to be sustained continuously throughout the season of ice and snow, itself an impossible proposition; and she could not, as she could in the aversiveness of labor, control and regulate the level of aversiveness by choosing when she would rest and when work; the timing of her actions would now be beyond her personal will and wholly dictated by the vagaries of the external world.

There is also a third, immediately apparent form of excess. The object may extend its reciprocating benefits to those wholly exempt form the process of projection. In hours when she sits by the fire, her brother, neighbor, or child can wear the coat; for what originated as a wholly interior counterfactual wish has been objectified into a sharable outcome. If material and verbal artifacts only reciprocated their specific makers rather than human-beings-as-makers-in-general, they would have almost the same absolute privacy as sentience itself. They are instead by nature social.

The fact that the object's reciprocating action includes but is not limited to its maker leads to a fourth form of excess: exchange. If the woman makes a second coat and trades it for the making of a wall (or alternatively, trades it for money with which she then buys a wall), then her projective action of coatmaking has brought her not only warmth but security. Although it seemed at first that she knew how to make one thing, it turns out that this set of concentrated gestures actually enables her to (indirectly) make many objects (a wall, a month's worth of food, an ointment to cure rashes, a pennywhistle), objects that in turn remake her in many different ways (she is now a warm, secure, well-enough-fed, and rash-free music-maker). Just as persons are not locked into the private boundaries of fixed sentient attributes, so made objects are not locked into the concrete boundaries of their sensuous attributes. As the human being may transform herself from a creature forever experiencing herself as vulnerable to the cold, to one primarily experiencing herself as a gifted coat-and-music-maker, so the coat (because of its inherent freedom of reference—that is, because it refers to the temperature instability not only of its maker but of its maker's brother, neighbor, and child) is transformable from an object with one set of sensuous attributes (e.g., soft blue cloth; an irregular two-feet by three-feet trunklike shape; an opening at the bottom and again at the top; movability) to an object with a different set of attributes (e.g., hard regular surfaces; rectangular eight-feet by five-feet shape; no openings; unmovability). Thus when the woman invested her own powers of creating in the object, the object became capable not only of *recreating her* (as well as other persons) but of *recreating itself* to be whatever object the woman in that week most wanted it to become.

Any artifact will ordinarily be characterized by at least one of these four, and usually all four, forms of excess.[44] The most celebrated artifacts of civilization—the isolation and construction of sulphuric acid, a Beethoven sonata, the U.S. Constitution, the smallpox vaccine, Genesis, the telephone, Arabic numerals, and so forth—will, despite the agonized years (and in some cases, several lifetimes) of labor that may have been entailed in the act of projection, be so extravagantly excessive in their referential powers that the calculus of projection and reciprocation will seem almost funny: one may turn away from the contemplation of magnitude with something of the resignation with which one surrenders in the attempt to picture the magnitude of interstellar distance. It is no doubt for this reason that when people study one particularly spectacular instance of invention, they sometimes conclude that the general phenomenon of invention could not possibly originate in the perception of need, for the vast and unanticipatable benefits of the object bear no resemblance to anything conjured up by the narrow word "need." But what is at present most important to notice is this: not only the most celebrated but the most ordinary and routine artifacts are characterized by excess. This is true to such an extent that one may accurately say that an artifact *is* this capacity for excessive reciprocation; what the human being has *made* is not object *x* or *y* but this excessive power of reciprocation. Thus the normative model must be one in which the total arc of action has in its second half a largesse not present in the first half; the total act of creating contains an inherent movement toward self-amplifying generosity.

It is crucial to recall that what is being presented here as a descriptive model is not a model of the relations between persons but of the relation between persons and the realm of made objects. People perform acts on behalf of (and also give objects to) other people from whom they may or may not anticipate any reciprocation: though a wholly unconditional love may be as unusual as a wholly unrequited love is distressing, human acts toward others routinely occur within a wide and benign framing asymmetry. So, too, the number of persons has the same fluidity of direction; it is ordinary to see five persons acting on behalf of one, as it is also ordinary to see one acting on behalf of five. The issue of reciprocity between persons is a complex and important subject, but is emphatically not the subject under discussion here. Whatever its characteristics, they cannot be derived from the model of the relation between persons and objects.

The model introduced above is called "normative" because it is almost omnipresent in the artifacts of civilization, whether the given artifact happens to be an uncelebrated or celebrated object. This is not to say that the human action of projection never occurs without the consequent and amplified object action of reciprocation. For example, in situations of survival, projection and reciprocation may be agonizingly close to one another; and it is for this reason that there will often be in such a situation a question about whether to go on in the projective act—e.g., expending labor on trying to locate beneath the frozen ground the remnants of potatoes for the next meal—since the aversive expenditure of energy may be only equal to, or even less than, the caloric restoration that the potatoes will bring. This situation, however, in which the two events are near equivalents, is not the model for artifice: it is an emergency measure and is itself a moment of failed or failing artifice. Similarly, there are countless instances in which one must perform the projective labor, often over many years, without knowing whether the made thing (a new technological invention, a new medical cure, a new philosophic treatise, a new country) will be invested with amplified referential activity, or indeed with any referential activity at all. It may be that an individual who devotes his life to finding the cause of yellow fever will not find the cause but will have contributed to an eventual discovery that has a collective authorship. But it may also be that he is pursuing a path of investigation so in error (and perhaps even a path whose erroneousness is already known on other continents) that it will not have contributed to the collective outcome; or it may be that he will find the cause but will die before he has time to make his way back up the river and tell anyone. There are many positive, and even laudatory, descriptive terms that apply to this situation, for there is no question that the projective act of creation often requires great risk, great courage, great spirit, and does so regardless of outcome; in fact, it is precisely to the degree that the outcome is unknowable that courage will be personally required of him. But to say this is not to say that the situation is a model (or even, strictly speaking, an instance) of artifice, for it is again a moment of failed artifice or nonartifice. No one knows this better than the person himself, for although he has recreated himself to be courageous (one may even say he has transformed himself from the original human given of expecting reciprocation from acts of making to one who continually "makes" with the recognized risk of in the end never having made), it was of course not this, but the cause and eventual cure of yellow fever for which he labored.

As the normal arc of making is framed on one boundary by the nonnormative models of failure and survival, so it is framed on the other side by an equally nonnormative structure of making that belongs to the realm of extreme well-being and leisure. If, for example, three persons all labor for twelve hours to produce for one person a pastry whose pleasure-bearing power is considered by that person unremarkable and which in any event lasts only three minutes, and whose caloric value lasts only one hour, this event will again not be one in which the reciprocating action of the made object is in excess, or even nearly equal to, the projective labor. It may be that this, too, should be recognized as an instance of failed artifice. But even if (as seems often the case) it is taken as an acceptable, or normal, that is to say, "a successful," moment of artifice, it can be so taken only because of the wider frame of artifacts surrounding it. That is, up until now the discussion has focused on the relation between maker(s) and a single made object; but some objects can only be understood in the context of a multitude of objects.

As has often been noticed, the realm of objects (material, verbal, and mental) tends to be numerically excessive. Although this "numerical excessiveness of objects" is a very different characteristic from the "excess of reciprocating action *within* an object," the former may in part be a reflection of the latter. Because so many of the invisible attributes of creating are themselves objectified and made visible in the materialized structure of the object world, it may be that the inherent, self-amplifying largesse of creating also comes to have a visible (and very positive, though not unproblematic) registration in the tendency toward numerical excess. One of the most eloquent depictions of this tendency is Defoe's *Robinson Crusoe*. When one recalls this story from a distance, it seems to be a narrative of survival on a remote island outside civilization; when, therefore, one returns to read it again, it is startling to discover that Crusoe's act of world-building, his reconstruction of civilization, is from a very early moment characterized by surfeit. The objects that float to him on a wrecked vessel from an older civilization—like a colony, his own small country is the cultural offspring of a parent country—are, even in the midst of their scarcity, tokens of superfluity: he finds there, for example, amid many other things, not one but three Bibles. So, too, his own constructions increasingly display this surfeit: his shelter grows increasingly extravagant, soon there are two houses, two boats, and a fence that is rhythmically returned to, extended, fortified, thickened now with earth, now with twining branches. Again, his verbal, computational, and mental world of calendars, journals, dreams, moral categories, and optical perspectives on his dwelling place contain this same tendency toward multiplication. Though the resulting culture of goods can be summarized in pejorative terms (e.g., having and hoarding), or again in a neutral descriptive vocabulary (e.g., the Protestant ethic of work and individualism), there is also something about the nature of making, and the inherent thrust of the civilizing impulse, that Defoe works to expose. Crusoe begins by being a person who "makes" either as a result of acute need (where willed artifice is the only available strategy of self-rescue) or as a result of accident (where artifice entails the human genius of observing a wholly unintended outcome), but increasingly becomes one who willfully "makes" merely to make. That is, in addition to transforming his external world, Crusoe has transformed the nature of *the act of creating itself*; he has remade making; he has remade the human maker from one who creates out of pain to one who creates out of sheer pleasure.

His story is relevant to the three-person twelve-hour pastry. At the point where the object world is characterized by abundance, an object may be invented in which projection is in excess of reciprocation merely to demonstrate and luxuriate in the fact that the structure of creating itself may be remade to be free of its ordinary requirement that reciprocation be in excess of projection: the new inequality of projection over reciprocation (which ordinarily signals emergency, survival, and failure) comes to be the vehicle of announcing one's very distance and immunity from the realm of fear, death, and failure.

That sense of immunity has been, of course, brought about not by the pastry itself but by the abundance of artifacts in which the ordinary ratio of narrow projection and wide reciprocation is firmly in place: the pastry expresses rather than itself creating the immunity. Through objects, human makers recreate themselves, and now this *newly* recreated self finds that it is no longer expressed in the existing object world, and thus goes on to project and objectify its new self in new objects (which will, in turn, recreate the maker, and so again necessitate new forms of objectification). Thus the continual *multiplication* of the realm of objects expresses the continual *excess of self-revision* that is occurring at the original sentient site of all creation.

This brief excursion into the subject of multiplicity calls attention to the fact that just as a single object has an identifiable structure, so the inclusive realm of abundant objects may have an identifiable structure, though the difficulty and complexity of this subject place it beyond the frame of the present discussion.

If made culture consisted of a handful of objects—a god, an altar, a blanket, and a song—it might be possible to articulate the relations between the objects. Indeed, as was seen earlier, in the Old Testament the number of artifacts in a given passage may be so small, and the chronological sequence in which they come into being so clearly etched, that not only the isolated structure of each but the structure of relations among them can be apprehended.[45] Once, however, one resides in a deep sea of artifacts, that task becomes much more difficult, and very possibly impossible. What becomes strikingly apparent, however, is the multiplicity of paths by which existing objects sponsor new objects (and in this picture of multiple paths, we gain some glimpse of the massive front on which the imagination is constantly at work, patrolling the dikes of made culture, repairing, filling gaps, extending, reinforcing).

First and most important, as was described above, an existing object, by recreating the maker, itself necessitates a new act of objectified projection: the human being, troubled by weight, creates a chair; the chair recreates him to be weightless; and now he projects this new weightless self into new objects, the image of an angel, the design for a flying machine. Second, just as the sentient needs and acuities are projected into objects, so objects themselves contain both capacities and needs that sponsor additional artifacts. An invention may have a latent *power* that suggests a new application and so requires a new modification of the original invention: as the bodily lens of the eye is projected in a camera, eventually a new kind of camera that can enter into the interior of the human body (and film the events of conception, the passage of blood through the heart, or the action of the retina) comes into being. Conversely, an invented object may have a *need* that now requires the introduction of a new object: once Crusoe successfully "makes" a crop, he must go on to make an object that protects the crop;[46] once Benjamin Franklin makes a glass harmonica, he must go on to create a case to ensure the longevity of the delicate instrument;[47] once a "new" virus has been isolated, a new medium may have to be created in which its growth can be observed; once a constitution is in place, many laws and customs arise to protect the constitutional privileges.

Third, as became clear in the previous chapter, a given attribute of the sentient creator (e.g., the capacity for creation itself) may be first projected into an extremely sublimated objectification (e.g., God) which then invites the invention of less sublimated, more materialized objectifications (altars, narratives, temple, ark, branching candelabra, rainbow-as-sign) to mediate between the human maker and the original Artifact. Or instead (as the writings of both Marx and Freud tend to suggest), invention may first occur in a freestanding artifact close to the body (dream object, individual craft object, patterns of family interaction) that then gives rise to successively more sublimated artifacts (market structures, civil structures, ideologies, philosophies, religion). The existence of multiple expressions of a given attribute affects the arc of projection and reciprocation within any solitary artifact. The movement in either direction will amplify the size of the margin by which object reciprocation exceeds human projection. A materialized objectification may invite a more extended, dematerialized, or verbalized objectification in order to extend the "sharability" or referential breadth of the original thing. Conversely, a verbalized objectification may necessitate the introduction of a more concrete version of itself, which by investing the verbalized abstraction with sensuous immediacy, contracts the projective labor. So, for example, the sense of protective unity sensorially available in a concrete shelter and in the lived patterns of family life may be greatly extended in the concept of "polis"; and, in turn, the projective act of apprehending and holding steadily available to the mind the remote concept of polis (or one's own existence as "citizen") may be assisted and relieved by the comparatively freestanding existence of maps, colored squares of cloth, courthouses, and verbal pledges.

The chronological sequence of object appearance may occur in either direction. If, for example, it were appropriate to think of any one sentient attribute (e.g., seeing) as objectified at a hundred sites of successive dematerialization or sublimation (e.g., crystal ball, eyeglasses, a fiction about a superman who sees through walls, microscope, satellite, prism, mechanism for objectifying invisible parts of light spectrum, medical procedures for eye transplants, speculative accounts of how color vision works, calendars and other objects for visualizing the passage of time, theories of knowledge, phenomenological descriptions of seeing, theological concepts of a providential overseer, astronomical calculations that make visible events that will only actually come to pass in the future—all these would belong not to the same but to many different sites), it would be noticeable that the creation of the various objects does not begin with level one and proceed through one hundred, nor begin with level one hundred and proceed back through one. Instead an object at level three might help to press into existence an object at level ninety-six, and this in turn might occasion the introduction of objects at levels forty-three through forty-nine. Further, a vast array of objects would come into being to express *the relation* between any two (or three, or thirty) levels of sublimation: for example, a fourth level projection of desire (e.g., a verbal recitation of a dream about a vulture) and a sixty-fifth level projection of desire (e.g., the Mona Lisa) might sponsor the creation of a third artifact (e.g., a psychological romance bearing the title "Leonardo da Vinci and a Memory of His Childhood"[48]) for the single purpose of expressing the relation between the first two artifacts. Finally, this interaction between existing sites of objectification would occur not just *within* any *one* sentient attribute such as "seeing" or "desiring" but among the entire constellation of attributes. Rather than speaking only of a twelfth-level projection of vision sponsoring a thirtieth-level projection of vision, one would also have to be speaking of a twelfth-level projection of vision sponsoring a twelfth- (or thirtieth-) level projection of touch (e.g., the existence of print might occasion the invention of braille, just as, conversely, the tactile qualities of another planet's surface might be translated and relayed back to earth as visual information).

This schematic description is introduced here only to recall the relentless tendency toward self-amplifying objectification that is omnipresent in everyday life. Because the interaction between successive sites of any one tier, or again across tiers, is so habitual, one comes to expect the introduction of a material artifact that incorporates and conflates the disparate interior actions of pump and computer, or instead computer and vaccine,[49] just as one also comes to expect the introduction of a verbal artifact that articulates the influence of optics on the Miltonic poetics of paradise, or instead demonstrates Walt Disney's unconscious adoption of the biological principle of neoteny in his invention and evolving conception of a famous cartoon mouse.[50] The introduction of new objects takes place within the frame of already existing artifacts. A particular molecular structure may be dreamed before it is seen; a period of breakthrough in quantum physics may first require as prelude the resolution of an argument about the legitimacy of visual metaphor;[51] and if we complain that the inventor of the relativity theory was, in his devout belief in God, guilty of an inconsistency, we are perhaps allowing a very local conception of "inconsistency" to deflect attention away from what, within the overall strategy of human making, has nothing inconsistent about it.

As in the previous section of this chapter, the interior of the object world has been entered in order to apprehend the invisible interior of the human action of making that is itself recorded in the object. The solitary artifact has been described here as a "lever" because it is only the midpoint in the total arc of action, and because the second half of that arcing action is ordinarily vastly in excess of the first half. It is this total, self-amplifying arc of action, rather than the discrete object, that the human maker makes: the made object is simply the made-locus across which the power of creation is magnified and redirected back onto its human agents who are now caught up in the cascade of self-revision they have themselves authored.

As the object attributes examined in the earlier section worked to expose some of the invisible attributes of the imagination (its attempt to invest the nonsentient world with the responsibilities of sentience; its ethical monotony; its original inseparability from compassion), so here the identification of the object-as-lever exposes additional attributes coexisting with those others. First, the imagination is large-spirited or, at least, has an inherent, incontrovertible tendency toward excess, amplitude, and abundance. Perhaps because it originally comes into being in the midst of acute deprivation, it continues to be, even long after that original "given" has disappeared, a shameless exponent of surfeit. This inherent largesse may manifest itself in a wholly benign form (e.g., the excessive reciprocating action within the single object) or instead in a form (e.g., the numerical excessiveness of objects) that, though essentially benign, is also problematic, and hence must itself be subjected to the problem-solving strategies of imagining. The source of the problem is also the source of the solution; for, as was observed earlier, the principle of excess as it occurs within a solitary object expresses itself *both* in the degree of the object's revisionary powers (the degree to which any one person is disembodied) *and* in its referential breadth (the number of persons who are the potential recipient of its actions; the object's nonspecificity of reference). Object-surfeit and object-sharability are related phenomena (just as, conversely, the original pain against which objectification works is characterized by both "acute *deprivation*" and "acute *privacy*"). Thus any problematic manifestation of surfeit, such as the numerical excessiveness of objects, can be eliminated by the translation of surfeit into sharability, by, that is, the distribution of the objects to a larger number of persons. Even if (as sometimes argued) the moral generosity of a people is a late flower of civilization, it *is* a flower of civilization: the element of largesse is from the beginning contained in the human action of imagining; it is already embedded in the ontological status of human beings as creators, a status that they seem (by most accounts) to have acquired at the first moment they became human beings.

The object-as-lever also exposes a second attribute of the imagination, its nonimmunity from its own action. The imagination's object is not simply to alter the external world, or to alter the human being in his or her full array of capacities and needs, but also and more specifically, to alter the power of alteration itself, to act on and continually revise the nature of creating. This was earlier apparent in the changing circumstances out of which, or on behalf of which, creating arises: the human being who creates on behalf of the pain in her own body may remake herself to be one who creates on behalf of the pain originating in another's body; so, too, the human beings who create out of pain (whether their own or others') may remake themselves to be those who create out of pleasure (whether their own or others'). This continual self-revision visible in the changing circumstances and ends is equally visible in the means, as was apparent in the transformation of weapon into tool and tool into freestanding artifact. Throughout this succession of displacements, the power of magnification remains; but the first object (weapon) acts on sentience, increasing its aversiveness and decreasing its acuities; the second object (tool) eliminates the problematic character of the first by moving away from a sentient surface altogether, and acting on a nonsentient one; now, finally, the third object (artifact) returns to the sentient surface but acts on it in a way opposite to the way it was acted on by the original object, for the artifact works to diminish the aversiveness of sentience and to amplify its acuities. Multiple artifacts collectively continue this same work: culture is the made-lever back across which human evolution occurs.

The recognition of the nonimmunity, or self-revising character, of the imagination, leads to the recognition of a third attribute that is a specific form of self-revision: the imagination tends to be self-effacing. Though the human power of creating is relentlessly at work in the multiplying realm of verbal and material artifacts, these multiple objects appear to interact with, sponsor, modify, and substantiate one another, thereby eliminating from attention the overtness and omnipresence of the fictionalizing process, and thereby diminishing the recognizability of the "madeness" of the made world, permitting one to enter it as though a natural given. Although all the material and verbal artifacts the imagination creates are created on behalf of the small sentient circle of living matter in the thick midst of which it itself resides, it is not in the end surprising to find that the imagination has no objection to the ease with which those artifacts shed their overt referentiality to sentience and become referential to one another. The imagination is not hostile to this activity because through this very activity it perpetuates its own. Although the capacity of artifacts to disembody us only comes about by their being themselves fictional extensions of sentience and so containing within themselves pictures of our own bodies, in their tendency to give rise to successively sublimated versions of themselves they systematically eliminate from their interior the picture of the human body, make progressively more unrecognizable their resemblance to the site of their own creation. Though it is by their being externalized images of the body that they derive the power to disembody, the recognizability of that resemblance would diminish the very work of disembodiment they exist to bring about (for they would exist as ongoing announcements of the problematic character of the body, a problem whose *intensity* would be everywhere signaled in the colossal *scale* of the culture required to accommodate the problem). Though the objects are projected fictions of the responsibilities, responsiveness, and reciprocating powers of sentience, they characteristically perform this mimesis more successfully if not framed by their fictionality or surrounded by self-conscious issues of reality and unreality.

This closing chapter has attempted to provide—as a postscript to the three-part structure of the mental and materialized action of making—a very partial list of the secondary attributes of creating. Artifacts themselves contain and expose some of those attributes, suggesting that the imagination works to distribute the facts and responsibilities of sentience out onto the external world; that the imagination tends to be ethically uniform on the issue of sentience; that the imagination is bound up with compassion; that the imagination has an inherent tendency toward largesse and excess; that the work of the imagination is not here and there, now on, now off, but massive, continuous, and ongoing, like a watchman patrolling the dikes of culture by day and by night; that the imagination forfeits its own immunity and is self-revising; and that, finally, the imagination is self-effacing, and often completes its work by disguising its own activity.

But the nature of creation, however self-effacing, must also be conceptually available and susceptible to description so that the periodic dislocations within its overall structure of action can be recognized and repaired. The collective effort to understand making, already very old, will always be ongoing. Like the work of making, it keeps itself going: "The craftsman encourages the goldsmith, / and he who smooths with the hammer / him who strikes the anvil" (Isaiah 41:7). Directed against the isolating aversiveness of pain, mental and material culture assumes the sharability of sentience. It holds within itself the universal salutation of Amnesty's whispered "Corragio!" It passes on the password of Isaiah's ancient artisans—"Take Courage!" (41:6).

NOTES

Notes to Introduction

1. Timothy Ferris, "Crucibles of the Cosmos," *New York Times Magazine*, 14 January 1979.
2. Walter Sullivan, "Masses of Matter Discovered That May Help Bind Universe," *New York Times*, 11 July 1977.
3. Virginia Woolf, "On Being Ill," in *Collected Essays*, Vol. 4 (New York: Harcourt, 1967), 194.
4. The emphasis here on "externality" and "sharability" does not mean that any assumption has been made about the reality of the object; for even an only imaginary object (e.g., ghost, unicorn) is usually experienced by the imaginer as existing outside the boundaries of the body; and though it may be less sharable than a real object, it is of course more sharable (nameable, describable) than objectlessness.

Though we may say, "The ghost she speaks of exists only in her own mind," the very fact that she has gotten us to speak that sentence means that the object, though unreal, is externalizable and sharable: she has made visible to those *outside* her own physical boundaries the therefore no longer wholly private and invisible content of her mind. What is remarkable is not that one person should enable another person to see a ghost (for this seldom happens), but that one person should routinely enable another person to see the inside of his or her consciousness.

5. Ronald Melzack, *The Puzzle of Pain* (New York: Basic, 1973), 41. See also revised edition, co-authored with Patrick D. Wall, *The Challenge of Pain* (New York: Basic, 1983).

This analogy is a striking and provocative one: its fertility is manifest in the very fact that it has led Melzack and others to important insights about attributes of pain other than intensity. Strictly speaking, however, it is almost certainly not the case that intensity is to the felt-experience of pain what light flux is to vision, since pain (however variable and multidimensional) is much closer to being one-dimensional than is vision, and much of its aversive and terrifying character arises from that one-dimensionality. In fact, when we attribute "intensity" to something (which we consistently do with pain and only occasionally do with objects of vision or hearing or taste), we usually in part mean that one dimension has become dominant (e.g., the redness of the red, the loudness of the siren, the pain of the pain). That is, it is in the nature of intensity to be wholly self-isolating, to so obsess attention that it breaks apart from any context against which it might be qualified or measured. By this perceptual process the intense becomes the absolute.

While the preceding paragraph calls into question the literal accuracy of Melzack's analogy, it simultaneously confirms the generous work of that analogy. One might say that if pain had a goal, it would be to be felt and known exclusively in its intensity. Those people working to make recognizable its other attributes are working against its insistent, self-isolating intensity, and therefore against pain itself.

327

6. Descriptions of the McGill Pain Questionnaire are also found in medical journals such as *Pain: The Journal of the International Association for the Study of Pain* (1975–present). Such articles, though mainly reporting on the accuracy of the questionnaire's specific diagnostic powers, sometimes describe the patients' reactions to being given the list of words. Those administering the questionnaire have been struck by the ease with which patients recognize what they consider the "right" word, by the "certainty" with which they differentiate appropriate and inappropriate adjectives, and by the "relief" and even "happiness" they express at having been given words to choose from at a time when they cannot formulate those words.

7. Conversation with Ronald Melzack, McGill University, Montreal, 9 June 1977.

8. The former director of A.I.'s Campaign to Abolish Torture explains that in *any one instance* of suspected torture, A.I. might feel the pressure to err on the side of assuming it is happening (and therefore immediately beginning to work for its elimination) rather than erring on the side of delaying action while confirmation is sought (and therefore permitting it to continue if it is indeed happening). But this pressure must be resisted, since a false report of torture in this one instance would then diminish A.I.'s ability to stop torture in all subsequent instances. (Conversation with Sherman Carroll, Director of Campaign to Abolish Torture, International Secretariat of Amnesty International, London, August 1977).

A.I.'s "Urgent Action Network" issues explicit instructions to those writing letters about whether the word "torture" can or cannot be used in an appeal on behalf of a specific prisoner.

9. J. K. Huysmans, *Against Nature (À Rebours)*, trans. Robert Baldick (Baltimore: Penguin, 1959), 92.

10. Friedrich Nietzsche, *The Gay Science*, trans. and introd. Walter Kaufmann (New York: Vintage, 1974), 249.

11. V. C. Medvei, *The Mental and Physical Effects of Pain* (Edinburgh: Livingstone, 1949), 40.

12. Homer, *The Iliad*, trans. A.T. Murray (Cambridge, Mass.: Harvard University Press; London: William Heinemann, 1924), Vol. I, iv, 117.

13. In one modern edition of Margery Kempe's writings, the editor found this use of the adjective "boisterous" so odd that he replaced it with the adjective "coarse." See *The Book of Margery Kempe*, ed. W. Butler-Bowdon and introd. R.W. Chambers (New York: Devin-Adair, 1944), xxv.

14. Ludwig Wittgenstein, *Philosophical Investigations*, trans. G.E.M. Anscombe (New York: Macmillan, 1953), 312.

15. Michael Walzer, *Just and Unjust Wars: A Moral Argument With Historical Illustrations* (New York: Basic, 1977), 24.

16. "Interview with Peter Benenson," *New Review* 4 (February 1978), 29.

17. William Safire cited in William Porter, *Assault on the Media: The Nixon Years* (Ann Arbor: University of Michigan Press, 1976), 196.

18. Interview with George Wallace, *Sixty Minutes*, C.B.S. broadcast, 17 April 1977.

19. "La rencontre Mitterrand-Reagan aura lieu les 18 et 19 octobre," *Le Monde*, 15 August 1981, 1. See also *Time Magazine*, 24 August 1981.

Notes to Chapter 1:
The Structure of Torture

1. The compensatory nature of torture, which gives rise to its continual recurrence in unstable countries, is especially visible in its use in situations of international conflict. Until 1940, according to Bruno Bettelheim, periods in which Jewish prisoners in Germany were released for emigration alternated with periods in which they were executed in large numbers; which of these two events took place depended on the strength and self-image of the Nazi government. He writes, "So there were the Jewish prisoners: wishing ardently for the destruction of the enemy; wishing in the same breath (up to 1940) that it would remain strong until they could emigrate, or (later on) remain safe forever to ward off their own mass destruction and the slaughter of their families" (*The Informed Heart: Automony in a Mass Age* [New York: Free Press, 1960], 202). Again in the present era, an Amnesty International mission to the Middle East to investigate allegations of torture connected with

the October 1973 war concluded that "In war, there is normally a severe hostility towards the captured enemy soldiers. This hostility is particularly strong during the most intensive periods of warfare, and the behaviour towards the captured enemy soldier is in many cases to be explained as an (inappropriate) revenge against military activity carried out by the enemy. For example, when one of the parties carries out bombing raids against towns of the other party, captured soldiers of the former party may be subjected to misdirected ill-treatment" (*Report of an Amnesty International Mission to Israel and the Syrian Arab Republic to Investigate Allegations of Ill Treatment and Torture* [London: Amnesty International Publications, 1975], 6).

2. Amnesty International's unpublished transcript of the 1975 trial in Greece of those torturers who served the Colonels' Regime; translated and collated from both verbatim and summary accounts of the daily court proceedings carried in the Greek newspapers *Vima, Kathimerini, Ta Nea,* and *Avghi*; 13, 65, 100, 77; hereafter cited as "Transcript of the Torturers' Trial." Many details for this chapter are taken from Amnesty International publications and from unpublished materials (trial transcript, medical reports, depositions of former prisoners) located in the research department of the International Secretariat of Amnesty International in London. As Amnesty International often warns, there is an unfortunate paradox in the fact that a thorough report of the brutalities in a particular country at least means that country has permitted an A.I. investigation, whereas there are other countries that refuse such investigations or that regularly kill the men and women they torture, thereby eliminating the possibility of making the violence known. A disproportionate number of details in this chapter will be based on torture in Greece since the 1975 trial makes accessible information about aspects of the process that could not otherwise be known.

3. *Report of an Amnesty International Mission to the Republic of the Philippines: 22 November– 5 December 1975,* 2nd ed. (London: A.I. Publications, 1976), 37.

4. "South Vietnam," in *Amnesty International Report on Torture* (New York: Farrar, Straus and Giroux, 1975), 167.

5. "Appendix: Special Report on Chile by Rose Styron," in *Amnesty International Report on Torture,* 257.

6. D. D. Kosambi, *The Culture and Civilization of Ancient India in Historical Outline* (London: Routledge and Kegan Paul, 1965), 110.

7. Although there is, of course, no way to demonstrate conclusively that the need for information is a fictitious motive for the interrogation, instance after instance can be cited of irrelevant questions. In 1974 an Irishman was asked such questions as "What would you do if the Prods were raping your daughter? . . . Stand by and take pleasure in it, I suspect" (Unpublished letter from a minister to various government and human rights groups in the archives of Amnesty International). In 1978 many Ethiopian schoolchildren were killed after first being made to provide the names of other schoolchildren who might be part of the anti-Government underground: " 'We're talking about kids,' said a diplomat, 'They're 10, 11, 13 years old. They pull a kid in and he gives them names of other kids—it could be someone that he had a fight with at school that day' " (John Darnton, "Ethiopia Uses Terror to Control Capital," *New York Times,* Late City Edition, 9 February 1978, Sec. 1, 1,9). That the information elicited in Ethiopia or Vietnam or Chile sometimes, in fact, determines the sequence of arrests and torture may only mean that governments sometimes depend on their opponents to provide an arbitrary structure for their brutality. Even those who accept that the need for information is the motive tend to call this into question in indirect ways. Amnesty International publications, for example, sometimes call attention to the fact that torture is an extremely inefficient means of intelligence-gathering, though its inefficiency is certainly not the grounds on which it is condemned. Again, in his discussion of torture in Algiers, Alistair Horne seems to assume that it is being used in order to elicit information, but he observes that the collating services of a country that uses torture are "overwhelmed by a mountain of false information extorted from victims desperate to save themselves further agony" (*A Savage War of Peace: Algeria 1954–1962* [London: Macmillan, 1977], 205). In his article, "Torture" (*Philosophy and Public Affairs,* 7 [Winter 1978], 124–143) Henry Shue analyzes and assesses both "terroristic torture" (where it is obvious that the purpose is not to extract information) and "interrogational torture" (where it may appear that the purpose is to extract information). But he prefaces his discussion of the second category with the comment,

"It is hardly necessary to point out that very few actual instances of torture are likely to fall entirely within the category of interrogational torture" (134).

8. "Transcript of the Torturers' Trial," 155.

9. Unpublished deposition of a former Chilean prisoner, Amnesty International Archives.

10. While those who withstand torture without confessing should be honored, those who do confess are not dishonored by and should not be dishonored for their act. Many contexts within our culture reveal our derisive attitude toward confession. So profound and powerful a play as Shelley's *Cenci* is seriously marred by Beatrice's sententious and ignorant condemnation of her brother and mother for confessing under torture. Intended to ennoble the heroine, the scene should embarrass the audience. Popular culture reveals the same attitude. Movies in the 1940s and 1950s like *Purple Heart* and *Bataan* so regularly showed the hero remaining tough and silent that the American public was shocked when Gary Powers "talked" rather than kill himself, and military law regarding confession was only very slowly changed in the next twenty years. That the civilian public is unintentionally allying itself with the torturers by this attitude was made very clear during the 1975 trial of the torturers in Greece where the torturers revealed in smug comments their assumption that their one remaining power over the former prisoners who were now testifying against them was their knowledge of who did and who did not confess. Similarly, an Israeli judge, Judge Etzioni, speaking at the Israeli Embassy in London, answered the London *Times* June 1977 reports of torture on the West Bank by saying, "We don't need to torture Arabs. It's in their nature to confess." While the speakers in both incidents were immediately corrected through public reprimand (in the courtroom in the first case and the news media in the second), there is a reason why the Greek torturers and the Israeli judge thought their comments would be appropriate. The most unfortunate sign of the negative public attitude toward confession is the embarrassment of former prisoners when they acknowledge having confessed. Although Amnesty International's medical form includes a section entitled "Interrogation and Torture," members of their missions never ask the prisoners what they were required to confess or whether they did confess. However, depositions are sometimes taken by people outside Amnesty and in these cases questions about confession are sometimes asked.

11. This relation between physical pain and irony will again be visible in the description of torture that follows. For now it should be noted that the place of irony in Sartre's short story is also the place it occupies in a great deal of the literature of the holocaust. The narrative texture of this literature is made up of endless ironies (see, for example, any page of Borowski's *This Way for the Gas, Ladies and Gentlemen*) none of which are experienced as ironies because of the overwhelming fact at the center. In Elie Wiesel's *Night*, a father and child on a train taking them to a concentration camp are in the presence of a woman who keeps seeing furnaces and fires. The others think her mad though she will turn out, of course, to have been a visionary. The moment when the actual furnaces are seen and the destiny of everyone is suddenly clear would, in the normal structure of experience, be the moment revealing the ironic position of the visionary and all those who misjudged her; but beside the fact of the furnace itself, it seems irrelevant in what way the end was or was not anticipated. Beside the horrible central fact, there are no peripheral facts, no relation between periphery and center, no irony, because in every direction there is only irony.

12. See, for example, Mircea Eliade, *Shamanism: Archaic Techniques of Ecstasy*, trans. Willard Trask, Bollingten Series, Vol. 76 (Princeton: Princeton University Press, 1964), 33.

13. Jean-Paul Sartre, "The Wall," in *The Wall (Intimacy) and Other Stories*, trans. Lloyd Alexander (New York: New Directions Paperback, 1969), 5. All subsequent references are to this edition; page numbers will be given in the text.

14. What from the inside is experienced as an increasingly insubstantial world may look from the outside as though the world is intact but the person is growing insubstantial, and so the experience is often represented as solid world ground on which the person no longer has a place. That is, from the inside, Ibbieta's bench feels thin and porous; but the way this same phenomenon is perceived by an outside observer would be more accurately represented (and is, of course, much more frequently represented) as Ibbieta being forbidden to sit on the bench. As one's world is obliterated, one's externalized self and therefore one's visibility is obliterated. This is in part why "property" and "personhood" become so easily confused.

15. Naturally Cordelia dies, as perhaps she always must. Though her death is horribly cruel, it

does not seem to be the unusual and gratuitous piece of authorial cruelty that it is sometimes claimed to be by people discussing the play. We usually see death represented from the outside; we see it as survivors; we see Van Gogh's chair empty of Van Gogh rather than as he, on the edge of death, would have experienced it, not the chair or world empty of himself but himself empty of chair and world. Traditional deathbed scenes with the young at the side of the dying old are representations from the outside, but from the inside . . . ? Perhaps the dying father experiences not the loss of himself, for he is still with himself, but the loss of his child. We grasp the agony of death only in the survivor's acute sense of loss. When we see the dying one respond to the death of his own daughter we see the cost to the self of the self's own death.

16. Lear, especially, understands this power of the voice. The play opens with his demand that each of his children, already physical extensions of himself, make him the sole content of their speech, thereby increasing still further his extension into the world. A great deal of the swirling energy of this play turns around his changing conception of the voice. This is clear in a juxtaposition of the first and final scenes: a father standing above a daughter demanding that she heave her heart into her mouth, and a father kneeling beside her, pleading for the smallest metonymous part of her, not her voice full of his name but her breath, her own small work of self-extension.

17. Karl Marx, *Herr Vogt*, as quoted and translated by Stanley Edgar Hyman, *The Tangled Bank: Darwin, Marx, Frazer, and Freud as Imaginative Writers* (New York: Grosset and Dunlap, Universal Library Edition, 1966), 118.

18. François Truffaut attributes this aphorism to Wilde in *Jules et Jim* (Paris: Seuil, 1971). Truffaut's French formulation is given in several slightly different versions in the English translation of the screenplay and the subtitling for the film itself.

19. George Eliot, *Adam Bede* (New York: Signet, 1961), 140.

20. Emile Zola, *Germinal*, trans. L. W. Tancock (Baltimore: Penguin, 1966), 338.

21. This is not to say that our sympathies are primarily with the torturer, for they are of course still with the prisoner; but to have moved even a small amount in the direction of the torturer in this most clear-cut of moral situations is a remarkable and appalling sign of the seductiveness of even the most debased forms of power, and indicates how wrong our moral responses might be in situations that have more complexity.

22. This same structure is present in many situations. A privileged group of people, for example, in this or any previous century, would probably never overtly justify its privileges on the greater pain, hunger, or frustration of the lower classes but rather on the mediating fact that the latter have less world (property, knowledge, ambition, talent, style, professionalism, and so forth). It is in many cases, of course, their pain and hunger that dissolves their world-extension, and their lack of world-extension that obscures their pain and hunger.

23. "Greek Nr. 5," Unpublished medical report, Amnesty International Archives.

24. *Philippines*, 24, 26, 31, 38.

25. *Report of an Amnesty International Mission to Spain: July 1975* (London: A. I. Publications, 1975), 7, 9.

26. *Workshop on Human Rights: Report and Recommendations, Nov. 29–1 Dec. 1975* (London: A. I. Publications, 1975), 4.

27. Alexsandr I. Solzhenitsyn, *The First Circle*, trans. Thomas P. Whitney (New York: Bantam, 1969), 614, 615.

28. *Philippines*, 51.

29. *Israel and the Syrian Arab Republic*, 12.

30. "Transcript of the Torturers' Trial," 11, 156, 32, 11, 64, 109.

31. *Philippines*, 24, 49, 28, 38, 27.

32. *Philippines*, 23, 39, 27, 28.

33. *Political Prisoners in South Vietnam* (London: A. I. Publications, n.d. [after January 1974]), 27, 28. The connections and reversals between the trial and torture are too immediately apparent to require elaboration; the "Transcript of the Torturers' Trial" becomes in many places a sustained counterpoint of the two.

34. Alexsandr I. Solzhenitsyn, *The Gulag Archipelago 1918-1956: An Experiment in Literary Investigation*, Vol. 1, 2, trans. Harry Willetts (New York: Harper & Row, 1974), 208.

35. "Transcript of Torturers' Trial," 70, 76, 115.

36. *Philippines*, 39.

37. Leonard A. Sagan and Albert Jonsen, "Medical Ethics and Torture," in *The New England Journal of Medicine*, Vol. 294, No. 26 (1976), 1428. See also *Chile: An Amnesty International Report* (London: A. I. Publications, 1974), 63.

38. Sagan, 1428.

39. *Report on Allegations of Torture in Brazil* (London: A. I. Publications, 1972, 1976), 25, 65.

40. *Israel and the Syrian Arab Republic*, 20-27.

41. "Repression en Uruguay," *Le Monde*, 20 June 1978.

42. This possibility has always been present. According to Jack Lindsay in *Blastpower and Ballistics: Concepts of Force and Energy in the Ancient World* (New York: Harper & Row, 1974), 346, one ancient writer of a medical treatise on articulations warns that devices developed to treat dislocations make a terrible force available to anyone who wants to misuse it; and the torture instruments of the Renaissance "were direct imitations of the Hippokratic apparatus."

43. So omnipresent is the willing or forced participation of doctors that various international medical associations meeting between 1975 and 1977 passed resolutions urging doctors of all countries to avoid both active and passive forms of assisting the torturers. Although everything possible to eliminate such assistance should certainly be done, the deconstruction of medicine in torture is probably no more dependent on the participation of actual doctors than the inversion of the trial is dependent on the presence of judges and lawyers.

44. Horne, 201.

45. Letter from a former Paraguayan prisoner, Amnesty International Archives. This man's description is somewhat unusual among former prisoners' accounts of torture in its inclusion of the internal sensations he was experiencing.

46. *Brazil*, 64; *Vietnam*, 28; "Transcript of the Torturers' Trial," 53; *Philippines*, 23.

47. *Report of an Amnesty International Mission to Argentina: 6–15 November 1976* (London: A. I. Publications, 1977), 24; *Philippines*, 29, "Transcript of the Torturers' Trial," 9, 48.

48. "Transcript of the Torturers' Trial," 46; A. I. Handout on Uruguay; *Brazil*, 64. The names given throughout this paragraph are the names of specific forms of torture. Where there is a more complete record of the torturers' language patterns, as there is for those who served the Greek Junta, it is clear that the torturers' incidental vocabulary is also drawn from these three realms: the prisoner might be held as though in a game of football, his interrogators accompanying each blow with shouts of "off sides," "goal," "foul"; the process as a whole with its many subordinate acts and with the prisoner gradually moving toward complete collapse was repeatedly referred to as "ripening"; pronouncements were made such as "whoever enters EAT must come out white as a butterfly" (30, 11, 45, 97, 156, 241).

49. The use of the terms "fraudulent," "false," and "fictitious" to describe the torturer's power will be clarified in the fourth section of this chapter and will be further clarified in Chapter 2 which compares the place of verbal fictions in torture and in war. While the torturer's physical power over the prisoner is as "real" as the pain it brings about, what is not in the same sense "real" is the translation of the attributes of pain into the cultural insignia of a regime, a regime whose absence of legitimate forms of authority and substantiation has occasioned the fictional display. Thus a distinction is being made here between the individual man torturing and his role as representative of a particular set of political and cultural constructs. (It should also be stressed that while the physical power of the torturer-as-individual may be accurately identified as "real," the scale of that power cannot be inferred from the scale of the prisoner's pain: that is, it requires neither strength nor skill to inflict hurt on a *wholly defenseless* human body; a weak child would be physically capable of inflicting similar hurt on human tissue had he the weapons or the impulse to do so.)

50. "Transcript of the Torturers' Trial," 48.

51. "Transcript of the Torturers' Trial," 6, 62.

52. *Vietnam*, passim; *Spain*, 9; *Philippines*, 25, 28; *Argentina*, 21; "Transcript of the Torturers' Trial," 68.

53. W. K. Livingston, *Pain Mechanisms* (New York: Macmillan, 1943), 2.

54. *The First Circle*, 640.

55. *Workshop on Human Rights*, 5.

56. "Transcript of the Torturers' Trial," 34.

57. *Argentina*, 38; *Philippines*, 45, 49, 50; "Transcript of the Torturers' Trial," 65; Unpublished medical report of former Greek prisoner; Unpublished medical report from former Chilean prisoner.

58. "Transcript of the Torturers' Trial," 54, 70, 77; Unpublished medical report of a former Chilean prisoner; *Workshop on Human Rights*, 4 (describing Portugal). Former prisoners from country after country describe being made to watch or hear another person being hurt. Instances of this are described by Reza Baraheni in "Terror in Iran," *The New York Review of Books*, 28 October 1976, 24.

59. The title of the Amnesty International newspaper *Matchbox* is derived from this incident.

60. Letter of a former Uruguayan prisoner, Amnesty International Archives.

61. Sheila Cassidy, "The Ordeal of Sheila Cassidy," *The Observer* [London], 26 August 1977.

62. For example, Ronald Melzack, a leading theoretician on the physiology of pain, writes, "If injury or any other noxious input fails to evoke negative affect and aversive drive (as in the cases described earlier of the football player, the soldier at the battlefront, or Pavlov's dogs) the experience cannot be called pain" (*The Puzzle of Pain*) [New York: Basic Books, 1973], 47). See also J. S. Brown's "A Behavioural Analysis of Masochism" in *Punishment*, eds. Richard H. Walters, J. Allen Cheyne, R. K. Banks (Harmondsworth: Penguin, 1972), 230–239, esp. 231. In David Bakan's brilliant little book, *Disease, Pain and Sacrifice: Toward a Psychology of Suffering* (Chicago: University of Chicago, 1968), 77, 79, he notes that pain's aversiveness in some situations has a beneficent effect since it is the only thing that can make tolerable the otherwise intolerable separation of a human being from his limb and, possibly, a woman from her baby.

63. *The Chronicle* [Willimantic, Conn.], UPI, 10 Feb. 1977, 7; UPI, 21 June 1976, 10. The articles themselves had none of the crudity of their titles.

64. Melzack, 19, 20, 72.

65. Melzack, 93; see also 76.

66. S. W. Mitchell, *Injuries of Nerves and Their Consequences* (Philadephia: Lippincott, 1872), 196.

67. Antonin Artaud, *The Theatre and Its Double*, trans. Mary Caroline Richards (New York: Grove, 1958), 23.

68. Albert Camus, "Reflections on the Guillotine," in *Resistance, Rebellion, and Death*, trans. Justin O'Brien (1960; rpt. New York: Random-Vintage, 1974), 180; and Hannah Arendt, *Totalitarianism* (New York: Harcourt-Harvest, 1951), 45, 108, 120.

69. Hannah Arendt, *Eichmann in Jerusalem: A Report on the Banality of Evil*, 2nd ed. (1965; rpt. Harmondsworth: Penguin, 1977), 106.

70. Bettelheim, 241.

Notes to Chapter 2:
The Structure of War

1. The use of the adverb "symbolically" here and later does not mean that a substitute for the human body has been made, nor a substitute for real pain. Rather, it refers to the fact that one person stands for many. Here Nelson Goodman's use of the word "sample" for one kind of "symbol" may be helpful (*Ways of Worldmaking* [Indiana: Hackett, 1978], 63–70).

2. For example, Article 5 of the United Nations Universal Declaration of Human Rights states unequivocally, "No one shall be subjected to torture or to cruel, inhuman or degrading treatment or punishment." Again, the International Covenants on Civil and Political Rights are unequivocal: though some "rights" may be suspended during a state emergency, no such qualification extends to torture. The unequivocal prohibition is again manifest in the fact that it has, as a result of the court decision in *Filartiga* v. *Peña-Irala*, 630 F. 2d 876 (1980), come to be considered a crime warranting extraterritorial jurisdiction: the crime of torture can be tried in the United States even if the act of torture did not occur within U.S. boundaries, and neither the victim nor the torturer is a

U.S. citizen. Harold J. Berman writes that this is "perhaps the most daring extension of extraterritorial jurisdiction" and that the extension was made on the basis of the universality of its condemnation. The "court held that since such torture was condemned in all countries of the world and was a violation of the International Covenants on Human Rights, the offender could be prosecuted by any state which had him in its custody" ("The Extraterritorial Reach of United States Laws," Report to the Legal Committee of the U.S.-U.S.S.R. Trade and Economic Council, Moscow, 17 November 1982, 12).

In contrast, prohibitions of war are often characterized by equivocation. On reservations, both within specific peace treaties and within international peace covenants, see below pp. 141–42.

3. As this distinction implies, the structure of torture is one that becomes visible in the experienced event itself, and thus it is appropriate to an analysis of that structure to invoke first person accounts, both actual and fictionalized. Because the structure of war does not itself reside in the immediately experienceable reality of war, there can be less reliance in this chapter on personal accounts, actual or fictionalized, even though the suffering of war is always occurring within individual sentient experience.

Within war, there may be an event that is torture (a soldier may literally torture an enemy prisoner) or one kind of event that more closely approximates torture than does another kind (the lone bomber above the city and the people below, to whom he is joined across the vertical expanse of the colossal weapon; for in the work of a bomb, the two ends of the weapon are the place from which it falls and the place where it lands). An event in war will approximate torture the more the injury is one-directional and nonreciprocal. Though the pilot in the example is himself susceptible to injury, the ratio of his power to inflict injury to his own risk is much greater than that of soldiers on the battlefield, where the injury-inflicting and injury-receiving possibilities are close to one another. However, his act is still occurring within the overall framework of reciprocity, a framework whose significance is elaborated below. For now, the critical point is that while within either peace or war) a single experiential event can occur which can be correctly identified as "torture," there does not exist any single experiential event which can be identified as "war." Its structure takes shape in a massive number of (hence individually non-experienceable) interactions.

4. Stockholm International Peace Research Institute, *Incendiary Weapons*, by Malvern Lumsden (Cambridge, Mass.: M.I.T. Press, 1975). This particular study is a complex analysis of political, commercial, and military facts, though the center of the book is a description and analysis of the form of wounding involved. Other works with the same benign intent do not always accomplish the overview so successfully as this one.

5. Carl von Clausewitz, *On War*, ed. and trans. Michael Howard and Peter Paret, commentary Bernard Brodie (Princeton: Princeton University Press, 1976), 91, 93. To "increase the suffering of" the enemy and to "destroy" or "annihilate the enemy's forces" are repeated *passim*. They refer to an action explicitly identified as having a structural centrality: "It follows that the destruction of the enemy's force underlies all military actions; all plans are ultimately based on it, *resting on it like an arch on its abutment*" (97, italics added). Though the object of injuring may go beyond the physical to the moral, it always includes the physical even when not limited to it (97).

6. Clausewitz, 98.

7. Conversation with Major Frederick Tweed, John Pennman Wood Library of National Defense, University of Pennsylvania, August 1984.

8. *Incendiary Weapons*, 82. According to this SIPRI study, the object of these massive droppings was to create fire-storms; but the study notes that U.S. Defense Department spokesman J. W. Friedheim denied this and said the object was to clear foliage (*New York Times*, 24 July 1972).

9. "Kamikaze: Flower-of-Death," Episode 6 in *World War II: G.I. Diary* (New York: Time-Life Films, 1978), North Carolina P.B.S. broadcast, Spring 1980.

10. Robert Whymant, "The Brutal Truth about Japan," *Manchester Guardian Weekly*, 22 August 1982.

11. For a description of the battle and its casualties, see Nicholas Golovine, *The Russian Campaign of 1914: The Beginning of the War and Operations in East Prussia*, trans. A.G.S. Muntz, Introd. Marshal Foch (London: Hugh Rees, 1933), especially Chapter 9, "The Death Throes of the Center Corps of the Second Army" (290–327).

12. On the difference between altering the surface of sentient and nonsentient surfaces, see this chapter, pages 147–48 and the next, pages 173–74.

13. For example, after the September 1982 massacres in the Shatila and Sabra camps in Beirut, Lebanon, newspaper accounts attempting to assess the extent of Israeli responsibility for the killings by Lebanese Christian Militiamen quoted Israeli Defense Minister Ariel Sharon as explaining his opposition to a multinational force in Lebanon on the basis that it would impede "mopping up" operations after the departure from Beirut of the P.L.O; and quoted army chief of staff Lt. Gen. Rafael Eytan as saying, "We will identify the terrorists and all their commanders. The area will be clean" (David Shipler, "Israelis Disclaim Any Responsibility," *New York Times*, 19 September 1982.) Similarly, communications between American generals at the end of World War II referred to the Ruhr pocket campaign as "mopping up the Ruhr" (D.D. Eisenhower, telegram, quoted in John Strawson, *The Battle for Berlin* [New York: Scribners, 1974], 105). Again, Churchill, urging the British population to sustain their support for the troops in the final months of war when Japan was not yet conquered, says in his 13 May 1945 radio broadcast, "I told you hard things at the beginning of these last five years; you did not shrink, and I should be unworthy of your confidence and generosity if I did not still cry: Forward, unflinching, unswerving, indomitable, till the whole task is done and the whole world is safe and clean" (*Victory: War Speeches by the Right Hon. Winston S. Churchill O.M., C.H., M.P.*, comp. Charles Eade [Boston: Little, Brown, 1946], 117). Hitler's use of this idiom in his racial programs is too familiar to require recitation here.

14. See this phrase, for example, in Winston Churchill's 26 March 1944 World Broadcast, "The Hour of Our Greatest Effort is Approaching" (*The Dawn of Liberation: War Speeches by the Right Hon. Winston S. Churchill, C.H., M.P.*, comp. Charles Eade [Boston: Little, Brown, 1945], 135). Specific sources are given here only as examples of general use. Unless stated in the text of the chapter, it should not be assumed that this language is especially pronounced in the specific author cited. Though Churchill, for example, often uses "neutral" language and often "redescribes" the act of injuring, he at other times states directly that physically hurting the enemy is the goal. In his 6 June 1944 address to the House of Commons on the liberation of Rome, he celebrates that event, but then goes on to specify that "General Alexander's prime object has never been the liberation of Rome. . . . The destruction of the enemy army has been, throughout, the single aim, and they are now being engaged at the same time along the whole length of the line as they attempt to escape to the North" (135). Again in the 26 March 1944 broadcast he says, "We were all confident of victory, but we did not know that in less than two months the enemy would be driven with heavy slaughter from the African continent, leaving at one stroke 335,000 prisoners and dead in our hands" (51); and, speaking in the same broadcast of Burma and the Pacific theatre, he says, "It is too soon to proclaim results in this vast area of mountain and jungle, but in nearly every combat we are able to count three or four times more Japanese dead, and that is what matters, than we have ourselves suffered in killed, wounded, and missing" (56).

15. Henry A. Kissinger, "Editor's Introduction," *Problems of National Strategy: A Book of Readings* (New York: Praeger, 1965), 4.

16. Arthur Waskow, "The Theory and Practice of Deterrence," in *Problems of National Strategy*, 67. Waskow, who also says "H-bombs will kill cities" (80), is writing on the dangers and complications of the balance of terror. He argues here against the Wohlstetter school assumption that "balance of terror" acts as a deterrence, and tries to show that it instead increases the possibility of war. That his odd use of language should invade even this passionate attempt to clarify thinking about the arms race signals its omnipresence. Again, this example shows that unless indicated in the text of the chapter, it is not possible to infer an author's position from his use of the language. Citations are given only to exemplify its formal use.

Sometimes the oddness of using the word "kill" for an inanimate object is partially acknowledged by setting the word in quotation marks: for example, "Harriers achieved most 'kills' of Argentine aircraft" and "A weapon such as the long-range air defense weapon Sea Dart can be effective if the enemy avoids it out of respect—but this will reduce the 'kills' to its credit" (Lawrence Freedman, "The War of the Falkland Islands, 1982," *Foreign Affairs* 16 [1982], 206, 207, 208).

17. That the centrality of injury often slips from view is also indicated by fictional accounts of war that so often have scenes in which the reality of an injured body is presented both to the reader

and to the fictional characters as a "surprise." Paul Fussell in *The Great War and Modern Memory* (London: Oxford University Press, 1975) writes that a "primal scene" in fictional accounts of World War I is one in which "a terribly injured man is 'comforted' by a friend unaware of the real ghastliness of the friend's wounds" (33, 35), and invokes as examples Act III, sc. 3 of R. C. Sherriff's *Journey's End* and the searing scene in Joseph Heller's *Catch-22* where Yossarian believes that Snowden is suffering from the huge and horrifying wound in his thigh, but then in opening Snowden's flak suit sees Snowden's insides begin to fall to the floor. Other works that contain a scene in which the hero is suddenly "surprised" by the nature of injury include Remarque's *All Quiet on the Western Front*, Crane's *The Red Badge of Courage*, and Stendhal's *The Charterhouse of Parma*. While the recurrence of such scenes suggests that the centrality of injuring in war has been lost sight of (otherwise it could not have the force of "surprise"), it may also be argued that the visual immediacy of an injured body is itself so overwhelming that it has the capacity to "surprise" and "shock," even if the observer is already acutely conscious of the fact that injuring is the central activity of war.

Fictional works that hold steadily visible the centrality of injury include Homer's *The Iliad*, Zola's *The Debacle*, Tolstoy's *War and Peace*, and Mailer's *The Naked and the Dead*.

18. When, in both formal and informal discussions of war, an occasion arises that requires the illustration of "typical" activity, the type of activity often chosen is "disarming." This tendency, which occurs in many different contexts (e.g., formal strategy on the one hand, selection for television scenarios on the other) can be represented here with game theory. In, for example, Anatol Rapoport's discussion of witholding information in situations of mixed strategy, he uses the following military situation: "A truck carrying munitions passes over either of two roads every day. Road #1 is good, and Road #2 is poor. The enemy sends a detachment to ambush the truck. The detachment has a choice of blocking Road #1 or Road #2. Each of the opponents has two strategies . . . " and so forth (*Fights, Games, and Debates* [Ann Arbor: U. of Michigan Press, 1960], 159ff). It is not that such situations do not occur in war, for they certainly do; and it might well be that the intriguing and thought-provoking character of the discussion would be less pleasurable if the illustrating example involved one group of men trying to kill sixty men passing on either Road #1 or Road #2. If, however, this is the only kind of model invoked rather than one based on injuring, it leads, through many repetitions, to an understanding of disarming as the central activity. Often the model activity used is one of taking forts or taking passes. See, for example, discussions of a classic dilemma in military strategy called "Colonel Blotto" (John McDonald and John W. Tukey, "Colonel Blotto: A Problem of Military Strategy," in *Readings in Game Theory and Political Behavior*, ed. Martin Shubik [Garden City, N.Y.: Doubleday, 1954], 56–58). The concept of a "fort" is the concept of an "armed space" and so is itself a model for the action of "disarming." Although the taking and holding of territory requires "out-injuring," and although the held territory is itself a visual objectification of discrepancies in the capacity to injure, neither of these descriptions tend to be emphasized. For an analysis of types of military models (mathematical, verbal, maps, analogues, computer simulation), as well as many specific models (e.g., TEMPER, CARMONETTE, TIN SOLDIER), as well as analysis of the relation between scenario models and historical accounts, see Garry D. Brewer and Martin Shubik, *The War Game: A Critique of Military Problem Solving* (Cambridge, Mass.: Harvard University Press, 1979). Injuring is in none of these identified as the central event, although again it can of course be argued that it is being assumed. The mixing of idioms appropriate to sentient and nonsentient materials occurs often here as in the summary of output for the CARMONETTE VI simulation of land forces and helicopter battles: "Output is in the form of computer printout listing all events assessed, with a summary of all *casualty events*, and summation of *kills* by target type and weapon types. Also available are summaries of weapon engagements (firings) shown by target type, rounds fired, *personnel and vehicles killed* for each of the selected range brackets" (136, italics added).

19. B. H. Liddell Hart in *Strategy* (New York: Praeger, 1954) observes that Clausewitz's term "disarming" is often used without recognition of the bloodshed that is entailed (73). Clausewitz himself uses the term almost as a synonym, seldom introducing it without also introducing the acts of killing and injuring; and he has a much more sustained articulation of the place of injuring than does Liddell Hart. Nevertheless, the latter's point is an important one for two reasons: first, Clausewitz

is (as Liddell Hart notes) so often quoted out of context; second, despite Clausewitz's merging of "disarming" and "injuring," there is a major ambiguity in his use of the first term, as elaborated below, pages 97ff. and in notes 82, 90.

20. For example, historian and government advisor Richard Pipes concluded a PBS interview by William Buckley with the observation, "We are building a deterrent which, in the event of Soviet aggression, would go not after Soviet cities and Soviet civilians but after Soviet weapons" ("Is Communism Evolving?" *Firing Line*, 9 December 1982).

21. Just as the conflation of animate and inanimate materials leads to the more sustained perceptual confusion of renaming injuring "disarming," so the connotations of protection embedded in "disarming" may in turn help to sponsor the confusion involving the concepts of "offense" and "defense." Clausewitz calls attention to the major error that arises by conflating the idea of defense (protection) with the technical distinction between offensive and defensive positions in battle (392). Thomas Nagel implicitly questions the moral license accorded by the term "defense" when in "War and Massacre" (in *War and Moral Responsibility*, eds. Marshall Cohen, Thomas Nagel, Thomas Scanlon [Princeton: Princeton University Press, 1974], 21, n. 11) he writes, "I am not at all sure why we are justified in trying to kill those who are trying to kill us (rather than merely in trying to stop them with force which may also result in their deaths)." It might be that it would be impossible to invent an incapacitating-but-relatively-noninjuring defense weapon, but since no attempt has been made to do so, its possibility or impossibility cannot be discussed. The benign connotations of the "defense" idiom have in any event not motivated the search for and discovery of such a weapon.

22. Or, as Henry Kissinger once called them, the "so called 'kill-ratios' " (*American Foreign Policy: Three Essays* [New York: Norton, 1969], 105). The qualification is an interesting example of the loss of reality of injuring. The term "kill ratio" is an accurate and literal description of the ongoing comparisons of the body counts on the two sides in any war. To preface the term with the qualification "so called" seems an apology for the brutality of language, as though the term itself, rather than the phenomenon it literally describes, required perceptual crudity. That is, since killing has disappeared from our understanding of the event, its sudden intrusive appearance in language ("kill-ratios") seems a distasteful barbarity only introduced at the moment of description.

23. Omar N. Bradley, *A Soldier's Story* (New York: Holt, 1951), 519.

24. Bradley, 495.

25. Liddell Hart, 212. See also 187, 210, 228, 232, 243, 244, 358, 359.

26. Liddell Hart, 291. Clausewitz, too, has passages that include this massive bodily dance, though they tend to be more submerged (e.g., 293, 391).

27. Liddell Hart, 212, 187.

28. Liddell Hart, *History of the First World War* (London: Cassell, 1970; rpt-London: Pan Books, 1972), 183, 185.

29. Churchill, *Dawn of Liberation*, 135.

30. The assigning of the language of injury to a geographical territory or to the overall armed forces need not *necessarily*, of course, deflect attention from the actuality of human injury. It may occur in a way wholly compatible with, even encouraging of, a perception of the nature of actual injury, as when at the end of *The Guns of August* (New York: Macmillan, 1962), Barbara Tuchman summarizes the long period of trench warfare: "Not Mons or the Marne but Ypres was the real monument to British valor, as well as the grave of four-fifths of the original BEF. After it, with the advent of winter, came the slow deadly sinking into the stalemate of trench warfare. Running from Switzerland to the Channel like a gangrenous wound across French and Belgian territory, the trenches determined the war of position and attrition, the brutal, mud-filled, murderous insanity known as the Western Front that was to last for four more years" (438).

Although the convention of single combatants is almost universal in strategic, political, and historical writings—at the very least, the army will be referred to as having "flanks" and "wings"—it becomes most pronounced in war propaganda. That is, in the rhetorical extremity that occurs in the day-by-day immediacy of war, the convention is lifted out of the analytic texture where its effects are partially absorbed by the stylistic density of the writing, and now becomes exaggerated and caricatured in political cartoons, journalistic accounts, and political speeches. For example, in his 1 January 1945 broadcast to resistance groups in Denmark, Churchill urges his listeners to hold on, saying, "The Nazi beast is cornered. . . . The wounds inflicted by the armed might of the Grand Alliance are mortal" (*Victory*, 117). Although no one would argue that this is the appropriate moment to ask the population to recognize that though the Nazi colossus has Hitler's face, it is made up of immensely ordinary human beings, it is also clear that the convention, in its most extreme rhetorical form, not only eclipses the perception of existing injury but assists the actual infliction of the injury (because of the place it often has in persuading a population to engage in renewed war acts).

31. Here as well as in the discussion that follows, whenever the external issue of war is invoked, "freedom" will be used as an example. Obviously there are many issues in wars that have neither the sublimity nor the clarity of freedom. However, it is being used here to give, in effect, the imagined participants of war "the benefit of the doubt." The logic here might be summarized as follows: imagine a war, one arising out of an issue that most people would agree would be an issue worth fighting for, and now ask what is the relation between that issue and the act of injuring. If the answer to this is not happy, how much less happy it will be if the issues themselves are equivocal or even (as happens) unequivocally tawdry. It is an understanding of the relation between the activity of injuring and the external issue (regardless of content) toward which we move.

32. Although a "use" for a given by-product may be found, it originally has connotations of "uselessness" and even "waste." On the use of the word "wasted" for "killed," see Michael Walzer, *Just and Unjust Wars: A Moral Argument with Historical Illustrations* (New York: Basic, 1977), 109.

33. Theodore C. Sorensen, *Kennedy* (New York: Harper, 1965), 684, 687.

34. Michael Walzer, "World War II: Why Was this War Different," in *War and Moral Responsibility*, 97.

35. The idea of indirect targets was, for example, advocated by Liddell Hart in *Paris, or the Future of War* (New York: Dutton, 1925) and in a number of articles preceding the book. He explains in *Strategy* that he was led to this advocacy by the spectacle of bloodshed and stalemate in the trenches. When he later realized that "indirect" civil and economic objectives also entailed endless human catastrophe, he found that he was less successful in persuading the Air Staff of his error than he had been in persuading them of his original position (363f). However, even in later works such as the one in which this explanation occurs, he sometimes speaks of civilian and economic targets as "indirect" and as though they entailed something other than the infliction of human suffering (357).

36. In isolation, any one of these phrases may appear sensible, as with Clausewitz's use of the "blood is the cost of slaughter" phrase: "It is always true that the character of battle, like its name, is slaughter [*schlact*], and its price is blood. As a human being the commander will recoil from it" (259).

37. The production and road metaphors elaborated earlier may themselves be understood as instances of description that work by "extension." The work of extension can, however, take many other forms.

38. Like any definition that is quoted widely, Clausewitz's is translated and interpreted in many different ways. Usually it is understood to be asserting that politics or national policy continues to be enacted through war, that war is the continuation of "policy" through other means (e.g., Liddell Hart, *Strategy*, 366). At other times, it is understood as primarily raising a question about whether civilian political authority or instead military authority is predominant during war: for example, drawing on the writings of Engels and Lenin, V. D. Sokolovskiy argues that military strategy may sometimes be determined by a country's politics, while at other times its political and economic situation may follow from strategic necessity, as when a nation during war makes an alliance with another nation that during peacetime had been perceived to exist in the enemy camp (*Soviet Military Strategy*, ed. Harriet Fast Scott, trans. from the Russian [3rd ed.; New York: Crane, Russak, 1968, 1975], 19, 20). Sometimes, the emphasis of the definition is understood to be the perception that war is a continuation of less overtly aggressive forms of conflict, as in Anatol Rapoport's paraphrase, war is the continuation of debate by other means (vii). In still other contexts, its emphasis is taken to be that politics is always operative, even in arenas where it appears to be absent, as for example in literary criticism (W. J. Mitchell, "Editors Introduction: The Politics of Interpretation," *Critical Inquiry* 9 [Summer 1982], iii). All of these interpretations are compatible with Clausewitz's own

introduction of the definition, though his emphasis is the first. War is, he argues, a phenomenon composed of three tendencies—first, primordial violence; second, the play of chance and probability in interaction with the creative spirit; third, a partially controlling element of rationality—and it is his elucidation of the third tendency that occasions his discussion of war's partial subordination to national policy. (Though the tripartite division is only overtly announced at the end of I, i, the whole chapter is a very clear and systematic progression through the three).

39. Liddell Hart, *Strategy*, 339.

40. Liddell Hart calls attention to a series of strategic mottos, all of which register the fact that by one side creating a situation in which it has two alternative paths of victory, the other side is left with no alternative but surrender: Bourcet's prescription, "strategy must have branches," one or the other of which cannot fail; Napoleon's "faire son thème en deux façons," Sherman's "putting the enemy on the horns of a dilemma" (343).

41. Clausewitz, 230. An historical moment illustrating Clausewitz's point is Lee's retreat at the battle of Antietam. Lincoln telegraphs McClellan, "Please do not let him get off without being hurt" (Lord Charnwood, *Abraham Lincoln* [New York: Holt, 1917], 306). Lincoln interpreted McClellan's refusal to pursue and injure Confederate troops during the retreat as motivated by a lack of sympathy with the Union, and dismissed McClellan from his command (307–309).

42. Paul Kecskemeti, *Strategic Surrender: The Politics of Victory and Defeat* (Stanford, Ca.: Stanford University Press, 1958), 8.

43. See the Brookings study of the 215 incidents between 1946 and 1975 in which the display of arms or armed forces was used by the United States for political purposes (Barry M. Blechman, Stephen S. Kaplan, *Force Without War: U.S. Armed Forces as a Political Instrument* [Washington, D.C.: The Brookings Institute, 1978]). Case studies include Laotian War, 1962; Indo-Pakistani War, 1971; Lebanon, 1958; Jordan, 1970; Dominican Republic, 1961–66; Berlin Crisis, 1958–59, 61; Yugoslavia, 1951; Czechoslovakia, 1968. Although the study analyzes the different incidents for many different factors (whether the weapons were nuclear or conventional, whether the purpose was "to assure, compel, deter, or induce," what the U.S. president's popularity rating was at the time, and many others), it does not explicitly confront the question raised here of whether the display of force counters another's display, whether it is one-directional or reciprocating, and whether it should be understood as a positive alternative to war or instead as a negative alternative to nonaggression. To some extent, however, the answers to the question can be determined from context.

44. Bertrand Russell, *Has Man a Future?* (New York: Simon and Schuster, 1962), 78.

45. Mouloud Feraoun, *Journal 1955–62* (Paris, 1962) as quoted in Alistair Horne, *A Savage War of Peace: Algeria 1954–62* (London: Macmillan, 1977), 208.

46. Walzer, *Just and Unjust Wars*, 109. Walzer's voice is here conflated with Randall Jarrell's; the perception and the sad weariness of the "voice" belong to both writers at this moment.

47. Louis Simpson, *Air With Armed Men* (London: London Magazine Editions, 1972), 114.

48. Leonardo da Vinci, "The Way to Represent a Battle," in *The Notebooks of Leonardo da Vinci*, trans. and introd. Edward MacCurdy (New York: Reynal and Hitchcock, 1938), Vol. 2, 269–71.

49. Work and war are nonsymmetrical because one is world-building and the other world-destroying. Just as the deconstruction of world-building in torture is sometimes registered in the language used in the process itself, so again the terminology in war sometimes registers this same relation to ordinary projects of construction: for example, one type of M113 tank is called "bulldozer" and another "bridgelayer," just as there is a complex launcher that (perhaps because of its array of gadgets) has the official nickname "kitchen" (Christopher F. Foss, *Jane's World Armoured Fighting Vehicles* [New York: St. Martins, 1976], 294–305). That war entails the systematic unmaking of the made-world tends to be especially striking to those who witness war within a city. For example, reporting back to Soviet News from Berlin in 1945, Red Army correspondent Lt. Col. Pavel Troyanovsky describes with astonishment the deconversion of each house, garden, street, and square into a fortified space, and comments that Berlin has ceased to resemble Berlin: it is no longer a city but a "nightmare of fire and steel" (quoted in Strawson, 152, 153). Similarly, Victor Hugo gives a haunting description of the slow-motion, object by object deconstruction of Paris in the 1848 revolution as random fragments of domestic life—doors, grilles, screens, beds, pots, pans, rags,

window frames, roofing ridges, fireplaces, carts, tables—one-by-one go into "the making" of the three-story high, seven-hundred-foot long Saint-Antoine barricade (*Les Misérables*, trans. Norman Denny [Harmondsworth: Penguin, 1980], Vol. 2, 292–94).

50. It is on the basis of this particular problem that Quincy Wright, for example, objects to the identification of war as a game. See his discussion in *A Study of War* (Chicago: University of Chicago, 1942), Vol. 2, 1146f., especially note 2 where he demonstrates the connection between the idea of "struggle" and that of "progress" in writings by Walter Bagehot, Ernest Renan, Karl Pearson, Herbert Spencer, as well as the German social Darwinists, Gumplowicz, Ratzenhoffer, Treitschke, and Steinmetz.

51. Fussell, 23–29.

52. Even the fact that American soldiers were able to fly out of combat areas and spend the evening in recreational centers that included showers, ping-pong, and so forth was unsettling to many Americans at home. In other words, the "work" of the soldier, as described above, was in this war partially suspended since, as in ordinary forms of work, the gestures and conditions could be temporarily abandoned at the end of each day.

53. Liddell Hart, *Strategy*, 358. Similarly, when Robert E. Lee was asked what he would have done had the North taken Richmond, he responded, "We would swap queens," meaning the South would have taken Washington (Charnwood, 302).

54. For example, Arthur M. Schlesinger Jr. summarizes how the Kennedy administration viewed Nikita Khrushchev's motives during the Cuban missile crisis: "With one roll of the nuclear dice, Khrushchev might redress the strategic imbalance, humiliate the Americans, rescue the Cubans, silence the Stalinists and the generals, confound the Chinese and acquire a potent bargaining counter when he chose to replay Berlin. The risks seemed medium; the rewards colossal" (Leslie H. Gelb, "20 Years After Missile Crisis, Riddles Remain," *New York Times*, 23 October 1982).

55. Churchill, *Dawn of Liberation*, 138.

56. Tuchman, 295.

57. Alexander Haig, "Peace and Deterence," delivered at the Center for Strategic and International Studies, Georgetown University, 6 April 1982.

58. Kissinger, *American Foreign Policy*, 103.

59. Churchill, *Victory*, 240. It is interesting to notice that as the "contest" language is used by soldiers, then strategists, then politicians, it becomes successively more disembodied (the first tend to cite games that require embodied players, the second use board games with symbolic players like checker and chess pieces, the third more often use only abstract structural language). These successive levels of disembodiment thus correspond to the speakers' distance from embodied participation in the war.

60. Hugo Grotius, *The Rights of War and Peace*, trans. A. C. Campbell, introd. David J. Hill (Washington, D.C.: M. Walter Dunne, 1901), 18.

61. Wright, 1, Table 41, 646.

62. Such as studies that examine either the "legality" (e.g., writings by Wright) or instead the "justness" and "moralness" (e.g., writings by Nagel and Walzer) of various aspects of war.

63. Sokolovskiy, 33, 12.

64. Kecskemeti, for example, suggests classifying war according to three criteria: (1) symmetry or asymmetry of military outcome; (2) degree of totality; (3) symmetry or asymmetry of political outcome (17 and passim).

65. The special peculiarity of this insistence on the "binary" during a civil war struck Lincoln. Charnwood writes, "Lincoln writhed at a phrase in Meade's general orders about 'driving the invader from our soil.' 'Will our generals,' he exlaimed in private, 'never get that idea out of their heads? The whole country is our soil' " (358).

66. Wright, 1, Table 41, 646.

67. Carl Schmitt, *The Concept of the Political* (1928), trans. and introd. George Schwab, afterword Leo Strauss (New Brunswick, N.J.: Rutgers University Press, 1976), 26ff.

68. Fussell, 75–80.

69. Even if there are more than two contestants, and the multiple contestants each enter individually rather than being arranged into two sides, they may still be thought of as conforming to the dual

structure. In a talent contest, for example, the group of participants may first be thought of as divided into two rough groups, those who certainly do not qualify and those who may qualify for the prize; that second group now in turn is divided into two, those who no longer qualify and those who may still qualify, and so on through successive refinements until at last the judgment must be made between the two individuals who finally remain, the winner and the one who was almost the winner but is instead the loser (or second place winner). This would apply as well to other contests, such as a swimming race.

70. Clausewitz, 77. An example of a war whose outcome did not depend on the relative levels of physical injury but on the relative levels of unacceptable injury is Vietnam. Henry Kissinger writes that the kill-ratios of United States to North Vietnamese "casualties became highly unreliable indicators. They were falsified further because the level of what was 'unacceptable' to Americans fighting thousands of miles from home turned out to be much lower than that of Hanoi fighting on Vietnamese soil" (*Foreign Policy*, 105).

71. The ratio of participants to the total population is much higher in the twentieth century than earlier (Quincy Wright gives the figures of 1 to 5 percent for the armies of the seventeenth and eighteenth centuries, 14 percent for World War I, and almost 100 percent participation for World War II [232 n. 32; 234, 242–44, 570, and see 1965 edition for World War II figures]); but even in centuries when the percent of the population participating was much lower, it would be, of course, much greater than in other forms of contest.

72. There are, in these various kinds of contests, two basic ways in which the judgment is arrived at. In one form (which would include dance, singing, mechanical invention, scientific discovery), the judgment is rendered by a third person and nonparticipant. In the second form (chess, swimming, running; war would also be included here) the determinants for winning are built into the internal structure of the contest itself; the winner stands revealed at the end of the contest, or (perhaps more accurately) the contest is over at the moment the winner stands revealed. There is no external judgment or decision needed. In fact, it may be that one reason why it is so difficult to replace war with arbitration and mediation is that the second way of arriving at a judgment (in which judgment is built into the structure of the event) is being replaced by the first form (requiring an external assessment).

Within the second form, in which the verdict is achieved within the contest process itself, there is a division between those forms in which it is the winner who is identified and the identification of the loser follows (as in a race where the first runner is the "winner" and the designation of "losers" logically follows from the fact that the designation of "winner" has already been appropriated by someone else) and those forms in which it is the loser who is first identified and the designation of winner that follows from it (as in war). Although war seems unusual in conforming to the second model, it is not exceptional: boxing and a fatal duel would also conform to this model, which suggests that injuring contests of any scale tend to belong here. This would also be applicable to less immediately recognizable forms of injuring such as endurance contests: in the dance marathons of the 1930s, for example, the "winner" was the *last* to drop out; that is, the successive identification of the "losers" eventually led to the identification of the winner.

73. Kecskemeti, 23.

74. It is of course also the case that just as these imaginary contests would lead to war, so war itself leads to more war; war tends to be self-amplifying. Each side's verbalized narrative of the dispute begins to take as its explanation of its own aggression the war acts carried out by the other side: they bomb London; thus, we bomb Berlin. Each successive event becomes a retaliation for the last event initiated by the opponent, as in the concept of "escalation." So, too, the original initiation of war may have the anticipated initiation by the other side as its justification: had the U.S. started a war at the time of the Cuban Missile Crisis, the reason for the war would be the fact that the opponent had created the conditions of war. So, too, it has been observed that "mineral wars" would be tautological, because the very minerals over which the war would be fought are acutely necessary to a nation's welfare only because those minerals are needed in weapons systems and engines. A similar argument can be made about territorial wars, each side fighting for territory perceived to be strategically crucial in protecting itself from future wars.

75. The prize in an ordinary contest may have one of two relations to the moment of winning.

First, a person who wins in a contest the title of "best singer" may receive as a prize an invitation to sing with the Berlin opera company or to become a professor of voice at Juilliard. In this case, there is an intrinsic relation between the contested attribute and the prize, so much so that the "contest" may in fact have been an "audition" for the prize; it is a formal prelude to the position awarded. Second, the "best singer" may instead receive as a prize some money or a vacation. The vacation is irrelevant to the title except that it objectifies and enables the winner to experience for the duration of a week the victory that actually occurred in moments (for example, the announcement of the winner and the applause may have lasted five minutes). Does the relation between a nation's victory at the end of the war and its right to determine certain postwar issues conform to the first model or the second: is the right to determine the issues an extension of the intrinsic attributes manifest in victory (like the position in the opera company) or is it instead a temporal prolongation and objectification of the moment of victory (like the money and vacation)? It would seem to be almost always the second.

76. This is not to say that Clausewitz or other exponents of the power of enforcement argument explicitly frame the question as it is framed here. But the framing context is implicit: that is, there is an attempt to describe what war is and thus (implicitly) to describe what differentiates it from other phenomena that might work in its place; there is also the intuition that whatever it is that differentiates it has something to do with the way in which war ends.

77. Like Thamyris the Thracian in Book 2 of *The Iliad*.

78. Kecskemeti, 13, 22.

79. Clausewitz, 91.

80. Clausewitz, 230. The notion of "equality" here troubles Clausewitz specifically because of the equality of casualties. It should also be noted that the concepts of "equality" and "inequality" tend in general to be problematic ones in discussions of war. Because the ongoing activity of war assumes an at least temporary equality of military power, and the end (in theory) only comes about by arriving at an at least marginal inequality, many discussions about the prevention of war assume that an anticipated equality of damage will eliminate the outbreak of war. This assumption is prominent in the "balance of power" or "balance of terror" conception of preventing war, just as M.A.D. ("mutually assured destruction") was spoken of for many years as an anticipated equality of damage that would prevent nuclear war. A parallel assumption was widespread in the period prior to World War I. Norman Angell's *The Great Illusion*—which "proved" that war was now impossible because the financial interdependence of countries meant that the economic and commercial damage of war would be suffered equally by all participants—was, according to Barbara Tuchman (24, 25), a cult book translated into eleven languages. That this assumption is wrong is suggested both by historical example (despite predictions, World War I did occur) and by general observations (such as those by Clausewitz) on the near-equality of casualties. Achieving an inequality of damage may at least be designated the goal of war (if not the actual outcome), yet there is anthropological evidence that certain peoples explicitly seek to achieve "parity" through armed conflict; their war ceases when the two sides are satisfied that they have achieved an equal number of casualties (Irenäus Eibl-Eibesfeldt, *The Biology of Peace and War: Men, Animals, and Aggression*, trans. Eric Mosbacher [New York: Viking, 1979], 176).

On the problematic concept of equality, see also Quincy Wright on the *"juristic* equality and inequality of belligerents" which has tended to correspond to the relative physical equality or inequality of participants (II, 1393, 981) and Michael Walzer on the *"moral* equality" of soldiers (*Just and Unjust Wars*, 35–41, 127, 137).

81. Clausewitz, 80. Bernard Brodie in "A Guide to the Reading of *On War*" points out that this was true of Clausewitz's Prussia, "virtually annihilated as a military power" in the Jena campaign of 1806 but back strong in the campaigns of 1813, 1814, and 1815 (in *On War*, 644). This would also be true of France's rapid, three year recovery from the staggering reparations that had been imposed on her in 1871. It would again be relevant to the defeat of Germany at the end of World War I and her return in World War II (some descriptions attribute her return to the incompleteness of the defeat in World War I while others to the extremity of her former defeat and punishment). The armed conflict between Israel and Arab countries is sometimes seen as a sequence of wars (e.g., five in thirty years) and is at other times seen as one continuous war. Eric Rouleau calls it a "permanent

war,'' and one U.S. economist, Oscar Gass, predicts it will continue until at least the end of the century (André Fontaine, "La 'pax hebraica,' " *Le Monde*, 14 June 1982).

These occurrences, however, do not alter the accuracy of differentiating war from other contests on the basis of the duration of its outcome: its outcome is "abiding" if not "eternal."

82. Clausewitz's categories of "real" and "ideal," or "Real" and "Absolute," are preserved in general commentaries on his work and are analyzed by W. B. Gallie in *Philosophers of Peace and War: Kant, Clausewitz, Marx, Engels and Tolstoy* (Cambridge: Cambridge University Press, 1979), 37–66.

The tension between unlimited (absolute) and limited (real) war, which Clausewitz openly addresses at many points, may itself arise from a conceptually prior ambiguity that he does not so overtly address. That is, the conflicting tendencies of "total" and "partial" destruction of the enemy themselves reflect an ambiguity in the concept they qualify, "destruction of the enemy." At an early moment in the book, Clausewitz writes: "The fighting forces must be *destroyed*: that is, they must be *put in such a condition that they can no longer carry on the fight*. Whenever we use the phrase 'destruction of the enemy's forces' this alone is what we mean" (90). This announcement wrongly suggests that "the destruction of the enemy's forces," or "the destruction of the enemy's capacity to injure" (whether partial or total) can be understood as something distinct from "the destruction of the enemy itself" (whether partial or total). But this is not the case since a live enemy population has, as a basic characteristic of the condition of aliveness, the capacity to inflict injury (it may, for example, if divested of its weapons, go on to invent new ones). Although Clausewitz tries to maintain the qualification—rarely using the phrase "destruction (or annihilation) of the enemy" without also in close proximity reintroducing the alternative phrase "destruction (or annihilation) of the enemy's forces" (see 92, 95)—he never explains what the second, as distinct from the first, would mean. For example, he writes, "What do we mean by the defeat of the enemy? Simply the destruction of his forces, whether by death, injury, or any other means" (227): the "means" other than death or injury is asserted but left unspecified.

83. J. F. C. Fuller, *A Military History of the Western World*, Vol. 3 (New York: Funk and Wagnalls, 1956).

84. Kecskemeti, 1, 216, 237, 239. Thus we tend to think of the extremity of defeat of World War II as typical, while it may in fact be anomalous, and conversely, tend to think of the lack of extreme defeat as in Vietnam as anomalous while it may be normal or an exaggeration of the normal.

Kecskemeti's important study is unusual in the literature of war for its sustained attention to the nature of war's ending.

85. Harry S. Truman, *The President's Message to the Congress: A Program for United States Aid to European Recovery* (19 December 1947) and George C. Marshall, *Assistance to European Economic Recovery: Statement before Senate Committee on Foreign Relations* (8 January 1948), Department of State: Publication 3022, Economic Cooperation Series 2 (Washington, D.C.: GPO, 1948).

86. Truman, 16. In fact, so far from punitive was the U.S. in this matter that Marshall at one moment in a different address acknowledges that "The charge has frequently been made that the United States in its policy has sought to give priority to restoration of Germany ahead of those other countries of Europe" (*The Problems of European Revival and German and Austrian Peace Settlements*, address 18 November 1947 to meeting sponsored by Chicago Council on Foreign Relations and the Chicago Chamber of Commerce and nationally broadcast, Dept. of State, Pub. 2990, Eur. Series 31 [Washington, D.C.: GPO, 1947], 13). Although what is at issue here is economic revival and not military revival, it is always recognized that the first can be relatively quickly translated into the second. Churchill once observed that the defeated country might have the advantage in any war that followed: "The nation that is defeated and disarmed in this world war will stride into the next with an advantage, for it will develop the new arms while we attempt to make the old ones do" (as cited in Bradley, 497). The argument against punishing the Axis populations was made by British as well as U.S. leaders: in his 13 May 1945 radio broadcast, Churchill says, "There would be little use in punishing the Hitlerites for their crimes if law and justice did not rule, and if totalitarian or police governments were to take the place of the German invaders," an argument that can be extended to Germany itself (*Victory*, 179).

87. Truman, 12. The only other country specified is Great Britain, but here the reference is much vaguer and carries much less successfully the idea of compelling recovery. Immediately after citing Germany's increase from 230,000 to 290,000 tons, he adds, "Similarly, coal production in the United Kingdom has risen markedly in recent weeks."

The importance of coal to the recovery of Europe is stressed not only in the technical reports on resources ("Appendix C: Summaries of Technical Committee Reports" of the *Committee of European Economic Co-Operation General Report* [Paris, 12 September 1947], Vol. 1, 80–86) but also in the "Historical Introduction" to the *General Report*: the recovery of Europe began impressively immediately after the war but then underwent a severe setback in the winter of 1946–47 because of coal shortages, especially damaging because of the intensity of the cold weather that winter (7).

Throughout the various reports and speeches surrounding the Marshall Plan, it becomes clear that preventing Germany's rapid economic growth is perceived as harmful to the United States in three ways. First, it will prevent Germany from becoming self-supporting and thereby necessitate larger financial contributions from America (Marshall, "Problems of European Revival," 13). Second, it will prevent Europe from achieving full recovery, since Germany's contribution is greatly needed for that recovery; and Europe's well-being is in a very deep sense required for the United States' well-being. Third, Germany's economic and industrial recovery comes to be understood as a way of preventing German militarism. In Marshall's original 5 June 1947 speech at Harvard, political stability and economic well-being are equated: "It is logical that the United States should do whatever it is able to do to assist the return of normal economic health in the world, without which there can be no political stability and no assured peace. Our policy is directed not against any country or doctrine but against hunger, poverty, desperation, and chaos" (*European Initiative Essential to Economic Recovery*, Dept. of State, Publication 2882, Eur. Series [Washington, D.C.: GPO, 1947], 4). In later articulations of this point, Germany's political stability (and by extension her military harmlessness) is still attributed to her economic well-being, but to an economic well-being that has an internal check on it because it is being shared with the rest of Europe. In *The Problems of European Revival and German and Austrian Peace Settlements*, Marshall specifies two ways of assuring that Germany will be peaceful in the future: first, disarmament; second, the production and sharing of her Ruhr valley coal with other communities so that she cannot again become an economic colossus (12, 13). So, too, in the *General Report*, the contribution of the Ruhr coalfields to the rehabilitation of Europe is explicitly designated a way of not allowing the German economy to "develop to the detriment of other European countries" (69). Viewed within the context of these considerations, Truman's recitation of Germany's leap from 230,000 tons to 290,000 tons a day is a celebration of the probability of Germany's revival, a celebration of the probability of the revival of the European community as a whole, and a celebration of the probable absence from the continent of any country with a natural dominion over her neighbors. That the western allies would have to assure themselves about the future harmlessness of Germany is unremarkable; what is remarkable is that they discovered a form of self-assurance that not only included Germany's well-being but was in fact premised on it.

88. For the most part, the various writings on the reconstruction of Europe do not even speak of Germany as a country separate from the rest of Europe; that is, the "us-them" language of necessity so operative in the war has by now disappeared. Because Germany was not represented at the Paris meetings, the *General Report* does recognize the distinction by using binary phrases such as "the participants and Western Germany." Marshall on the other hand (even in *The Problems of European Revival and German and Austrian Peace Settlements*) tends to speak simply and collectively of Europe. Insofar as there is "us-them" language in these writings, the "them" is not Germany but the Soviet Union. (See Arthur Schlesinger, Jr.'s classic article, "Origins of the Cold War," *Foreign Affairs* 46 [October 1967], 22–52.) What is most important about this disunity from the point of view of the issues raised in the present analysis is that the split between the U.S. and the U.S.S.R. *is represented* in these writings as occurring precisely on the two countries' different conceptions of the structure of war—on, that is, the Soviet Union's assumption that war carries the power of its own enforcement and that, as almost a structural necessity, the defeated must be without the power of self-renewal. Even in the original Harvard speech, Marshall says, "Any government which maneuvers to block the recovery of other countries cannot expect help from us. Furthermore gov-

ernments, political parties, or groups which seek to perpetuate human misery in order to profit therefrom politically or otherwise will encounter the opposition of the United States'' (4). In his 8 January 1948 speech before the Senate Committee on Foreign Relations, he is more explicit, presenting Russia as encouraging ''economic distress'' rather than recovery (7); and similar descriptions occur in ''The Problems of European Revival'' (4, 8, 9, 10).

89. On the convention of single combatants, see above 70–72, and on the binary, see above 87–88. Although the concepts of ''two'' and the ''binary'' are related, they are distinct, the concept of ''two'' having attributes that the ''binary'' does not have. Although the binary is a necessary psychological and structural characteristic of war, ''two'' is emphatically not, and the conflation of the terms is an unfortunate one.

In connection with the convention of single combatants, it is interesting that Grotius notes the etymology of ''*bellum*'' in ''*duellum*'' in his description of war entailing the fundamental condition of disunity, and peace the fundamental condition of unity (18).

90. Sigmund Freud, ''Why War?'' in *Character and Culture*, ed. and introd. Philip Rieff (New York: Collier-Macmillan, 1963), 136, 138.

Clausewitz's *On War* opens with an invocation of the two-person model: ''I shall not begin by formulating a crude, journalistic definition of war, but go straight to the heart of the matter, to the duel. War is nothing but a duel on a larger scale. Countless duels go to make up war, but a picture of it as a whole can be formed by imagining a pair of wrestlers'' (75). His use of the two-person model is complicated by his alternation here between one version of the model in which the opponent is killed (the duel) and one version in which the opponent is divested of his power but not killed (the wrestling match), thus showing the ambivalence about whether the ''annihilation of the enemy's injuring power'' ultimately requires the ''annihilation of the enemy'' present throughout (see n. 82). That is, it is as though Clausewitz is aware of the problem of the translation of the two-person model into the two-peoples model and retroactively builds the ambiguity into his modified two-person model in this most carefully revised and rewritten chapter of the work.

91. Alexander Haig, ''Peace and Deterrence,'' cited in Theodore Draper, ''How Not to Think About Nuclear War,'' *New York Review of Books*, 15 July 1982, 38.

92. Draper, 38.

93. See, for example, Walter Millis, ''Truman and MacArthur,'' and Morton H. Halperin, ''The Limiting Process in the Korean War,'' in *Korea: Cold War and Limited War*, ed. and introd. Allen Guttmann (Lexington, Mass.: D.C. Heath, 1967), 69–78, 181–201.

94. Kissinger, *American Foreign Policy*, 193f. For example, U.S. maps of Vietnam were, according to Kissinger, neatly divided into three color zones for government, contested, and Viet Cong held territories, while such territorial divisions were irrelevant to Hanoi and the strategy of guerilla warfare; further, their validity was also undercut by the actual *temporal* division of territories that for a while escaped U.S. attention, Saigon holding control of villages in the daylight hours, Hanoi holding sway after dark and throughout the night.

95. Douglas MacArthur, ''No Substitute for Victory,'' Letter to Representative Joseph W. Martin, 5 April 1951, in Guttmann, 20.

96. Liddell Hart, *Paris, or the Future of War* (New York: Dutton, 1925), 41–43, and passim. See also *Strategy*, 363.

97. Kecskemeti, 192–206. (Kecskemeti's conclusions are based on an analysis of many factors such as the postwar interviews with Japanese policymakers conducted by the United States Strategic Bombing Service team.)

98. Walzer, ''World War II: Why Was this War Different,'' 101. See also David Irving, *The Destruction of Dresden*, introd. Ira C. Eaker (New York: Holt, 1964).

99. For example, Liddell Hart writes that ''Soldiers universally concede that general truth of Napoleon's [words]'' (*Strategy*, 24); and Sokolovskiy cites both Engels and Lenin as designating the ''moral'' element decisive, but he then goes on to cite numerous sources from the West and concludes that ''modern bourgeois military theoreticians'' are in their writings inclined to overestimate the significance of this element (33–35).

100. Throughout both his speeches and his memoirs, Montgomery repeatedly designates morale ''the single most important factor'' (*Forward From Victory: Speeches and Addresses* [London:

346 NOTES

Hutchinson, 1948], 76, 97, 204, 237, 270, 273; and *The Memoirs of Field-Marshal Montgomery* [Cleveland: World Publishing, 1958], 77, 81, 112, 388). Though this repeated observation is sometimes occasioned by very humane considerations (such as when he is advocating better medical care for the soldiers), it is at many times occurring in the midst of the circular argument that morale is the greatest single factor in victory, and that victory is the most important factor in creating high morale.

101. The truth of this last point is debatable: it can certainly be argued that the ability to go on injuring and to bring about the surrender of the opponent is (every bit as much as nursing another while oneself in pain) a manifestation of the "spirit." This is, for example, a central thesis in Hegel's analysis of the master-slave relation. But the central and most crucial point here is that, whether or not this is a manifestation of the spirit, it is in no sense separate from the activity of injuring and out-injuring, even though it is invoked in military descriptions as something distinct from those physical actions.

102. The visual image of "high morale" is sometimes derived from images of elated soldiers after "victory" has been achieved, or virtually achieved—the exuberance of American and British soldiers in the Ruhr valley seeing hanging from house after house the white bed sheets of surrender (e.g., Bradley, 494), or the elation of Zhukov and the Russian soldiers in Berlin seeing the red flag run up on the Reichstag (Strawson, 155).

For visual images of the terror, fear, and exhaustion in battle, see da Vinci, 269–71, and also E.V. Walter, "Theories of Terrorism and the Classical Tradition," in *Political Theory and Social Change*, ed. and introd. David Spitz (New York: Atherton, 1967), 133–60, as well as Clausewitz's 1, 4, "On Danger in War."

103. While the first three objections are specific to the "moral-morale" argument, the last is structural and would apply to other explanations that insert into war's ending some activity other than injuring. It may even be that this would be applicable to Clausewitz's explanation of how problematic wars sometimes end by one side assessing the conditions and judging that victory is unachievable and the damages of trying too great. Here the activity of "injuring" is replaced by the activity of "thinking"; but if it is to be replaced by thinking, why only make the substitution in the last moments rather than substituting it for injuring altogether?

104. Pierre Bourdieu, *Outline of a Theory of Practice*, trans. Richard Nice (Cambridge: Cambridge University Press, 1977), 95.

105. Mark Zborowski, "Cultural Components in Response to Pain," *Journal of Social Issues* 8 (1952) 16–30; and M. K. Opler, *Culture and Mental Health* (New York: Macmillan, 1959), and "Ethnic Differences in Behavior and Health Practices," in *The Family: A Focal Point for Health Education*, ed. I. Galdston (New York: New York Academy of Medicine, 1961).

106. François Jacob, *The Logic of Life: A History of Heredity*, trans. Betty E. Spillmann (New York: Pantheon, 1973), 75–81 and passim.

107. On the involvement of the antibody system in war and invasion, see William H. McNeill, *Plagues and Peoples* (New York: Anchor-Doubleday, 1977); and for conscious use of this, see literature on germ warfare. On the conscious alteration of the gene pool in political situations, see, for example, studies of the policies regarding intermarriage held by different colonizing European countries in Africa.

108. This point has been made by many people both in fiction and nonfiction. For example, the immunity of the body to the state, even when all other aspects of personhood have become alterable, is persistently visible in the plays of Bertolt Brecht whose characters, even when they have surrendered all aspects of consciousness to some political entity outside themselves, continue to have bodies incapable of taking orders: "Stop limping," shouts a sergeant to an otherwise obedient private moving over the mountains of *Caucasian Chalk Circle*, "I order you to stop limping!"—an order that is unsuccessful. So, too, in the more uniformly harsh idiom of *A Man's A Man*, where the military characters are interchangeable in name, uniform, and verbal acts of self-nullification (a character named Galy Gay becomes a character named Jeriah Jip and then a character named Bloody Five), the body continues to be (from the point of view of a perfect military utopia) grotesquely individuating: Jeriah Jip is recognizable by his vomit, Bloody Five by his uncontrollable erections,

Galy Gay by his large appetite for food. These bodily attributes remain outside the sphere that can be remade by the military, and preserve the fact that there are three discrete individuals involved.

So, too, physical pain, either naturally occurring or self-inflicted, is often represented as an individual's last hold on personal identity before the surrender to an external force or system: a girl in Ionesco's *The Lesson* has only the blaring pain of a toothache to make possible the act of resistance to her teacher-dictator, just as in popular movies like *Ipcress File* or *Thirty-Six Hours*, it is again physical pain—the gash of a nail in one case, a small paper cut in the other—that makes resistance possible. Of course, if the state itself inflicts and controls the pain, as in torture and war, it then controls the body as well as all aspects of consciousness.

What is represented in literature is especially important insofar as it reflects what actually occurs in historical reality where the loyalty of the body to its own impulses and origins is even more hauntingly visible. Bruno Bettelheim, for example, describes a solitary moment of resistance at one of the concentration camps when a German guard recognized in a line of women entering the showers a woman who had been a dancer. He ordered her to step out of the line and dance for him. She did so, and as she moved into the habitual bodily rhythms and movements from which she had been cut off, she became reacquainted with the person (herself) from whom she had lost contact; recalling herself in her own mimesis of herself, she remembered who she was, danced up to the officer, moved her hand with grace for his gun, took it, and shot him. Though she was of course herself moments later killed, the story is cited by Bettelheim because of the courage displayed there, and because it constituted an exceptional moment of resistance, and exceptional precisely because for the most part the human body was so successfully appropriated by the state in the camps. It is not coincidental that the most precious survival advice Bettelheim himself received in the camps was to retain as much control over his body as possible, by determining the time of day in which he would take food into his mouth (or even cloth, anything that would allow the mime of chewing food and permitting the entry of the external world into the body), and so also the time of day given to excretion, preserving by these apparently modest acts his own autonomy over an intimate sphere beyond the reach of the state (*The Informed Heart: Autonomy in a Mass Age* [New York: Free Press, 1960], 264, 265, 132, 133, 147, 148).

109. Christopher S. Wren, "China's Birth Goals Meet Regional Resistance," *New York Times*, 15 May 1982.

110. Although many objections to school integration were raised out of racist impulses, others were made on the basis that it was unfair to make children responsible for revisionary justice that should be carried out by adults. On this basis Hannah Arendt, for example, objected in her "Reflections on Little Rock" (Elisabeth Young-Bruehl, *Hannah Arendt: For Love of the World* [New Haven: Yale University Press, 1982], 309–13). Significantly, Arendt thought integration should have as its primary focus the repeal of miscegenation laws: that is, the adult human body, through intermarriage, was the most important locus of learning racial equality.

111. Bourdieu, 94.

112. American Law Institute, *Second Restatement of the Law of Torts* (St. Paul, Minn.: American Law Institute Publishers, 1966), Vol. 2, Sec. 402A (Reporter: William L. Prosser), 349f. and Appendix, Vol. 3, 1f.

113. Vincent Bugliosi with Curt Gentry, *Helter Skelter: The True Story of the Manson Murders* (New York: Norton, 1974; rpt—New York: Bantam, 1975), 383. Bugliosi, the prosecutor of the Tate-LaBianca trials, is here describing the requisitioning of handwriting samples from the defendants.

114. Interview with David Ogden, Clerk to Supreme Court Justice H. A. Blackmun, 5 August 1982.

115. An understanding of the kind of consent process entailed in war requires a model wholly different from that which would include (for example) voting. The outcome of a vote is an outcome consented to but it is itself an objectification of what was *wanted* as an outcome: thus the acceptance of the outcome of voting is unproblematic; only its nonacceptance would be peculiar. The outcome of war, in contrast, is accepted when it is deeply antithetical to what was wanted by a population.

116. For example, Churchill in March 1944 assesses the war in the Pacific as one in which three or four Japanese are being injured for every one Allied soldier (*Dawn*, 56), or again, President

Lyndon Johnson in his 18 June 1966 "News Conference Statement" about Vietnam, notes, "Since January 1, 1966 we have lost 2,200 of our men; the South Vietnamese have lost 4,300 of their men; our allies have lost 250 of their men. But the Viet Cong and the North Vietnamese have lost three times our combined losses. They have lost 22,500 of their men" (in *Documents on American Foreign Relations: 1966*, ed. Richard P. Stebbins with Elaine P. Adam [New York: Harper for Council on Foreign Relations, 1967], 225). The citation of kill-ratios is too ubiquitous and familiar to require the recitation of additional instances here.

117. For the meaning of "out-injures," see above p. 128.

118. William H. McNeill, *The Pursuit of Power: Technology, Armed Force, and Society since A.D. 1000* (Chicago: University of Chicago Press, 1982), 133.

119. McNeill, 138n., 132, 129.

120. McNeill describes the spread of technology as well as, for example, drill books (135). Strategy, too, tends to be international. While this is of course true of Clausewitz (whose work was as familiar to the Allied as to the Axis leaders in World War II, as it is now equally familiar to contemporary military theoreticians in the U.S. and U.S.S.R.), it is also true of less classic texts. Israeli General Yigael Yadin describes his use of Liddell Hart's *Strategy of Indirect Approach* in the 1948–49 Arab-Israel War, and mentions that among documents belonging to a captured Egyptian commander was this same book, which became a cherished souvenir. He adds in a footnote that the Egyptians "did not grasp the essence of the book" and could therefore be surprised by the Israelis' strategic use of its concepts (Yigael Yadin, " 'For By Wise Counsel Thou Shalt Make Thy War.' A Strategical Analysis of the Arab-Israel War," in *Strategy*, 386–404, see 396).

The culture of war is a shared culture. So Erwin Rommel approaches a Rumanian battery in World War I and finds that they are "Krupp guns! German workmanship!" as he writes in his strategic analysis of that war, *Infantry Attacks* (trans. G. E. Kiddé [Potsdam: Ludwig Voggenreiter Verlag, 1937; Washington, D.C.: The Infantry Journal, 1944], 93), and in turn American General Patton reads Rommel's book immediately before the Saar Campaign of World War II ("Diary, 8 November 1944," in Martin Blumenson, *The Patton Papers: 1940–45* [Boston: Houghton Mifflin, 1974], 571), an event memorialized for the American public in the film, *Patton*, when George C. Scott, his voice thick with excitement, looks through field glasses at his imagined opponent and shouts triumphantly, "Rommel!! You magnificent bastard!!! *I READ YOUR BOOK!!!!*"

121. Here and wherever the discussion refers to the loss of national beliefs, the only "beliefs" that are being referred to are the "disputed beliefs," the ones that are at issue in the war. In most cases this represents only a small fraction of national self-belief (though large enough to have warranted war).

122. Homer's consistent inclusion of some benign attribute of civilization in the midst of a description of a man's death is often interpreted as a way of crediting the soldier at the moment of his loss; and because it occurs equally in his depiction of Greeks and Trojans, is seen as indicative of his deep fairness. See, for example, Simone Weil's discussion of Homer's "extraordinary sense of equity" in *The Iliad, or The Poem of Force*, trans. Mary McCarthy (Wallingford, Pa.: Pendle Hill, 1956) 32f; as well as many other commentaries on epic objectivity. The interpretation given here is meant as an additional rather than an alternative explanation of Homer's inclusion of these details in the moment of the warrior's death.

123. Homer, *The Iliad*, trans. A. T. Murray (Cambridge, Mass.: Harvard University Press; London: William Heinemann, 1924), Vol. 1, 5, 70f; 59f; 6, 13f; 12, 378f.

124. As was stressed earlier, and as will be returned to in later chapters, this process of transfer is facilitated by and occurs across the sign of the weapon.

125. These different forms of substantiation are elaborated briefly at the end of this chapter, and developed more fully in the second half of the book.

126. This distinction is described more fully in Chapter 1, 38–40, 57–58, and in Chapters 4 and 5 in Part Two.

127. D. D. Kosambi, *The Culture and Civilization of Ancient India in Historical Outline* (London: Routledge & Kegan Paul, 1965), 102.

128. The confusion between invincibility achieved by the nonmortality of death on the one hand

and by the nonmortality of immortality on the other perhaps has as a parallel the rich history of confusion between atemporality and eternity, a confusion that occurred because both are nontemporal.

It is interesting to notice in this connection that the architecture and technology of "protection" in war tends to go in two different directions. One type (as can be seen in helmets, air raid shelters, radical city) is a materialized image of decreased sentience: it tends (whether in something the size of a helmet or instead a cement block shelter) to have thick uniform surfaces unbroken by any opening into the outer world; they look defensive, full of fear, and (from the outside) totalitarian. The other type of protective device is "supersentient"; that is, the object has many instruments of extension that magnify the work of the senses, through devices exaggeratedly sensitive to heat, vision, and smell (see, for example, the visual images in Keith Mallory and Arvid Ottar, *The Architecture of War* [New York: Pantheon-Random, 1973], 215, 216, 229, 219, 275).

129. Genesis 24:2–9; see also 47:29.

130. Although the turning inside out of the body is most extreme and literal in wounding, it is also at least mimetically present in other kinds of rituals: the Angami Naga taking an oath wears his clothes inside-out, the seams displayed to the outside world (John Henry Hutton, *The Angami Nagas* [London: Oxford University Press, 1969], 144). It should be stressed that what is important in all these examples is that the body is made an emphatic and unignorable object of perception. The literal fact of "opening" or exposing the body is just one way of accomplishing that: thus the phrase often used here, "open body," may be more generally understood as the "unignorable body."

131. These and many parallel instances are found in A. E. Crawley, "Oaths" in *Encyclopedia of Religion and Ethics*, ed. James Hastings with John A. Selbie and Louis H. Gray (New York: Scribners, 1928), Vol. 9, 431, and have been collated from many sources.

132. *Bureau of American Ethnology Annual Report* (Washington, D.C.: GPO, 1881–1933), Vol. 23, 485, 511, 513, 514. Dennis Tedlock interprets this act as indicating "honesty" or "speaking from the heart," since the lethal end of the weapon is in the interior of the body, and the wind moving through the feathered end emanating from the mouth represents the voice ("In Search of the Miraculous at Zuni" in *The Realm of the Extra-Human: Ideas and Actions*, ed. Agehananda Bharati [The Hague: Mouton, 1976], 273–83).

133. J. H. Hutton, *The Sema Nagas* (London: Oxford University Press, 1968), 165. Serious oaths among the Lhota Nagas are sometimes accompanied by the act of making a mixture of materials that includes material from the latrine, and thus from the inside of the human body (J. P. Mills, *The Lhota Nagas* [London: Macmillan, 1922], 102); and the most serious oath among the Ao Nagas was taken on a human skull (Mills, *The Ao Nagas* [London: Macmillan, 1926], 126).

Both in these works on contemporary peoples, and in Crawley's account of oaths among ancient peoples, there are instances in which no actual body, human or animal, is cut open; but the image of the open body is presented in an accompanying verbal image, "If my words are not true may I be torn open by a wild animal" and so forth. Thus here the work of material substantiation is accomplished imagistically.

134. These and other differences in national self-description in the period preceding World War I are given by Tuchman, 17–43.

135. Often the dispute over which country's political reality will be recognized becomes a dispute over which country's physical reality will be recognized—that is, over which side's suffering will be recognized. So, for example, the duration of the Arab-Israeli dispute has sometimes been attributed to the fact that each side perceives itself as the victim and insists on being so recognized internationally. Similarly, whether a nation's action in an international conflict is condemned or endorsed by other nations often turns on whether it is perceived as acting out of strength or vulnerability: while, for example, Russia's invasion of Afghanistan was universally recognized as an act of aggression rather than defense, it was less strongly condemned by countries that saw it as a manifestation of Russia's long-standing anxiety about the defenselessness of her own borders. Because the claim to "suffering" and even "physical suffering" has such an important place in the structure of dispute, the U.S.-Iran confrontation might be understood as a model of this aspect of conflict; on one side was the reality of individual suffering of the dying Shah; on the other side was the

reality of suffering of many tortured Iranians; and the suffering of the hostages became an intermediate third term (intermediate both numerically and nationally) through which each side argued the legitimacy of its position.

136. That each side credited the reality of the other's claim was visible not only in the central events (the transportation of the dying Shah away from the American mainland; the eventual return of the United States citizens) but in many incidental events as well. The fact that the United States was willing to televise each day the Iranians' claims was itself a major act of crediting the Iranian right of self-description (more typically in a dispute, communication with the other side is cut off). The Iranian acknowledgment of U.S. cultural and political reality was visible in their obviously stress-filled attempts to find a spokesman who would have a compelling "believability" to Americans, not a religiously gowned mullah, not even the partially westernized Ghotbzadeh, but the wholly westernized, elegant European figure of Bani-Sadr. Another way of phrasing this is to say that each side realized that certain aspects of itself would have "unreality" within the other's borders, that a religious spokesman would not "fly" on American television sets, just as American technology (helicopters) would not fly in the sands of Iran.

137. On the relation between the psychological act of "believing," which must be renewed each day, and the material objectification of belief that relieves the believers of having to renew the action of "believing," see below, Chapter 3, pp. 171–72 and Chapter 4, pp. 219–43.

As one may be relieved of the work of believing by materialization, so conversely the dissolution of the external objectification may necessitate a return to the work of belief. Thus, feelings of national belief, or "patriotism," tend to become most acute at the moment when the external forms are jeopardized.

138. Thus it is possible to back up for a moment to the metaphors used for injury and understand why each of them asserts the transformation of the hurt body into the external issue, why (if one is including the external issue in the metaphor) injury is the intermediate product that will in its final formation turn into freedom, why injury is the cost or money that will in the end be exchanged for, traded in for, or transformed into freedom; why injury is the road at the end of which there suddenly appears (as though an inevitable extension of the road itself) the town of freedom. Insofar as each of the metaphors calls attention to a phenomenon of transformation or transference, it calls attention to something that literally occurs in war; for the attributes of the hurt body are "transferred" to the issues, the attributes of the hurt body are "transformed" into attributes of the issues."

But the unanchoredness of the external object of war makes it clear that the metaphors are often used inaccurately. For example, attempts to assess the issues of war by means of the "cost" metaphor often take the form of the sentence, "Is issue X worth it?" (a form that can be used in either a justification of or a lament for the injuries). But the phrase misstates the relation between the "X" and the "it." Issue X is not worth the death and damage; issue X is not worth anything—or at least its worth has been seriously called into question by one large group of people (the opponent). The exponents of issue X are trying to invest it with worth through the process of war. Thus, the issue is not worth the injury; rather, the injury confers worth on the (otherwise worthless) issue.

139. Just as in dispute, each side tends to perceive its own description as "real" and the other side's as an invention, so too within the actual moment of conflict, each side is openly aware that the other side is lying but does not describe its own strategic acts as lies. In, for example, the account of the 1958, 1959, 1961 Berlin Crisis given in The Brookings Institute's *Force Without War*, the authors explain that the situation was complicated by the fact that Khrushchev had in 1958 hinted that Soviets were "producing ICBMs on a regular basis." The United States knew this was a lie, a "bluff," but could not openly say so because its information was gained by the "high-level U-2 flights" begun in 1956. Only when these flights were "inadvertently exposed in May 1960 when the U.S.S.R. finally succeeded in bringing one down" could the United States alert the public to the hollowness of Khrushchev's boast. Throughout the description, the actions of the two countries are presented as though they are nonsymmetrical: it is the Soviets who are "lying," and the United States is presented as being at the mercy of its opponent's dishonesty despite the fact that it is smart enough to know the opponent is lying. But on the issue of "truthfulness" the two countries are in a parallel position: the U.S. cannot call Khrushchev's bluff because it is itself lying, engaged in secret activity of surveillance unknown to the opponent or to its own population. Thus two kinds

of distortion occurred, the U.S.S.R. claiming more than they had; the U.S. acknowledging less than they had (348).

140. Liddell Hart, *History of the First World War*, 259.

141. Clausewitz, 202, 215, 218, 233.

142. Niccolò Machiavelli, *The Prince*, trans. Luigi Ricci, revised E.R.P. Vincent (London: Oxford University Press, 1925; rpt—New York: Mentor, 1952), 42, 57, 92, 100.

143. Arthur Schopenhauer, *The World as Will and Idea*, trans. R. B. Haldane and J. Kemp (New York: Dolphin-Doubleday, 1961, 346–51. Schopenhauer writes: "In all cases in which I have a right of compulsion, a complete right to use *violence* against another, I may, according to the circumstances, just as well oppose the violence of the other with *craft* without doing any wrong, and accordingly I have an actual *right to lie precisely so far as I have a right of compulsion....* Whoever would deny this must still more deny the justifiableness of strategem in war, which is just an acted lie (350, 351).

144. Summaries of the contents of these various peace plans are given in Sylvester John Hemleben, *Plans for World Peace through Six Centuries* (Chicago: University of Chicago, 1943), 84, 70, 93.

145. Hemleben, 163, 158, 165, 172–74.

146. Article III, "Major Provisions of the Treaty on the Non-Proliferation of Nuclear Weapons," (Text of Treaty signed at Washington, London, and Moscow, 1 July 1968) in *Progress in Arms Control?*, introd. Bruce M. Russett and Bruce G. Blair (San Francisco: W. H. Freeman, 1969), 230.

147. Fletcher Pratt, *The Battles That Changed History* (New York: Doubleday, 1956), 316.

148. Pratt, 336.

149. Kecskemeti, 104, 105.

150. See Nicholas Harman, *Dunkirk: The Patriotic Myth* (New York: Simon and Schuster, 1980).

151. For example, Omar Bradley emphasizes in his memoirs the precision of the campaign west of the Rhine that made it a textbook model of maneuvers (506), and also assesses positively the strategic logic of staying in the Ruhr, though he also acknowledges that belief in the myth of German Redoubt played some part in this decision (536, 537). Field-Marshal Albert Kesselring in contrast writes in his memoirs that the Germans assumed that from a strategic point of view "the Ruhr had no interest for Eisenhower; his objective lay to the east;" and thus when Eisenhower decided to concentrate on the Ruhr it was the fulfillment of "the hope I had hardly dared conceive—that strong American forces would allow themselves to be drawn into the mountains by our weak troops" rather than progressing to Berlin (*Kesselring: A Soldier's Record*, trans. Lynton Hudson, introd. S.L.A. Marshall [New York: Morrow, 1954], 300, 303, 313, 314, 315, 317). Thus, from Kesselring's point of view, the decision to stay in the Ruhr assisted German rather than Allied objectives (328, 330; for example, "The Russians had broken through and were closing in on Berlin towards the end of April. While the decisive battle of the war was awaited there, the British and American forces were astonishingly passive. One had the impression they had packed up"). In September 1944, Eisenhower had thought of Berlin as the most important objective (Strawson, 103) but by the spring of 1945 he can radio Montgomery, "That place has become, so far as I am concerned nothing but a geographical location, and I have never been interested in these" (quoted in John Tolland, *The Last 100 Days*, [New York: Random House, 1965], 325). On Eisenhower's announcement to Stalin (rather than to the British or American authorities) that he would stay in the Ruhr, on Stalin's pleased reaction, on the American high command's at first startled but then supportive reaction, and on Churchill's shock, anger, and incomprehension, see Tolland 327f., Strawson 85, 111, 119, 160. On Stalin's assessment of the critical strategic importance of Berlin (despite his disclaimer to Eisenhower) see these same works; and on the competition between Soviet Marshals Zhukov and Konev to take Berlin, see also Cornelius Ryan, *The Last Battle* (New York: Simon and Schuster, 1966), 249f.

152. Liddell Hart, *Strategy*, 396.

153. Kecskemeti, 81.

154. *War and Peace*, Bks. 3 and 4; second epilogue. For a summary and analysis of Tolstoy's arguments against "the great men theory," see Gallie, Ch. 5.

155. Sigmund Freud, "Reflections upon War and Death," in *Character and Culture*, 112.

156. Marc Bloch, *Memoirs of War*, trans. and introd. Carole Fink (Ithaca, N.Y.: Cornell University Press, 1980), 126.

157. Bloch, 34.

158. Bloch, "Réflexions d'un historien sur les fausses nouvelles de la guerre," in *Revue de Synthese Historique* 33 (1921): 2-35.

159. A critical element differentiating "torture" (as in the concentration camps) and the second function of injuring in all war is elaborated in Section IV.

160. For an analysis of arguments about torture, see Henry Shue, "Torture," *Philosophy and Public Affairs* 7 (Winter 1978), 124–43. Especially pernicious in discussions of torture is the argument that a hypothetical case can be imagined in which, for example, saving a city from a nuclear bomb might depend on torturing the madman who placed it there and knew where it was hidden (see Shue, 141; and for an example of the invocation of this argument see Michael Levin, "My Turn: The Case for Torture," *Newsweek*, 7 June 1982, 13). Introducing an "imaginable" occasion for torture that has no correspondence with the thousands of cases that actually occur has the effect of seeming to change torture to a sanctionable act. As Shue points out, the absolute prohibition against torture must be kept in place; and should the unlikely "imaginable" instance actually ever occur, the torturer would have to rely on convincing a jury of peers that the context for his act was exceptional (55). One may go further than this and point out that surely anyone who had the choice between on the one hand torturing and saving-the-city, and on the other hand not torturing and not saving-the-city, would almost certainly choose the first; but so, too, anyone confronted with the choice between on the one hand saving-the-city and being himself imprisoned and possibly executed, or on the other hand not saving-the-city and not being imprisoned or executed, would almost certainly also choose the first. That is, torturing should be perceived with the same acute aversion with which one's own legal culpability and one's own death are perceived; and while it is certainly possible and desirable that a jury would exonerate anyone in this situation, it does not follow that any such guarantee should be provided before the fact. That one might *have to* do something one day that is wrong does not mean that the act has ceased to be "wrong" and punishable. It is unlikely that any saviour of the city would actually be inhibited by the lack of pre-existing moral and legal assurances of immunity.

It is a peculiar characteristic of such hypothetical arguments on behalf of torture that the arguer can always "imagine" someone large-spirited enough to overcome (on behalf of a city's population) his aversion to torture, but not so large-spirited that he or she can also accept his or her legal culpability.

161. See Quincy Wright's discussion of the connection between armed strength and sovereignty in Machiavelli, Bodin, Grotius and others (Ch. 24, "Sovereignty and War," 895–922), and see Walzer, *Just and Unjust Wars*, 60f., and 98f.

Sometimes the concept of "sovereignty" is even understood to absolve a country from treaties and agreements formed to assure peace. Henry Kissinger, for example, writes, "A government as subtle as Hanoi must have known that there are no 'unconditional' acts in the relations of sovereign states, if only because sovereignty implies the right to reassess changing conditions unilaterally" (*American Foreign Policy*, 119); and Carl Schmitt in his 1928 essay makes a similar point, "As long as a sovereign state exists, this state decides for itself, by virtue of its independence, whether or not such a reservation [self-defense, enemy aggression, violation of existing treaties . . .] is or is not given in the concrete case" (51).

162. See Alexander Hamilton, *Federalist Papers* 23, 24, 25, 26, 27, 28, 29 in which control of the army by states or instead by the federal government is analyzed, and throughout which the "original right of self-defense which is paramount to all positive forms of government" (180) is assumed (*The Federalist Papers*, McLean edition, indexed and introd. Clinton Rossiter [New York: Mentor-New American, 1961], 152–88.

163. Schmitt, 62.

164. For example, after the Vietnam War, of all the occurrences either at home or in Southeast Asia that might have caused self-questioning in the minds of the generation that resisted the war, perhaps nothing had as much power to cause self-doubt (at least momentarily) as the sudden widespread recognition (perhaps "recognizable" during the war but not actively "recognized" until after

the war) that those who had gone to war and those who had stayed at home in protest fell into two distinct economic groups.

The revelation that the protest itself had been in part made possible by the condition of economic privilege was especially stunning because central to the protest itself had been the principle that economic might should not be used to determine the political reality of a less economically privileged population.

165. Alexis de Tocqueville, *Democracy in America*, trans. Henry Reeve, rev. Francis Bowen, rev. and ed. Phillips Bradley (New York: Vintage-Random, 1945), Vol. 2, 265f, 269, 273, 283.

166. For example, Erik Erikson describes Adolf Hitler as a young boy, walking along the roads with schoolhood friend August Kubizek, mentally unbuilding and rebuilding each house and vista they approached, as well as the older Hitler in his bunker with a room full of books on the architecture of opera houses. Erikson concludes, very tentatively, "Maybe, maybe, if he had been permitted to build, he would not have destroyed" (*Young Man Luther: A Study in Psychoanalysis and History* [New York: Norton, 1958], 105, 107, 108). It is not clear here what Erikson means by the passive syntax "if he had been permitted to build"; it is part of any builder's or artist's work to create the conditions in which he or she can do that work; and Erikson's slightly problematic speculation would make immediate intuitive sense if it read, "Maybe, maybe, if he had found the strength, courage, and persistence to build, he would not have destroyed."

167. Whether or not this correspondence exists is debatable. If it does exist, there are at least two antithetical explanations of its existence: one, that a militarily and economically powerful country is also an artistically active culture (Ezra Pound is perhaps the most familiar exponent of this reading); two, that "artistic excellence" is an unstable and arbitrary category, and those who have other forms of political power will also have the power to designate their own form of art as the best of that era.

168. Siegfried Giedion, *Mechanization Takes Command: A Contribution to Anonymous History* (New York: Norton, 1969); Lewis Mumford, *Technics and Civilization* (New York: Harbinger-Harcourt, 1934); William McNeill, *The Pursuit of Power*.

169. Gallie, 63.

170. Leonard S. Woolf, *International Government: Two Reports, Prepared for the Fabian Research Department, Together with a Project by a Fabian Committee for a Supernational Authority that Will Prevent War*, introd. Bernard Shaw (New York: Brentano, 1916), 378, 379.

171. For example, Schmitt argues this about the Kellogg Pact of 1928 and the Geneva League of Nations (50, 51, 56).

172. This would at first appear to be inapplicable to the "defense" argument. But defense assumes that another country has initiated war; thus if a substitute is found for the two occasions that permit initiation, it will by eliminating initiation also eliminate defense.

173. This is obviously not to say that the content of the artifact is in itself an indifferent matter, for it is of course of crucial importance. It only ceases to matter if torture is used on its behalf: then the mode of substantiating the artifact overwhelms and makes irrelevant any considerations of content. The nature of a construct's content, however, will ordinarily greatly affect the method by which it is substantiated. A constitution that is benign in its interior is much less likely to occasion torture than a constitution that is not, since the first will naturally elicit the consent of the people and thus bring about benign forms of substantiation.

In this regard, it is relevant to notice that the content of the issues of war fall into two categories: the material (land, wealth) and immaterial (ideas, descriptions, religious beliefs, ideologies of cultural style, and so forth). This division is important because of the following peculiar fact. The object in the first category is fought over because it *cannot be shared*: each wants it and neither wants the other to have it. The opposite occurs in the second category: now the object tends to be fought over because there is *the insistence that it should be shared*; one side, for example, has a religious or ideological belief and insists that the other population should also have it. This suggests that successive levels of disembodiment or dematerialization of culture permit (encourage, even require) successively wider acts of sharing. That is, a material object itself allows the translation of an attribute of the unsharable body to exist in the external sharable world; but there are limits on the extent to which it can be shared (though more sharable than the body, it is less sharable than wholly disembodied artifacts, language, ideas, belief systems and so forth). So great is the impulse toward sharing the dematerialized realm that when it is not shared it can cause the catastrophe of war that was produced for exactly the opposite reason in the material realm.

174. See above, Ch. 1, 38–40, 57–58; Ch. 2, 126–27; and below Ch. 4–5.

175. See above Ch. 1, 38–45.

176. Hume's word "vivacity" (*A Treatise of Human Nature*, ed. L.A. Selby-Bigge [Oxford: Clarendon, 1896], 153, 154, 629) is, when applied to the present context, perhaps the most interesting. Although it is in some ways a problematic noun (see Mary Warnock's discussion in *Imagination* [Berkeley: University of California, 1976], 134–35), it is also extremely revealing because of its etymology in the verb "*vivere*," to live (as is also the adjective "vivid"). In describing a perceived tree as vivid or as having vivacity, what may actually be being described is the vivid or intense feeling state of "seeing"; in other words, one's own intense aliveness is at that moment experienced (just as dull or neutral sensory content prevents intense awareness of the perceptual experience). Thus it is not the tree's (or the gate-post's) aliveness that is experienced, but one's own aliveness that is experienced and then attributed to the object or content of the perceptual act, the tree. The importance of this point will become clear later where this ordinary form of substantiation is compared with forms of analogical substantiation. As will become apparent, the relation between the perceptual act and object, seeing and tree, has a crude analogue in the sensory condition of pain (hurting) and an imagined object. Just as in the first case the "aliveness" or "realness" is transferred to the object (as registered in the descriptive word "vivacity") so in the second case there is a transfer of the "aliveness" or "realness" of the experience of pain to the imagined object.

177. Jean-Paul Sartre, *The Psychology of Imagination* (New York: Philosophical Library, 1948), 177-212. Some of these same attributes are described by Karl Jaspers in *General Psychopathology*, trans. J. Hoenig and Marian W. Hamilton (Manchester: Manchester University Press, 1962), 69.

178. Jack Davis, lecture on "Conviction," University of Connecticut, February, 1974.

179. Warnock, 166. The competing perceptual content on waking is not only visual but somatic. Thus it has been observed that it is much easier to remember, or to "sustain," one's dream upon waking, if the awakened dreamer does not move his or her body in any way. If one rolls over, or simply extends an arm, the dream spills away with the gesture. This same phenomenon applies to daydreams as well. Those for whom a daydream begins to become too real or vivid, often say that they "shake" themselves out of it: in effect what they are at that moment doing by the brief gesture of "shaking" is introducing competing perceptual (somatic) content.

180. What is described here as a progression from human hurt to animal hurt to no hurt, could instead be described as a progression from human hurt to animal hurt to *plant hurt* if one chooses to argue that the "realization" of the invented chair requires the carving of wood and thus the "wounding of trees." The significance of this alternate description is elaborated below in Ch. 3.

181. Philippe Ariès, *Centuries of Childhood: A Social History of Family Life*, trans. Robert Baldick (New York: Vintage-Random, 1962).

182. Michel Foucault, *The History of Sexuality*, trans. Robert Hurley (New York: Pantheon, 1978).

183. Stephen J. Pyne, *Fire in America: A Cultural History of Wildland and Rural Fire* (Princeton: Princeton University Press, 1982).

184. Nelson Goodman in *Ways of Worldmaking* insists on the distinguishability of fictions and frauds (94 and passim).

185. Even a "pre-emptive" strike has its structural equivalent in conventional war, as Liddell Hart's "a strategy so superb as to eliminate battle" made clear. In both kinds of war, both one-directional and two-directional injuring are strategic possibilities.

186. Stendhal (Marie Henri Beyle), *The Charterhouse of Parma*, trans. Lowell Bair, introd. Harry Levin (New York: Bantam, 1960), 30.

187. Noel Perrin, *Giving Up the Gun: Japan's Reversion to the Sword, 1543–1879* (Boulder: Shambhala, 1980), 25.

188. McNeill, *Pursuit of Power*, 67, 68, 94.

189. McNeill, 167–70.

190. It is also possible that the opposite is the case, that the shift of skill away from the "using"

of weapons to the "making" of weapons means that they can more easily, and therefore more readily be used. Thus, rather then freeing people from war, it enslaves them to it. Whereas earlier one would have to choose between a lifetime devoted to war preparation or instead total abstention from warlike activity, one no longer has to make an either-or decision. Whether the change is seen as freeing or instead constraining depends on whether one imagines that in that earlier period one would have been a samurai-knight or instead a farmer.

191. John Locke in *Of Civil Government* (Ch. 5) observes that our fundamental idea of "property" arises from a person's perception of his own relation to his body. Even philosophies which reject the notion of "property" accept the notion of one's own body as one's own property. Thus Marx's essential objection to capitalism was on the basis that it appropriated the laborer's body.

192. The politics of the body in slavery are a less extreme instance of what occurs in torture. The closest equivalent for nuclear war is torture, not slavery. These differences turn on the nature of the relation between work and pain and will be elaborated in Chapters 3, 4, and 5.

Notes to Chapter 3:
Pain and Imagining

1. Gilbert Ryle, *The Concept of Mind* (London: Hutchinson, 1949), 267f.

2. Jean-Paul Sartre, *The Psychology of Imagination* (New York: Philosophical Library, 1948; rpt—Secaucus, N.J.: Citadel, 1972), 208.

3. Not only is there no form of sentience specific to "imagining," but it does not, unlike other forms of sensation, even seem to be anchored in a specific part of the body. Though "images" may typically be experienced as appearing in the head, it requires very little effort to "push" the image into some other part of the body: it is almost as easy to make an imagined blue flower arise in the interior of the calf of the leg as it is to make it arise in the head; just as the picture of a foot race can occur along the interior path of the forearm, with its starting point at the elbow and its finishing point at the wrist. The "natural" location within the head (which may occur in part because of the analogue with the objectified content of hearing and seeing) becomes habitual, but is a habit that is subject to alteration.

4. The analysis of it as *both* act and object may be only implicit in a given analysis or may instead be explicit, as in Edward S. Casey, *Imagining: A Phenomenological Study* (Bloomington: Indiana University Press, 1979). Of most importance is the fact that whether or not the language of intentionality is explicitly used, such discussions tend always to include the specification of an object. Casey (49) quotes Ryle (251, 254) as asserting that there is no object in imagining, but in context Ryle seems only to mean there is no perceptual or actually sensed object: Ryle's entire discussion of necessity proceeds through a series of invoked objects.

5. Sartre, 177–212.

6. In some traditions "pleasure" has been understood as the absence of pain rather than as itself an actively experienceable condition, while in others it has been understood as a discrete sensory phenomenon the experience of which does not depend on the prior presence (or even anticipated presence) of pain. Even in the second case, however, it has tended to be understood as a *bodily* state in which something other than *the body* is experienced: see for example, Eugene Minkowski's description of "pleasure" or "contentment" as the feeling that accompanies the expansive, outward movement into the world, as when one completes an act, or makes a decision ("Findings in a Case of Schizophrenic Depression," trans. Barbara Bliss, in *Existence: A New Dimension in Psychiatry and Psychology*, ed. Rollo May, Ernest Angel, Henri F. Ellenberger [New York: Basic, 1958; rpt—New York: Simon-Touchstone, n.d.], 134). Thus the two conceptions of pleasure are not as deeply at odds with one another as they may at first appear, for in each (overtly in the first case, less overtly but recoverably in the second) it is a condition associated with living beyond the physical body, or experiencing bodily sensations in terms of objectified content.

7. It may at first appear that this description would be inapplicable to an intentional state such as fear, since the state itself is elicited by, rather than eliminated by, its object. But, as has often been noted, objectless fear may have a much greater aversiveness than fear-and-object since, in the second case, the existence of the object gives the person a course of action: he may act to eliminate the object, to move away from the object, or to placate the object, all of which are ways of altering the state of fear itself (and are ways not available to him in objectless fear, which thus places him much closer to the aversive passivity of the person in pain).

This argument would not, however, wholly satisfy the objection, since objectless fear is an unusual condition (except in those descriptions of the ordinary modern state of anxiety as an ongoing state of diffuse, objectless fear). The analysis of the relation between state and object in ordinary intention as occurring within the framing relation of the boundary conditions of pain and objectified self-transformation must be understood as it would apply not only to the alternatives of objectless fear and fear-and-object (where it is clear that the first places the person closer to pain than the second) but to the more ordinary alternatives of no fear and fear-and-object.

To understand this, it is helpful to return for a moment to the relation between pain and its imagined object. The imagined object in pain may be one of two kinds. First, it may be an object (artifact or objectified condition) in which the hurt is eliminated: if one is hungry, imagining food; if one's back aches, imagining and longing for a chair when there is none actually present or perhaps even yet invented. Second, the envisioned object may instead be an imagined cause of the pain, as has been stressed in the many examples given here of the occurrence of agency language: so a person with an acute "stabbing" pain in the leg conceives of it as "stabbing," imagines a causal knife, and may even (as in imagistic therapy) work to diminish the pain by mentally moving the knife away; or, the person may instead conceive of the leg itself as the cause, and imagine himself existing in a world without the offending limb (itself a change in the realm of externalized objects); or a person with pains in his chest and stomach may conceive of God as the cause, and so work for forgiveness and the cessation of pain. Both categories of imagined object are paths toward the elimination of pain (whether effective or ineffective): in the first case, the object (food, chair) directly eliminates the sensation; in the second, the anterior cause of the sensation is imagined, and one then works to alter one's relation to that anterior object (pushing the knife away; having the leg removed; praying to God).

The second category of pain-and-imagined object makes possible an understanding of the nature of fear-and-object that explains how the object, even though responsible for producing the sentient condition of fear that is close to pain, can be itself understood as moving the person away from pain and toward the opposite boundary of objectified self-transformation. Here fear-and-object (or the object and the fear it has evoked) reverse the temporal relation between pain-and-imagined object: seeing a knife in the vicinity of one's leg, one fears it and acts to remove it, push it away, rather than waiting until one's leg is already hurt and trying (much less effectively) to alter the pain by mentally reversing the causal action; or again, fearing God's wrath, one works to heal the relation, preventing the anticipated infliction of hurt. Thus the external object occasions a modified and diminished form of sentient suffering (fear), in order to allow the object to be acted upon, in order to prevent the more extreme form of sentient suffering (physical pain). Fear-and-object can thus itself be understood as a partially objectified, hence halfway eliminated, form of pain.

8. Because ordinary forms of sentience are, when objectless, close to pain, an understanding of pain may eventually make possible a better understanding of the much more inclusive phenomenon of sentience. That is, it would be difficult to approach an understanding of sentience by attending to the sentient experiences of, for example, seeing or hearing or hungering because it is so difficult to hold visible these occurrences separate from their objectified content. Pain, in contrast, makes possible a recognition of the characteristics of sentience (whether that sentience occurs as hurting, seeing, touching, hungering) distinct from the characteristics it (i.e., seeing, touching, hungering) acquires in its habitual interaction with the realm of objects. Thus to some extent, the "language of physical pain" can be understood more broadly as "the language of physical sentience."

9. The class of intentional objects is, in other discursive contexts, commonly understood to include

both existing objects and nonexisting ones—hence the widespread use of the term "intentionality" to designate a relation between state and object, one feature of which is that the object "may or may not" exist. The discussion that follows assumes (and takes an interest in) the distinguishability of existing objects and imaginary ones, but does so in order to show the multiple ways by which they become implicated in one another (thus also, in effect, resulting in the "may or may not exist" condition of any one intentional object).

10. The two antithetical conceptions of work are, for example, described by Karl Löwith in his chapter on "The Problem of Work" in *From Hegel to Nietzsche: The Revolution in Nineteenth-Century Thought*, trans. David E. Green (Garden City, N.Y.: Anchor, 1967), 2, ii, 260–283.

11. The centrality of the categories of "act" and "object" in Marx's analyses of work in both *Grundrisse* and *Capital* will be returned to in Chapter 4. Hannah Arendt's important analysis of the distinction between "work" and "labor" as depending on the temporal stability (e.g., table) or instability (e.g., bread) of the object (*The Human Condition* [Chicago: University of Chicago, 1958; rpt—Garden City, N.Y.: Doubleday-Anchor, 1959], 72–88) may be derived from the much more elaborated account of the distinction between objectless and objectified work in Marx, whose entire political critique depends on it.

12. See, for example, the extensive literature on industrial disease and industrial accidents.

13. Both the benign and the deconstructed form of the relation are summarized in the introductory discussion of "pain and agency"; the deconstructed form is elaborated on in the chapters on torture and war.

14. At many points in the opening chapters of this study, it has been noted that a person attempting to objectify pain will invoke the image of the weapon. It is interesting that philosophic discussions of the imagination also invoke the sign of the weapon (Sartre's nail [84], Ryle's boxing match [260,261]), even though the ostensible subject is not pain, thus suggesting the accessibility of this sign, even among the full array of imaginable objects. (That is, the writers are attempting to invoke imaginary objects that the reader will simultaneously be able to imagine, and thus the fact that the image of the weapon is among those invoked suggests that it is presumed to be universally invocable.)

15. L. W. Sumner, *Abortion and Moral Theory* (Princeton: Princeton University Press, 1981).

16. Even if, for example, anthropologists were one day able to show that the "first" artifact had been a bowl rather than a hammer, this would not change the fact that the first artifact was a tool (or weapon) since in order to make the bowl, the hand had to first be "made" a tool—that is, in the making of the bowl, the hand had to be used as a shaping agent.

17. James Madison, "Number 51," *The Federalist Papers*, McLean edition, indexed and introd. Clinton Rossiter (New York: Mentor-New American, 1961), 322.

18. Alexander Hamilton, "Number 27," *The Federalist Papers*, 175.

19. Conversation with Henri Jann, National Humanities Center, Research Triangle Park, North Carolina, September 1979.

20. George C. Marshall, *Assistance to European Economic Recovery: Statement before Senate Committee on Foreign Relations* (8 January 1948), Department of State: Publication 3022, Economic Cooperation Series 2 (Washington, D.C.: GPO, 1948), 2.

21. Jacob Talmon, "Portrait of a Humanist and His Dilemmas: In Memory of Charles Frankel," National Humanities Center, Research Triangle Park, N.C., 25 September 1979.

Notes to Chapter 4:
The Structure of Belief

1. An inconsistent pronoun will be used in descriptions of God in order to hold visible the idea that God is an artifact (it), that God is an artifact invested with attributes of personhood (he), and that God is an artifact credited with greater reality and authority than human beings (It, He).

2. Passages are cited from the Revised Standard Version. For the most part, the ideas attended to in the discussion that follows do not depend on the specific language of a particular translation. Occasionally a discrepancy in phrasing between the Revised Standard Version (hereafter RSV) of

the Old Testament and the Jewish Publication Society (hereafter JPS) translation of The Torah, The Prophets, and The Writings will be noted, if the difference in phrasing appears to alter even very slightly the point under discussion.

3. On the narrative emphasis on multiple births, see also, for example, Genesis 29:32–35 and 30:22,23 as well as passages about twins, 25:24–26, 38:27–30.

4. Even the psychologically diverse moments cited in this paragraph affirm this point, for each is a moment in which something (laughter, tears, speech) suddenly emerges out of the body, and each is also bound up with the issue of parenting and hence with the more dramatic emergence from the interior of the body: for example, as has often been noted, Sarah's laughter and the birth of the baby are directly connected by the fact that the baby's name (Isaac) means "laughter,"

5. Another example of a passage in which the double movement is accomplished within a single sentence is Genesis 29:10.

6. Judith Wegner has called my attention to the fact that in the Talmud women are sometimes explicitly referred to as "pitchers" or "vessels," though such passages are frequently much more tonally problematic than in the scriptures (e.g., b.Sabb.152a).

7. The fact that the Hebrew word "adama" is used for the dust of the original creation, whereas the word "erez" tends to be used for the dust of the generational promise does not prevent them from being imagistically linked, any more than, for example, the varying words "soil" and "dirt" (or "stones" and "rocks") in an English poem would prevent the recognition of a possible connection between the lines in which those words occurred.

8. The twentieth-century mind tends to be fascinated by the *mechanism* for replication (e.g., DNA, RNA) rather than by the sheer *fact* of replication. However, the way in which the reality or aliveness of something may assert itself through its powers of self-replication, and the capacity of the modern mind to experience awe in the presence of such multiplication, are still visible in the literary genre of science fiction which very characteristically presents an entity—whether vegetable, animal, or simply some diffuse and nearly jellylike form of matter—that at an initial moment appears in one place, two hours later appears in twenty-six places, and three days later is everywhere. The fact that multiplication in this genre (as again in medical contexts: for example, the geometric increase of an organism in contagion) is perceived as negative and elicits fear, should not divert attention from the more central fact that it is a powerful occurrence and elicits a powerful response; and of course if it has positive connotations (as in the survival and spread of one's own species rather than that of another species, terrestrial or extraterrestrial) it will be perceived as "splendid" rather than "terrifying."

Alterability or self-alteration is not just a secondary feature of "aliveness" but perhaps its most central characteristic. Erwin Schrödinger writes, "What is the characteristic feature of life? When is a piece of matter said to be alive? When it goes on 'doing something', moving, exchanging material with its environment, and so forth, and that for a much longer period than we would expect an inanimate piece of matter to 'keep going' under similar circumstances" (*What is Life? And Other Scientific Essays* [Garden City, N.Y.: Doubleday-Anchor, 1956], 69).

9. The human tendency to count in precarious situations is exemplified by the phenomenon of "triage" used in wartime as well as in other situations in which large numbers of people are simultaneously injured. Similarly, what is know as "the rule of nines" enables those attempting to rescue someone who has been badly burned to determine very quickly the seriousness of the injury at a moment when the mind of the rescuer might shut down, or be incapable of subjectively computing the scale of damage on the maimed visage before him. Again, procedures for administering artificial respiration or cardio-pulmonary resuscitation require that the rescuer know by heart certain fixed sequences of counting, and that he or she be able to enact bodily those verbal sequences. Counting may also, in any given disaster, become an improvised, rather than a formally learned, strategy of survival: there are no doubt many people who—like Zola's Catherine in *Germinal* slowly emerging from a collapsing mine by devoting her mind and legs to the dogged, progressively more difficult feat of reciting numbers—only live because they counted. Newspaper headlines, too, make visible the deep mental reflex of counting when human life is jeopardized.

On the physical and the arithmetic, see also below pp. 268–70.

10. Although the discussion here and in section II focuses on the most central and frequent events

of generation and wounding, there are also other kinds of bodily events where God's anticipatory precision is used to elicit belief: God may, for example, predict the person one will in the next moment "see" (1 Samuel 10:1f) or a pattern of "eating" (2 Kings 19:29; Isaiah 37:30). He also predicts moments of disbelief, as does Jesus in the New Testament (the predicted denial of Peter, the predicted betrayal of Judas). This last occasion for prophecy is one of the most interesting since it allows "disbelief" to be subsumed into the procedures for eliciting "belief," thereby preventing the existence of any mental ground outside the religion since "rejection" is, through anticipation, converted into a source of confirmation.

11. The transference of the powers of the woman's body to God is a projection of female attributes onto a male persona. The repeated role of the husband's prayer in securing for the woman God's gift of fertility makes the male appropriation of female powers even more strikingly apparent.

12. The mental proximity of creating and wounding is embedded even within the fragile structure of the event cited here: the supremacy of Jacob over the animals is demonstrated not by his wounding of the animals (that is, by the act of imprinting himself on their bodies) but by wounding the tree (which is perceived either as nonsentient or nearly so) and then having those markings somehow incorporated into the animal body.

13. The RSV phrase "starting places" is translated in JPS as "starting points." The emphasis on the word "start" is slightly more emphatic in JPS, recurring three times rather than two in 33:1,2. Both translations then move through an extended sequence of parallel constructions, each opening with the phrase, "They set out from . . ." (33:3f).

14. The discrepancy between God's voice and the people's voicelessness may appear to be qualified by the framing fact that this difference is itself occurring within a verbal text that is the people's history and hence their voice. This framing qualification, however, is itself complicated by the fact that the verbal scriptures are, like the ten commandments, themselves understood within the Hebrew tradition to be authored by God (or written at his direction). The relation of the verbal form of the scriptures to their content will be elaborated in section III.

15. The repeated RSV word "murmuring" is in JPS translated as "railed" or "muttering." (On the dissolution of language in complaint, see also Chapters 1 and 2.)

16. In both RSV and JPS the repeated word "stiffnecked" is used in Exodus 32–34; the RSV phrasing of "hardened his heart" in Exodus 7-9 is in JPS given as "stiffened his heart" and sometimes, less graphically, as Pharaoh remaining "stubborn." Passages cited from Zechariah, Jeremiah, Proverbs, Psalms, Nehemiah, and Isaiah expressing the resistance to belief as the hardening of a specific body part are translated in JPS either identically or with minor variations: for example, "stubborn shoulder" may be "balky back" (Zechariah 7:11,12) or "forehead of brass" may be "forehead of bronze" (Isaiah 48:4).

17. That the reader's relation to the text, Isaac's relation to Abraham, and Abraham's relation to God are parallel to one another was brought to my attention by Michael McKeon, Spring 1980.

18. The insistence on "rest" rather than "work" in this commandment is also discussed by Göran Agrell, who gives a number of different interpretations (such as the possibility that the necessity of labor is here simply being assumed) in *Work, Toil and Sustenance: An Examination of the View of Work in the New Testament, Taking into Consideration Views Found in Old Testament, Intertestamental, and Early Rabbinic Writings*, trans. Stephen Westerholm (Lund, Sweden: Håkan Ohlssons, 1976), 16f.

19. There are varying traditions for the precise numbering of the commandments.

20. Probably the most familiar contemporary instance of this phenomenon is the emphatic inequality in the representation of female and male bodies in western art, film, and above all, magazine imagery. The newsstand in almost any city tends to present to all who pass on the street a proliferation of images of women unclothed, or effectively unclothed, which is distressful to at least half of the population who pass by each day. (It subverts women's autonomy over their own bodies; their power to determine the degree to which they will or will not reveal their own bodies is pre-empted by the prior existence of such images in the most public, most communal, of spaces.) Opposition to pornography is sometimes deflected into discussions of the particular content of the photographs or drawings, attempts to assess whether the images are themselves beautiful or ugly, dignified or degrading. But the content and tone of such images vary considerably (some are in themselves

beautiful; others comic; others unpleasant; others contorted) and much more crucial is the framing fact that, comparatively speaking, men have no bodies and women have emphatic bodies. That the very serious political problem here is independent of the tone and content of the images can be illustrated with section I of the Sunday *New York Times*: the steady presentation of the disembodied male voice of the news column side by side with the drawings and photographs of women announces, iconographically, a relation between the two halves of the population that makes any discussion of the isolated content of the images (they are often quite beautiful) irrelevant.

21. Examples of other passages in which the making of a graven image is followed by an intensification of the human body include Hosea 13:16 (where because of idol-worship, "pregnant women are ripped open"), Isaiah 47:1-4 (where women are stripped naked and made to grind meal without the protective covering of their clothing) and Jeremiah 13:22-27 (where because of attention to other gods, God lifts up the skirts of the people over their faces, again exposing the body).

22. Their earlier cultural act of naming (2:19) is initiated by God, performed at his direction; their making of clothing is the first act of making wholly independent of God.

23. It should be noticed that when men and women conflate or threaten the categories, they are sometimes *temporarily* effective. The making of the calf in Exodus *is* (temporarily) the shattering of God's voice; Moses breaks the tablets containing the ten commandments. Similarly, the making of the apron of leaves is immediately followed by the description of the sound of God "walking" or "moving about in" the garden: for a brief moment he has taken on a body. The effectiveness of men and women in revising the categorical separation is elaborated in section III.

24. In, however, 33:11 the Lord is described as speaking to Moses "face to face."

25. This does not mean that "healing" is never a "sign" in the Old Testament (e.g., Miriam, after being inflicted with leprosy as a sign [Numbers 12:10], is then restored to health [12:12-15]; Naaman's leprosy is cured [2 Kings 5:8-14]), but means that healing does not have nearly the same frequency nor the same centrality as the rhythmically invoked scenes of hurt. In contrast, the infliction of hurt on human beings is not only not central to the New Testament, but seems to have almost no place there at all.

26. See Chapter 3, 172-76.

27. In his infancy, Jesus tends to be represented without clothing or with the white cloth of the swaddling clothes; this same fragile and minimal band of white cloth recurs in one major form of the crucifixion picture. On "draped" (Jerusalem type) and "undraped" (Antioch type) images of Christ on the cross, as well as on the timing of the emergence of the crucifixion as a central image in the first ten centuries of Christianity, see Kenneth Clark, *The Nude: A Study in Ideal Form* (Garden City, N.Y.: Doubleday-Anchor, 1956), 306-309.

28. On crucial exceptions to the "predicatelessness" of God in the Old Testament, see section III.

29. Herbert Schneidau, *Sacred Discontent: The Bible and Western Tradition* (Baton Rouge: Louisiana State University, 1976), 5.

30. The ears of this writer can only hear the orphan-like tone of God's announcement; for the suggestion that God's voice may at such moments instead have the sound of a bureaucrat, I am indebted to Allen Grossman.

31. The final phrase of this passage is given more concretely in JPS as everyone "will hiss over all its wounds."

32. The conflation of creating and wounding also occurs in other ancient texts such as the *Iliad*. For example,

So the arrow grazed the outermost flesh of the warrior, and forthwith the dark blood flowed from the wound.

As when a woman staineth ivory with scarlet, some women of Maeonia or Caria, to make a cheekpiece for horses, and it lieth in a treasure-chamber, though many horsemen pray to wear it; . . . even in such wise, Menelaus, were thy thighs stained with blood, thy shapely thighs and thy legs and thy fair ankles beneath. (4, 11.137f., trans. A.T. Murray, Cambridge, Mass.: Harvard University Press, 1924)

Here the merging of creating and wounding works more in the direction of eliciting a compassion for, or an honoring of, the wounded (though it may therefore also work to glorify human acts of

·killing other persons since the wound is now an artifact and thus the warrior-who-wounds is a kind of craftsman).

33. I would like to thank Rabbi Jack Bemperod who in a conversation at the Hastings Institute, Spring 1980, first called my attention to the Isaiah passage in which God speaks as though in physical pain. (The phrase "woman in travail" is more concretely given in JPS as "woman in labor.")

34. Because of its beauty, its association with the cessation of the death-laden storm, and its explicit designation as a sign of the covenant between people and God, the benign connotations of the rain-bow may make it difficult to recover the fact that it is God's bow (or "My bow") set in the clouds (i.e., put aside, rather than used). The word "bow" (*keshet*) used as a noun elsewhere in the scriptures refers to the weapon: the curved or arched shape of the rainbow makes it a bow whose scale and splendor are appropriate to God; as a bow that is set aside, it becomes an image of the now benign (because unused) divine weapon.

35. See above, Chapter 3, 175.

36. The RSV "lips" is in JPS "pipes."

37. The repeated verb "passed over" in the RSV (also used in all these verses in the King James Version) is in JPS always given as "crossed over" (the Hebrew ʿabar, rather than passah) except for Joshua 4:7 where "passed over" is used for the movement of the Ark, the actual artifact that has made possible the parting of the waters and thus the passage of the people.

That the two events are counterparts of one another is suggested by many shared attributes: both the exodus and the entry entail the parting of the waters (the Red Sea in the first instance; the Jordan in the second); one of the people's first acts while encamped in Gilgal is to renew the original "passover" ritual (Joshua 5:10); in each event the circumcision ritual occurs and is a required prelude to one's participation in the passover observance (Exodus 12:48; Joshua 5:2), and so forth.

38. Except in each of the two, metals and materials are kept (Exodus 12:35; Joshua 6:18,24).

39. As is true of the Deuteronomy song, many verbal observances and rituals are, though substitutions for the substantiating work of the body, introduced as being recorded in the body, as though to signal (or recall) the original locus of substantiation. This is true of the rituals of cleansing and ordination, for example, in which blood is placed on the right ear, right hand, and right foot (Leviticus 8:22; 14:14). See also Exodus 13:9; Deuteronomy 6:6,8; 11:18; and Joshua 1:8.

40. Max Weber, *The Protestant Ethic and the Spirit of Capitalism*, trans. Talcott Parsons, introd. Anthony Giddens (New York: Scribner, 1958).

41. E. Digby Baltzell, *Puritan Boston and Quaker Philadelphia: Two Protestant Ethics and the Spirit of Class Authority and Leadership* (New York: Free Press, 1979; rpt—Boston: Beacon, 1982).

42. Marx is, of course, drawing on a very rich tradition of nineteenth-century philosophic materialism (especially the work of Hegel and Feuerbach) as well as on a complicated weave of political and economic writings. His sources are not the subject here and will not be attended to.

43. Karl Marx, "Preface to the First Edition" (1867), in *Capital: A Critique of Political Economy*, Vol. 1, trans. Ben Fowkes, introd. Ernest Mandel (New York: Random-Vintage, 1977), 90.

44. *Capital 1*, 727.

45. Karl Marx, *Grundrisse: Foundations of the Critique of Political Economy*, trans. and introd. Martin Nicolaus (New York: Random-Vintage, 1973), 661. Marx repeats the analogy on 670: "(In the human body, as with capital, the different elements are not exchanged at the same rate of reproduction, blood renews itself more rapidly than muscle, muscle than bone, which in this respect may be regarded as the fixed capital of the human body)." This same reliance on the body as an explanatory model recurs in the much more carefully thought-through writing of *Capital I*: for example, "Among the instruments of labour, those of a mechanical kind, which, taken as a whole, we may call the bones and muscles of production, offer much more decisive evidence of the character of a given social epoch of production than those which, like pipes, tubs, baskets, jars etc., serve only to hold the materials for labour, and may be given the general denotation of the vascular system of production" (286).

The use of the body-state metaphor in the fable of Menenius Agrippa to discourage workers from rebelling against the state on the grounds that they are rebelling against their own projected body is dismissed by Marx as absurd because the fable "presents man as a mere fragment of his own body" (481,2). He makes no attempt to explain how this use of the metaphor can be absurd when at the same time his own large structural dependence on the metaphor is legitimate: his political critique rests on the idea that the collective wealth of the state is the projected body (labor) of the workers, and that the appropriation of that wealth is thus an appropriation of the worker's own body. Though the political conclusions he derives from the metaphor are opposite to those of Menenius Agrippa, the metaphor itself is the same; and Marx unfairly pretends that the metaphor, rather than the interpretation, is intellectually weak.

46. On his own insistence upon the scientific character of his work see, for example, *Capital 1*, 433, as well as his letters quoted by Martin Nicolaus, "Foreword," *Grundrisse*, 56.

47. Marx himself invokes a similar explanation when, for example, he criticizes those who would dismiss money as a "mere symbol," noting that "In this sense every commodity is a symbol, since as value, it is only the material shell of the human labour expended on it" (*Capital 1*, 185).

48. *Capital 1*, 287.

49. *Capital 1*, 289.

50. *Capital 1*, 296. See 308.

51. This passage, as well as those that follow, are from Jack Cohen's translation of the Fourth and Fifth Notebooks of Marx's *Grundrisse*, first made available in English under the title *Karl Marx: Pre-Capitalist Economic Formations*, ed. and introd. E. J. Hobsbawm (New York: International Publishers, 1965), 67. The same passages in the Nicolaus translation of the full *Grundrisse* again convey the compelling character of the original relation between the worker's body and the land, though they tend to be slightly less graphic: where, for example, Cohen translates the land as "a prolongation of the body," Nicolaus translates it as "man's extended body"; what Cohen translates as the "intimate merging of the human body and the earth," Nicolaus translates as "the entwining" of body and earth. The passages cited are from the Cohen translation, but the page and line number for the equivalent phrasing in Nicolaus are indicated also (as in this instance, Nicolaus, 471, 1.21f.).

52. Cohen, 69; see Nicolaus, 473, 1.6f.

53. Cohen, 81; see Nicolaus, 485, 1.16f.

54. Cohen, 85; see Nicolaus, 488, 1.31f.

55. Cohen, 89; see Nicolaus, 491, 1.32f.

56. Cohen, 92; see Nicolaus, 493, 1.31f.

57. Cohen, 108; see Nicolaus, 505, 1.20f. Passages similar to these cited from the *Grundrisse* also occur in *Capital 1*. For example, "Thus nature becomes one of the organs of his activity, which he annexes to his own bodily organs, adding stature to himself in spite of the Bible. As the earth is his original larder, so too it is his original tool house" (285).

58. James J. Gibson, *The Senses Considered as Perceptual Systems* (Boston: Houghton, 1966), 112.

59. For an extended discussion of the physical continuity between the body of the worker and the raw materials of earth as it occurs in literature, see E. Scarry, "Work and the Body in Hardy and Other Nineteenth-Century Novelists," *Representations* 1, 3 (Summer 1983).

60. Although this confusion may sometimes be read into the text, at certain moments it seems to reside in the text itself.

61. See Mandel's discussion of strikes in "Introduction" to *Capital 1*, 48, 49.

62. For example, *Capital 1*, 951 and passim. John McMurtry in *The Structure of Marx's World-View* (Princeton: Princeton University Press, 1978), 64, also calls attention to Marx's original vocabulary and the idea of body damage by describing property as a dismembering of the worker's external organs. McMurtry makes the second important point that by conceiving of the earth as a prolongation of the human body, Marx's position "devalues nature by depriving it of any claim to independence," permitting it to have value only as it becomes an extension of the human being (65n). Marx's starting point is thus analogous to the starting point in Genesis where, as described earlier (see above, p. 222), the natural world is subverted and reconceived as the outcome of, or territory proper to, creating.

63. See, for example, Ernst Fisher's descriptions of tools as substitute body parts in "The Origins of Art," in *Marxism and Art: Writings in Aesthetics and Criticism*, ed. Berel Lang and Forrest Williams (1972; rpt—New York: Longman, 1978), 142: "Man, or the pre-human being, had orig-

inally discovered—while gathering objects—that for instance, a sharp edged stone can take the place of teeth and fingernails for tearing apart, cutting up, or crushing prey.'' See *Capital 1*, 493, n.4.

64. Cohen, 91; see Nicolaus, 492, 1.29f.

65. Cohen, 87; see Nicolaus, 489–90.

66. Cohen, 95; see Nicolaus, 495.

67. Cohen, 107n; see Nicolaus 504n.

68. In *Capital 1*, Marx repeatedly makes the analogy between "individual consumption," where the product or artifact is the individual's own body, and "productive (or social) consumption," where there come to be products separate from the bodies of the individuals (289).

69. Just as the recognition that a made thing (whether "cloth" or "the state") is a projection of the human body can be interpreted to have political implications consonant with Marx's own view or instead implications that are exactly the opposite (as in the fable cited in note 45), so the counterpart thesis, that the *body is itself an artifact* may lead to political conclusions either compatible with or instead sharply divergent from Marx's own. In the Sixth Notebook of the *Grundrisse*, for example, Marx records a passage from *The Principles of Political Economy*, whose author, John Ramsey MacCulloch, is almost never mentioned in *Capital* without anger:

Man is as much the *produce of labour* as any of the machines constructed by his agency; and it appears to us that in all economical investigations he ought to be considered in precisely the same point of view. Every individual who has arrived at maturity . . . may, with perfect propriety, be viewed as a machine which it has cost 20 years of assiduous attention and the expenditure of a considerable capital to construct. And if further sum is laid out for his education or qualification for the exercise of a business etc., his value is proportionally increased, just as a machine is made more valuable through the expenditure of additional capital or labour in its construction, in order to give it new powers. (London, 1825, 115, cited in *Grundrisse*, 615, 616)

This passage, which takes man's self-recreating nature as license to demote him to the object world, again finds its way into the Eighth Notebook (849), where it is greeted with equal astonishment.

The existence of divergent interpretations of the thesis strengthens, rather than undercuts, the thesis itself; for it suggests that the thesis is independent of, and conceptually prior to, any particular political ideology.

70. Cohen, 84, italics added; see Nicolaus, 487-8.

71. Stephen Jay Gould, "Posture Maketh the Man," in *Ever Since Darwin: Reflections in Natural History* (New York: Norton, 1979), 207ff. Gould often returns to the problem of the *a priori* assumption of "cerebral primacy" in his essays on natural history (see, for example, "Our Greatest Evolutionary Step," and "Piltdown Revisited," in *The Panda's Thumb: More Reflections in Natural History* [New York: Norton, 1982], 125–133, 108–124).

72. Frederick Engels, "The Part Played by Labour in the Transition From Ape to Man," in *Karl Marx and Frederick Engels: Selected Works* (New York: International Publishers, 1977), 359. Hegel also emphasized the importance of upright posture and the hand as the primary tool ("Philosophy of Mind," Sec. 411); Gould cites Ernst Haeckel as possibly Engels's immediate source.

73. Most notably, and widely noted, in the Russian Lysenko experiments (1934–64).

Although Engels's phrasing invites and justifies the general interpretation of his assertions as requiring a Lamarckian mechanism, it is also true that the Darwinian mechanism of natural selection alone allows for the appearance of the progressively more agile hand in very early ancestors.

The general issue raised here must be seen within the frame of the larger fact that human evolution has been primarily cultural rather than biological. Gould, for example, writes, "All that we have accomplished since [the time of the Cro-Magnon people] is the product of cultural evolution based on a brain of unvarying capacity" ("Natural Selection and the Human Brain: Darwin *vs.* Wallace," in *The Panda's Thumb*, 56). Like our contemporary science, our ancient myths also stress the independence of cultural progress from the limitations of the body, for the craftsman (e.g., Hephaestus, Philoctetes) is often physically handicapped. Insofar as evolution is cultural rather than biological, there is no question about humanity's self-conscious assumption of responsibility for its own evolution, nor is there any question about its ability to pass on benefits *acquired* during one generation's lifetime to the next generation.

74. The hand has this same kind of primacy in the scriptures as well, for it is referred to many hundreds of times, far more than any other bodily part, such as face, head, or heart.

75. The stipulation that this is a difference in degree is important, for animals have not only *verbal* or vocalized forms of communication but also the rudimentary equivalents of *materialized* objectification, artifacts or tools. Engels, for example, mentions in his essay the spider's weaving of a web, as does Marx as well (*Capital 1*, 284). The web is a tool that not only assists the catching of prey but does so by extending the range of the creature's sentience: that is, just as a person may literally feel different surfaces at the end of a walking stick or scissors (noted earlier), so it is today recognized that some spiders, by poising their legs on the threads of the weave, feel in the vibrations of the threads the approach of another creature; thus the threads act as a literal extension of their nervous systems. Frequently cited instances of *instinctual* animal construction include the arch of the termite, the habit of some birds to use one of their own feathers to fish with, and the fact that some fish have a growth on their bodies that acts as a "fishing lure" to attract other fish, either for food or reproductive processes. As an example of the last, see Gould's discussion of the *Lampsilis ventricosa* ("The Problem of Perfection, or How Can a Clam Mount a Fish on Its Rear End?" in *Ever Since Darwin*, 103f.). After saying that human beings are, in their capacity for objectification, differentiated from other animals by a difference in degree, one must then go on to acknowledge that the degree is itself so vast a one that the difference comes to seem almost as qualitatively great as if there had been no point of identification.

76. *Grundrisse*, 832.

77. Nelson Goodman, *Ways of Worldmaking* (Indiana: Hackett, 1978), 63f.

78. Marx traces in detail the changing path of internal referentiality. His very ability to identify and define discrete phases within the transformation of capital depends on it: what makes a single phase apprehensible as a phase, what enables him to move inside the vast artifact, the total economic system, and begin to map the glacial nuance of its gigantic activity, seems to be, more than anything else, the periodic changes in the flow of referential activity.

To designate a phase is to designate a place which, though only a fractional part of the total artifact, is nevertheless a part in which the total integrity of the artifact for a moment coalesces and coheres, a place about which one may say "whatever it is an artifact 'is' is here," "whatever it is an artifact 'does' is done here," a place where its essential "reality" or (in the shared idiom of economics and aesthetics) its "value" may be found to reside. But its value *is* its capacity for referential activity, and its capacity for referential activity is most easily apprehended at the moment when its direction suddenly shifts. In our reading of Marx's now completed map of the path of value, the fact of value is first a point, its referential direction is a line extending out from the point, and its changing direction is a fork in the line. But it seems probable that for the author the sequence of discovery ran in the opposite direction: the vague shape of a fork first catches Marx's attention, and in coaxing its outlines into heightened visibility he then determines the exact length of the line, and designates the precise location of the point. At any rate, whatever its role in the process of discovery, the fork—which will itself fork and refork many times until there is a gigantic tracery of forks that lies across the subject of Marx's investigation—comes to constitute the very center of each successive phase.

Marx is not a maker of line-drawings (or of maps or of lace or of nets) but a maker of texts and what is here described as the repeated marking of a fork is in *Capital* visible in the rhythmic reentrance onto the page of a set of double terms: material object/commodity, commodity/money, wage/profit, work/labor, use value/exchange value, relative term/equivalent term, absolute surplus value/relative surplus value, surplus value/profit, constant capital/variable capital, fixed capital/circulating capital, labor process/valorization process, production capital/capital of circulation, commodity capital/commercial capital, money capital/money-dealing capital, profit-bearing capital/interest-bearing capital, and so forth. Not in every instance but in many, many instances the companion terms identify something that is originally singular but that acquires a doubleness from its duality of referential direction. A material object, for example, is essentially one thing, both projected out of and reciprocating or referring back to what immediately preceded its own existence, the human being. The double terms "use value" and "exchange value" express the fact that, although the object continues to be singular in the circumstances out of which it arose (for the "value" of both its use value and its exchange value derives from its having been projected out of a human being), it no longer has a singular direction of reference: in its use value the material object retains its original obligation to refer back to its predecessor, while in its exchange value it acquires a freedom from that mandatory return to the site of its own making and refers instead to other material objects. This same pattern recurs across many other pairs of double terms. Eventually over the course of *Capital 1, 2,* and *3,* Marx distills a single structure of activity common to and repeated in each successive phase. This structure of activity can be applied retrospectively to the original human act of making that starts the sequence and, when so applied, works to clarify the relation between "creating" and "reference-breaking." Because the intricacies of this require an entry into Marx's complex and sometimes circular economic vocabulary, it will be attended to in an essay separate from the present study.

79. What is here the case is probably almost always the case. What is frequently talked about as an artifact's—a painting's, a poem's, a piece of philosophy's, a legal system's—internal referentiality (or autonomy) would probably in every instance be accurately understood to be a discussion of its "*temporary* internal referentiality" or "*eventual* freedom of external reference." An object that refuses to surrender its referentiality will be destroyed; if it both refuses to surrender its referentiality and cannot be destroyed, we then enter the nightmare situation of the sorcerer's apprentice. Certain objects having these two attributes exist both in historical reality (e.g., nuclear waste that cannot be disposed of) and in science fiction, but it is crucial to notice that no one is celebrating their "autonomy" in the way that the (only apparent) autonomy of art works is sometimes celebrated. When an artifact is genuinely autonomous, it is celebrated by no one. Thus an autonomous object cannot be taken as a model either for what an artifact normally is or what it should be.

80. Karl Marx, "Economic and Philosophical Manuscripts" (1844), in *Early Writings,* trans. Rodney Livingstone and Gregor Benton, introd. Lucio Colletti (New York: Random-Vintage, 1975), 284.

81. Although furniture does not have quite the primacy of shelter and food, it is included here as belonging to the first circle immediately beyond the edges of the body because the absence of furniture usually signals the near-absence of shelter and food. Furniture may be sold to pay for shelter, and shelter may be forfeited to pay for food, but, after that there are no other objects to pass through. Engels in his descriptions of the working poor (*The Condition of the Working Class in England* (1845), trans. and ed. W. O. Henderson and W. H. Chaloner [1958; rpt—Stanford: Stanford University Press, 1968]) includes precise details about furniture, which tends either to be wholly absent or, as though caught in a moment of arrested motion, to be in the state of disappearing before our eyes: for example, "the only furniture consisted of two rush-bottomed chairs with seats gone, a little table with two legs broken, one broken cup and one small dish. . . . and in one corner lay as many rags as would fill a woman's apron. It was on these rags that the whole family [seven persons] slept at night. . . . she had had to sell her bed during the previous year in order to buy food. She had pawned her bedding with the grocer for food. Indeed everything had been sold to get bread" (37, and passim).

82. *Economic and Philosophical Manuscripts,* 286. Similarly he writes that the "capitalist purchases labour . . . from the worker in order to capitalize a sum of money, and the worker sells his labour . . . in order to prolong his life" ("Appendix: Results of the Immediate Process of Production," in *Capital I,* 991).

83. *Grundrisse,* 891.

84. It can be argued that the particular rubrics "laborer" and "capitalist" belong to a specific, sharply etched historical moment in industrialism, and have ceased to be the terms through which problems in material distribution can today be best understood and repaired. Even if this is the case, Marx's analysis continues to be helpful in two ways: first, in his framing assumption that political and economic injustices can be approached and understood through an understanding of creating; second and more specifically, in his recognition that where an inequality in material distribution exists (whether the affected populations are "laborers and capitalists" or instead, "physical laborers and mental laborers," "makers and users," "users and owners," "first-world citizens and third-world citizens," etc.), there will be an inequality in embodiedness. Here the structure of conflict differs from what occurs, for example, in war, where there is an ongoing, shared intensity of

embodiedness to produce only as a final outcome an inequality of embodiedness (i.e., inequality in level of injury), carrying with it the inverse inequality in world-extension (i.e., the less injured is the winner, and thus has the greater "right" to determine the disposition of postwar issues).

85. As was visible in earlier chapters, a discrepancy in world-extension thus conceals the deeper discrepancy that is its inversion: the "having of objects" is the "have not" condition of pain, and the "not having of objects" is the "have" condition of pain. The inversion in the "have/have not" language that occurs when one moves from expressing the inequality in terms of objects to expressing it in terms of the body is not a matter of word-play, for which of the two formulations is used influences the way the inequality itself comes to be perceived and explained.

86. *Capital 1*, 433. All subsequent references to Marx's writings are exclusively to *Capital 1*; page numbers will be cited in the text.

87. At one point (298–300), the capitalist enters *Capital 1* long enough to have a three-page-long hypothetical debate with the author; but the capitalist, having been so introduced, is once again subtracted out, for the passage ends by saying that he would actually never have entered into such a conversation, that even the "function" of defending his own point of view is performed by someone else, in this case, by "professors of political economy" who are paid to describe economic conditions in a way sympathetic to the owners.

88. On the "capitalist" as one who is by definition exempt from the process and thus has a personal life, see 423, 667, 741. On the capitalist's potential complexity of personhood in his personal life in contrast to his vacuity of presence in the process of production, see, for example, 343. Marx is only very infrequently ambiguous on this point, as in "Appendix," 990, where he describes workers and owners as equally engulfed in the process.

89. The capitalist's body enters only in the form of a joke, as in Marx's occasionally repeated play on the idea of the worker "tanning" (or not tanning) the capitalist's hide, or giving him a "hiding." For example: "In tanning ... [the worker] deals with the skins as his simple object of labour. It is not the capitalist whose skin he tans" (425, and see 280, 1007, and elsewhere). Whether this is a *funny* joke is debatable, but insofar as it *is* a joke, the joke depends wholly on suddenly subverting the capitalist's state of physical exemption and, for a fleeting moment, imagining him as physically vulnerable to the process or to other persons in the process.

90. Unlike the continually differentiated workers, the capitalist tends to be referred to only by the general rubric, "capitalist." If a particular kind of capitalist is specified, that specification is almost immediately retracted. For example, in Part 8 of *Capital 1*, Marx speaks separately of the "agricultural capitalist" and the "industrial capitalist," but then adds in a footnote that the distinction is not a precise one: "In the strict sense the farmer is just as much an industrial capitalist as the manufacturer" (914).

91. Ernest Mandel, "Introduction" to "Appendix: Results of the Immediate Process of Production," 944.

92. Mandel, "Introduction," 944. In summarizing the changes in Marx's manuscript, Mandel cites Marx's 31 July 1865 letter to Engels in which he expresses his hope of making *Capital* a "dialectically articulated artistic whole." Marx also tells Engels, "I cannot make up my mind to send off anything before I have the whole thing before me. Whatever shortcomings they may have, the virtue of my writing is that they are an artistic entity, and that can be achieved only by my method of never having them printed until I have them before me in their *entirety*" (*Karl Marx–Friedrich Engels, Selected Letters: The Personal Correspondence, 1844–77*, ed. Fritz J. Raddatz, trans. Ewald Osers [Boston: Little, Brown, 1980], 112).

93. "Appendix: Results of the Immediate Process of Production," 950.

94. "Appendix," 949.

95. The tool has an important place in Marx's writing. It restores the referent because it mediates between worker and artifact, and thus when the image of the tool is held steadily visible, the original site of human projection is held visible as well. For this reason the tool is often taken as a summarizing sign of Marx's work.

In this connection, it is interesting to notice that a potentially profound change in the "signs" of nationhood has occurred in the twentieth century: for the first time, tools appear again and again on the flags of many countries, increasingly coming to displace weapons as the chosen sign of national self-identification. The overall occurrence is not itself attributable to Marx: although some countries whose national flags bear an image of the tool explicitly seek to identify themselves with Marx (e.g., U.S.S.R.), others have no such identification (e.g., Austria, India). There exist, of course, many sources and precedents: the banners of medieval guilds often contained very beautiful depictions of tools; later, the banners of some of the city companies of England did also; many of the United States' state flags, adopted primarily in the nineteenth century, included plows, mining tools, axes, scythes, sickles, anvils, and rakes (they also included bows, arrows, guns, and swords, but the tools outnumber the weapons); and so forth.

Despite the existence of many precedents and sources, the twentieth-century willingness to make the tool not simply the sign of a group (e.g., guild), city, or region (state), but the sign of the nation-state itself seems a significant change. Prior to the twentieth century, national flags and coats of arms do not include tools; in fact, it is unusual for them to include any man-made object other than swords, shields, and crowns. Two striking exceptions are the red stocking cap of liberation that occurs on the national flag or coat of arms in some Latin American countries (Cuba, El Salvador, Argentina, Nicaragua) and which had already begun to surface in the nineteenth century, and the Irish harp which, though not officially adopted until 1919, occurred earlier in regiment flags. Since this is the first century in which tools have emerged as a major sign of national identification, it is impossible to assess whether that appearance represents a change of very little, or instead, very great significance. Although persons living in the third century A.D. might notice the increasingly frequent appearance of the cross, it would not have been possible for them to guess the scale of the cumulative weight the sign was then in the midst of acquiring.

National flags that have, during at least *some period within* the twentieth century, depicted tools include the following. The hammer of industry and the sickle of farming was adopted by the U.S.S.R. in 1923: it occurs not only on its national flag but also on the flag of each of its fifteen constituent republics such as Georgian S.S.R. and Armenian S.S.R.; these two tools, or some variant of them, occur on the flags of the autonomous republics (the mattock and horsewhip of Eastern Mongolia, the sickle and rake of Tuva, the anchor and pick of the Far Eastern Republic). A hammer and a sickle are held by the eagle of the Austrian national flag. A hammer and a pair of dividers appear on the national flag of the East German Democratic Republic. A spinning wheel (which is at the same time a charka) is on the national flag of India. There are a hammer and a hoe on the national flag of the People's Republic of the Congo. There is a hoe on the flag of Upper Volta. A hammer and hoe appear on the national flag of Costa Rica. An armillary sphere appears on the flag of Portugal. Countries whose state arms (but not necessarily their flag) have at some point depicted tools include Liberia, Zambia, Tanzania, Namibia, Gambia, New Zealand, Trinidad, Honduras, and Panama. Countries whose state arms have included the word ''work'' or ''labor'' include the Central African Republic, Chad, Republic of Dahomey, Zaire, Upper Volta, the People's Republic of the Congo, and Barbados.

In addition to hand tools, flags sometimes include machine tools. The unity of agriculture and industrial work, for example, can be represented by hammer and sickle, or instead by a cogswheel and a sheaf of grain. A cogswheel appears on the national flag of Burma, of Mongolia, and of Bulgaria. It again appears in the state arms of the People's Republic of China, of Botswana, of Poland, of North Vietnam, and of Italy. Larger machine tools, such as a power station, have occurred on either the flag or the arms of Zambia, Romania, and North Korea. (Catalogues of flags consulted include Whitney Smith, *Flags: Through the Ages and Across the World* [Maidenhead, England: McGraw-Hill, 1975]; A. Guy Hope and Janet Hope, *Symbols of the Nations* [Washington, D.C.: Public Affairs Press, 1973]; and Terence Wise, *Military Flags of the World* [New York: Arco, 1978]).

Notes to Chapter 5:
The Interior Structure of the Artifact

1. On the meaning and use of this word, see Chapter 1, and below, p. 293–96.
2. Judgments about persons that are made on the basis of skin color are atavistic, since such

judgments cannot be made without mentally depriving people of their clothing, divesting them of the habit of self-recreation, and reconceiving of them as beings prior to culture.

3. Sigmund Freud, *Civilization and Its Discontents*, trans. and ed. James Strachey (New York: Norton, 1961), 41, 42.

4. Philip Fisher, noting the way words used to designate parts of the human body (e.g., hand, lips) are also used to designate parts of objects (e.g., a cup's handle and lip), writes, "Imagine that a cultural taboo existed such that no word for a part of the body could also apply to things. Jealous and timid, the human race could fear a contamination from the flow of resemblances and linkages between man and things. That we in fact do the opposite makes possible both the flooding of the world of matter with human meanings and the subsequent recovery of the human image from that world" ("The Recovery of the Body," *Humanities in Society* 1 [Spring 1978], 140).

5. Jonathan Miller, *The Body in Question* (New York: Random, 1978), 208.

6. Jeremy Bernstein, "Calculators: Self-Replications," in *Experiencing Science* (New York: Dutton, 1980), 237, 8.

7. John Fitch, *The Autobiography*, ed. Frank D. Prager (Philadelphia, 1976), 113, cited in Brooke Hindle, *Emulation and Invention* (1981; rpt—New York: Norton, 1983), 28.

8. Marx's attribution of "aliveness" to inanimate objects occurs in two forms, either as a straightforward attribution (see above, Chapter 4), or instead in the form of a complaint that a given object is characterized by "indifference" or "obliviousness" (to complain that a problematic or defective object is indifferent is to imply that a successful object would not be characterized by such unawareness).

9. Barry M. Blechman, Stephen S. Kaplan, *Force Without War: U.S. Armed Forces as a Political Instrument* (Washington, D.C.: The Brookings Institute, 1978), 2.

10. The word "literally" here refers to the plane of literal and overt events *within* the narrative.

11. See Chapter 1, 52f. and 62n., and Chapter 3, passim.

12. Though this theme surfaces in complex ways in many of Bergman's films, it is most simply and starkly presented in his late film, *Fanny and Alexander* (1983).

13. Sheila Cassidy, "The Ordeal of Sheila Cassidy," *The Observer* [London], 26 August 1977.

14. These two objects are cited in medical and torture reports (e.g., "Transcript of the Torturers' Trial," 42) read at the International Secretariat of Amnesty International, London, 1977.

15. Miguel Angel Asturias, *Strong Wind*, trans. Gregory Rabassa (New York: Dell-Laurel, 1975), 196, and see 7, 8, 9, 22 for similar use of objects.

16. Charles Dickens, *Bleak House*, ed. Norman Page, introd. J. Hillis Miller (Harmondsworth: Penguin, 1971), 690, and *Our Mutual Friend*, ed. and introd. Stephen Gill (Harmondsworth: Penguin, 1971), 379. I would like to thank Deidre Murphy for bringing the Dickens examples to my attention.

17. Plato, *Laws*, trans. A. E. Taylor, in *The Collected Dialogues of Plato Including the Letters*, ed. Edith Hamilton and Huntington Cairns, Bollingen Series LXXI (Princeton, N.J.: Princeton U. Press, 1961), 1432.

18. Oliver Wendell Holmes, *The Common Law*, ed. Mark DeWolfe Howe (Boston: Little, Brown, 1881, 1963), 23.

19. Holmes, 25.

20. Holmes, 33.

21. Chief Justice Marshall, as cited by Judge Story (*Malek Adhel*, 2 How. 210), as cited by Holmes, 27.

22. Holmes, 13.

23. The transcripts of cases alluded to in the discussion that follows, as well as other unpublished trial materials (e.g., depositions, closing arguments where not included in the transcript), were made available to me in 1979 through the generous research facilities of two Philadelphia law firms: La Brum and Doak; and Beasley, Hewson, Casey, Erbstein, and Thistle.

Because the subject of this discussion is "object failure," the analysis will draw primarily on cases in which, according to the jury's verdict, the object *did* fail—that is, cases in which the object (or the defendant company) *was* responsible for the bodily hurt suffered by the plaintiff. However, the three structural elements described here are not dependent on or limited to the point of view of the plaintiff, and need only be inverted to be applicable to a case in which the jury has ruled for the defendant. For example, the unifying function of "the path of the accident" would be equally characteristic of a trial that had as an outcome a verdict for the defendant, except that (according to the jury) the trial will have demonstrated that such a path did not exist, that the plaintiff and the object never converged on such a path (that the plaintiff was not hurt, or if hurt, was not hurt by the object).

24. Transcript of Proceedings, Janice, Salvatore, and Theresa Foresta v. Philadelphia Gas Works, Roper Corp., Roper Sales, Mars Wholesale, and Roberts Brass, No. 15038-10 (Pa. C.P., Nov. term 1974).

25. It might at first seem that a play and a trial would be differentiated by the fictional content of the first and the historical content of the second. But a play may, of course, have an actual historical action for its subject, just as, conversely, the lawyers in a trial may disagree about the degree to which the subject matter in front of them is fictional or historical. As will be clarified below, however, the play and trial *are*, in the end, distinguished by their respective "fictionality" and "reality," but this distinction applies to the audience's (or jury's) ability to act on the subject matter rather than to the subject matter itself.

26. In very exceptional instances, a work of literature may be intended to bring about actual social action, or may do so whether or not such action was intended (Stowe's *Uncle Tom's Cabin* is perhaps the most frequently cited example of this very small category). Further, it would probably be accurate to say that the more a literary work has, or is intended to have, this outcome, the more closely it will approximate a trial: thus, for example, Bertolt Brecht, who wanted his plays to have concrete social effects, repeatedly described them as trials, their themes as court pleas, and their audiences as juries.

27. If, of course, the case is one in which there is a question about whether the plaintiff actually suffered any hurt, then the defense will in its closing argument summarize these doubts rather than (as in the kind of case under discussion) accepting the indisputable physical damage as a given and arguing that it is irreversible. Here, the defense may suggest that not only would such juror action *not* undo the accident, but it might also bring harm to the defendant. Because the closing tends to discredit the juror's power to act on the accident itself, it works to credit and invite audience passivity and inaction.

28. Harry Lipsig, quoted by Alan Richman, "For the Afflicted, a Champion in Court," *New York Times*, 25 April 1979.

In *Foresta* v. *PGW*, Paul R. Anapol opened his closing argument for the plaintiff by comparing the jury's exceptional power to bring in a verdict with Congress's power to make war and peace (Transcript, Vol. 11, pp. 69–72), and throughout the closing he repeatedly returned to the subject of their authoritative capacity to act, as, for example, at the moment when he began to speak specifically about the physical suffering of the plaintiffs (p. 151f.).

This same approach is visible in the closing arguments of Jim Beasley, one of Philadelphia's leading plaintiff lawyers. Throughout his closing for the plaintiff in *Flores* v. *Lubbock Manufacturing Company* (a case presenting an accident in which the plaintiff had suffered unthinkable kinds of hurt), he repeatedly called on the words of figures like Oliver Wendell Holmes and Theodore Roosevelt to remind the jurors that their present role was perhaps the most important one they would be assigned in their lifetime. Toward the end of the closing, this power of action was increasingly presented in the language of counterfactual reversal: they were invited to transform the catastrophe into "a verdict which is noble" (Transcript of Closing, pp. 13, 15); the final sentences credit the jurors with almost cosmic powers of reversal— "Jimmy and with him, his family, in part can be delivered from this pit of bottomless affliction by your verdict. His sun has been darkened, his moon does not give light, and his star has fallen from the heavens . . . Now let your verdict come with much power and glory to give compensation for his unbearable losses" (21).

29. The plaintiff's lawyer can specify exact figures for medical costs, unemployment, and so forth, but cannot specify a figure, nor even a precise procedure for arriving at a figure, for the physical suffering (nor can the judge; this is left exclusively to the discretion and authority of the jurors). The plaintiff's lawyer will, however, address the subject of monetary compensation for pain. At one time it was permissible for the lawyer to say to the jurors, "How much would you pay *not* to have this happen to you; how much would you pay *not* to be subject to this degree and duration

of pain?'' Though no longer allowed in most states, this specific approach is cited here because it is such an overt articulation of the phenomenon of counterfactual reversal toward which the overall efforts of the plaintiff's lawyer are directed: the sentences cited explicitly place the jurors in a temporal position prior to the event and ask them to arrive at a figure that will *negate* the occurrence of the accident (as though the density of the object world acts as a buffer between the body and external agents of pain). This counterfactual reversal of the pain is, of course, present, even when the cited formulation is disallowed. Sometimes in their difficult task of attempting to translate degree and duration of suffering into a monetary form, the jurors will, on their own initiative, think through the translation in terms of a specific object that will enhance the life of the person (for example, the cost of a college education). Thus, world-extension is explicitly poised against the annihilation of world content that earlier occurred in the physical pain. The terms used for the financial award—''damages'' and ''recovery''—also suggest mimetic reversal.

30. In some cases, both these judgments are made together at the close of the trial; in other cases, the trial is subdivided into two parts, one on the question of liability (after which the jurors arrive at a verdict), then followed by a second part on the monetary question (after which the jurors arrive at a decision about the appropriate size of the award).

In the first arrangement, where the trial is unitary, there is an inconsistency built into the structure of the defense argument: the defense lawyer must argue, ''This object (or its maker) was not responsible, and if it was responsible, the award should be as follows,'' or ''We're not liable, but if liable, only for a small amount.'' Sometimes the judge's charge will take note of this inconsistency: for example, the judge in *Jenkins* v. *Pennsylvania Railroad* cautions, ''Let me say, ladies and gentlemen, prematurely, that because I talk now about damages, you should gain no implication from that that it is my will that you should bring in a verdict for the plaintiff'' (Transcript of Proceedings at 318, sec. 171a; No. 3774 [Pa. C. P., Sept. term 1964]; *rev'd*, 220 Pa. Super. 455, 289 A.2d 166 [1972]). The plaintiff lawyer, in contrast, has a structurally consistent argument: ''This object was responsible and the award should be as follows.'' The division of a trial into two distinct parts appears to eliminate the difficult structural inconsistency in the defense argument, since he or she need only move to the second position once the first position has already been lost, and thus the first position is not prematurely undercut by the necessity of simultaneously introducing the second.

31. Because of the consistent and overwhelmingly ''self-evident'' evidence of the hurt suffered by the plaintiffs, none of the defense lawyers openly disputed the fact or even the degree of hurt. But at one point the PGW attorney introduced a stove expert to testify against one of the co-defendants (not against the plaintiff). This witness had not earlier been present in court and, in making assessments about the stove, spoke somewhat cavalierly, or at least ignorantly, about the degree of injury. Interrupting the proceedings, the judge, as though struck, turned to the witness, and said with the quiet incredulity of one who is deeply offended, ''Didn't you know, didn't you know, that 75 percent of this little girl's body was burned?'' (It would later be explained that PGW had shown the witness photographs taken long after the healing process was underway and allowed him to misperceive them as pictures taken immediately after the explosion). Ordinarily, a witness's statements of ''fact'' are called into question or refuted by the attorney on cross-examination, or by the attorney's introduction and questioning of a different witness. Thus this occasion of judicial intervention was a riveting moment in the trial, and one later referred to in the closing argument for the plaintiff (Transcript, Vol. 11, 120, 121), and again referred to by the attorney for Mars (one of the co-defendants) in their closing against PGW (Vol. 12, 72). The judge had, in this moment, not only announced that the witness was in error but, in effect, announced that the freedom and fluidity of interpretation appropriate to so many courtroom subjects was, in some simple and absolute way, deeply inappropriate to this one.

This contrast between the fluidity of verbal constructs and the nonfluidity of certain bodily facts has also been evident in the nonlegal contexts encountered in earlier chapters (see above, 2, p. 127f., 133f.; and 4, 192, 268f.).

32. According to defense attorney Dan Ryan, many lawyers feel that in section 402A of *Restatement of Torts*, the American Law Institute acted to extend greatly (rather than merely to summarize) object expectation in the United States; though the *Restatement* does not carry the force

of law, it has worked to revolutionize the law in areas such as 402A, shifting the legal trend from the side of the property owner to the side of the consumer (Conversation, LaBrum and Doak, Philadelphia, July 1979).

33. Similarly, the issue of smellability was an important issue in *Hennigan* v. *Atlantic Refining Co.* (Transcript of Proceedings at 1230f., 1240, 1292f., 2477, and passim, 282 F. Supp. 667 [E.D. Pa. Nov. 1967]; *aff'd*, 400 F.2d 857 [Dec. 1968]), just as "visibility" was an issue in *Murphy* v. *Penn Fruit* (Transcript of Proceedings at Vol. 2, 484 and passim, No. 4172 [Pa. C.P., Apr. term 1973]; *aff'd*, 274 Pa. Super. 427, 418 A2d 480 [1980]).

34. Transcript of Proceedings at Vol. 2, 444–531, Murray v. Beloit Power Systems, 79 F.R.D. 590 (D.V.I. 1978).

35. The place of blame and its psychological counterpart, guilt, is difficult to formulate. The judge in such a case will often point out to the jury that there is no question of criminal guilt at issue; but (as defense lawyers have sometimes noted), a verdict against the defendant may carry with it a form of social stigma. In some cases there are, in addition to compensatory damages, punitive damages. In very exceptional circumstances criminal charges may be brought: by the middle of 1979, seventy-six lawsuits had been filed against Ford in connection with its Pinto; seventy-five of them were civil suits; the seventy-sixth was a criminal case including three counts of homicide (Reginald Stuart, "Year-Old Recall of Ford's Pinto Continues to Stir Deep Controversy," *Sunday New York Times*, 10 June 1979).

36. Melvin M. Belli cites the figure of ninety-eight percent in *"Ready for the Plaintiff!"* (1956; rpt—New York: Popular Library, 1965), 66.

37. See, for example, Embs v. Pepsi-Cola Bottling, 528, S.W.2d 703, 706 (Ky. Ct. App. 1975).

38. Harold J. Berman, "American and Soviet Perspectives on Human Rights," Congress of the International Political Science Association, Moscow, 16 August 1979, published in *Worldview*, November 1979, 20.

39. Berman, 16.

40. An analysis of an extreme historical moment of the failure of reciprocation is given in Chapter 4, iv; and an analysis of a more extreme instance in which persons are deprived of autonomy over the phenomenon of projection is given in Chapters 1 and 2, v.

The failure at either site is a deconstruction of the artifact (whether the made thing is a state or any other political or nonpolitical construct); and thus the emphasis here is on the importance of protecting both sites (as appears to be the growing tendency in the two countries cited). This, however, is not to say that if one could protect *only* one site or the other, one or the other would be equally good; for, as suggested earlier, the site of projection has a primacy. The privileging of this site does not depend on one's allegiance to the concept of democracy, since a democracy is an expression of that primacy rather than the vehicle by which the concept comes into being (see above, Chapter 2, section v). In ordinary circumstances, however, the two so entail one another that if one of the two actions is intact, the other will also be (though perhaps to a lesser degree) intact. If, for example, one enjoys the reciprocating benefits of a fiction, one will tend to enter willingly into creating it and sustaining it; if one is deprived of its reciprocating benefits, one will choose not to enter into it and may actively rebel against it.

41. See 3, 173–76; and 4, 213–21, and passim.

42. I. E. S. Edwards, *The Pyramids of Egypt*, illus. J. C. Rose, 3rd ed. (Harmondsworth: Penguin, 1976), 262.

43. Interviews with Craftsmen, *On the Road*, narr. Charles Kuralt, prod. Ross Bensley (New York: C.B.S. News, 1983), C.B.S. broadcast, 26 June 1983.

44. Thus, if reciprocation is received in a symbolic form (e.g., money) rather than in direct access to the completed object, it will be difficult to determine the appropriate amount of "compensation," for she should not be paid simply for the action of coatmaking, nor for the coat, but for the excessive reciprocating power of the coat (the thing she has actually made).

She might only be paid for the difficulty of the action of coatmaking (but this would be the same as if she had each day repeated the warming dance of labor without having ever made an object); she might be paid for the number of days she devoted to making the coat (but the action of the coat lasts not for days but for eighteen months); she might be paid an amount that would accommodate her own needs, rather than those of both herself and her children (but the coat's referential powers extend to more than one person, as is evident in its entry into the marketplace texture of exchange). If, in summary, she were paid an amount only *the equivalent* of the aversiveness she experienced, this would be the same as her never having engaged in an act of "making" at all, since an act of making, by definition, entails a nonequivalency that benefits the maker.

45. See above, Chapter 4, section iii.

46. Daniel Defoe, *The Life and Adventures of Robinson Crusoe*, ed. and introd. Angus Ross (Harmondsworth: Penguin, 1965), 128.

47. Letter "To Giambattista Beccaria," 13 July 1762, in L. Jesse Lemisch, ed., *Benjamin Franklin: The Autobiography and Other Writings*, Farrand text (New York: Signet, 1961), 248.

48. Sigmund Freud, *Leonardo da Vinci and a Memory of His Childhood*, trans. Alan Tyson, ed. James Strachey (New York: Norton, 1964).

49. For example, the mathematics of imaginary numbers assisted the discovery and work with electricity, just as DNA analysis has drawn on, among other things, Fourier mathematics and the linguistic analysis of the syntactical features of language (Horace Freeland Judson, *The Eighth Day of Creation: The Makers of the Revolution in Biology* [New York: Simon-Touchstone, 1980], 537f.); just as the descriptive model of the repressor mechanism in genes has drawn on the structural model of the computer (Philip J. Hilts, "On Divinity Avenue: Mark Ptashne and the Revolution in Biology," *Scientific Temperaments: Three Lives in Contemporary Science* [New York: Simon, 1982], 188).

50. Stephen Jay Gould, "A Biological Homage to Mickey Mouse," in *The Panda's Thumb: More Reflections in Natural History* (New York: Norton, 1982), 95–107.

51. Arthur I. Miller, "Visualization Lost and Regained: The Genesis of the Quantum Theory in the Period 1913–27," in Judith Wechsler, ed., *On Aesthetics in Science* (Cambridge, Mass.: MIT Press, 1979), 73–105.

INDEX

Aaron, 202, 209, 236
Abimelech, 194, 195
Abram/Abraham, 127, 186, 187, 188, 190, 191, 195, 196, 197, 204–5
Absence: of Abraham and Isaac in Nahor, 197; of capitalist from workplace, 264–66, 366n.89; of God, 197, 211, 214; of Sartre's Annie and Pierre, 163–64
Accident: in product liability trial, 297–98, 299; in war, 74–75
Adam, 185, 187, 209
Agency, 13, 27–28, 45, 47, 52–53, 54, 56–57, 84, 154, 180; language of, 13, 15–19, 27, 61, 172, 356n.7. *See also* Weapon
Aging, 32–33
Agrell, Göran, 359n.18
Agrippa, Menenius, 361
Algeria, 42
Algiers, 98, 329n.7
Aliveness, 22, 192, 230, 358n.8; conferred on mental or material object, 71, 147, 247, 264, 285–306, 354n.176, 368n.8; as weapon, 38, 343n.82
Altar, 189–90, 204, 211–12, 234, 238, 252, 284
Alteration: of body in childbearing, 192–93; of body by God, 194–95, 199, 237; of body through verbal and material making, 254–55; of body in war, 64, 112–13, 123; common to wounding and creating, 183, 204; of land, 199, 247; power of, in weapons and tools, 174–75, 315; of world and body in work, 82, 253; of world in perception and imagining, 168, 212
American Civil War, 98, 105, 138, 144, 339n.4, 340n.53

American Revolution, 138
Amnesty International, 9–10, 13, 17, 50, 51, 62, 130, 326, 328 nn. 1, 8, 329 nn. 2, 7, 330n.10, 333n.59
Analogical verification, 13–14, 21, 124–27, 205, 278, 353n.173; believer's *vs.* nonbeliever's body, 149; and benign forms of substantiation compared, 125, 139, 146, 147, 212–13, 215, 280, 354n.176; of God, 201–204; of issues in war, 119–20, 121, 124–39, 350n.138; need for displaced by artifacts, 14, 185, 222, 235–36, 238, 240–41, 242, 249n.133; not eliminated by material culture, 242, 257, 276; in torture and war compared, 143–48
Anapol, Paul R., 369n.28
Anesthetic, 174
Angell, Norman, 342n.80
Animals, 174; and artifacts, 364n.75; in Plato's homicide statutes, 293, 294; and verification, 126, 127, 138–39
Animism: attribution of pain to weapon, 16; in economic writings, 264, 286; in homicide statutes, 293–94; in maritime law, 294–95; by military, 67, 71, 286–87, 289, 336n.18; pathetic fallacy in art, 286, 305
Annunciation, 217
Antietam, 339
Arbitration, relation to structural attributes of war, 341n.72
Architecture: in Hitler's childhood, 353n.166; of war, 348–49n.128
Ardennes, 70
Arendt, Hannah, 58, 347n.110, 357n.11
Argentina, 44, 47, 335n.16, 367n.95
Aristotle, 297

373

and sparing, 204, 238; and parental body, 188, 192, 204
Childbearing: in China, 110; in Old Testament, 185–98, 199–200, 240; in U.S. law, 112
Chile, 28, 31, 42, 50, 291
China, People's Republic of, 110, 367n.95
Churchill, Winston, 12, 65, 72, 86, 335 nn. 13, 14, 337–38n.30, 343n.86, 347n.116, 351n.151
Circumcision, 204, 235, 361n.37
Civil war. *See* American, Spanish, Chinese
Clausewitz, Carl von: *On War*, 12, 20, 65–66, 77, 78, 96, 97–101, 103, 133, 141, 334n.5, 336n.19, 338nn. 36, 38, 342nn. 80, 81, 343n.82, 345n.90, 346n.102, 348n.120
Cleansing: ritual in Old Testament, 202, 361n.39; vocabulary of war, 66, 335n.13
Code, 133
Cohen, Jack, 361n.51
"Colonel Blotto," 336n.18
Colonels' Regime. *See* Greece
Colonial wars, 98
Committee of European Economic Cooperation, 99, 344n.87
Communion, 216
Compassion: and imagination, 304, 325; as a lending of the body, 197; as a lending of the voice, 50; materialized in artifacts, 288–93; punishment of objects lacking, 293–95; and reciprocal sentience, 233; subverted by agency, 58–59; willed refusal for, 37, 56, 58
Complaint, 54, 200–201, 276
Concentration camp, 41, 42, 58, 66, 137, 157, 328n.1, 330n.11, 347n.108
Confession: as "betrayal," 29–30, 35, 36, 47, 54, 330n.10; objectifies dissolution of prisoner's world, 20, 34–35, 36, 38; tape-recorded, 49; as wound, 46
Consent, 21, 112, 114, 115, 122, 144, 152–56, 347n.115, 353–54n.173
Constitution, U.S.S.R. and U.S. compared, 308–10
Contest: binary *vs.* two-person model, 87–88, 100–1, 340n.65, 345 nn. 89, 90; language of, 85–86; 332n.48, 340n.50; out-injuring, with other contests compared, 91–108, 114, 137–38; play, game, contest differentiated, 84–85; play, work, war differentiated, 82–84; prize, 86, 96, 341–42n.75; two forms of, based on source of judgment, 341n.72; war as, 81–90
Cost, idiom in war, 75–77, 80, 350n.138

Counterfactual perception, 22, 310, 315; materialized in craft, 289–91; materialized in trial, 299–301, 369nn. 28, 29
Counting, 192, 269–70, 358n.9. *See also* Body count; Kill-ratio
Craftsman: handicapped, 363n.73; warrior as, 360–61n.31
Crane, Stephen, *The Red Badge of Courage*, 336n.17
Creating: becomes conflated with wounding, 127, 180, 183, 184, 197–99, 206, 226, 257–58, 294; brings about disembodiment, 133–41, 219, 307–24; divests external world of inanimateness, 281–307; endows mental object with a freestanding form, 21, 146, 171–72, 177, 214, 217, 218–20, 222–23, 241, 280, 290–92; habit of, embedded in cultural framework, 177, 179–80; moral resonance of, 22, 222, 242, 244, 257–58, 281, 283, 301, 304, 306; nationbuilding as, 108, 140–42, 177–79, 308–10; pain and, 145–49, 197–98, 238–41, 244, 257, 277, 280, 289–92, 324; scale of territory, 222, 226, 249, 325, 362n.62; sites of projection and reciprocation, 231–33, 243–44, 249–51, 256, 258–60, 270, 276, 307–24
Creativity, as argument for war, 140–42
Cross, 19, 213, 214, 216, 217, 219, 367n.95
Crossbow, 151
Crucifixion, 34, 215
Cuban Missile Crisis, 75, 340n.54

Dance: of labor, 290–91, 316; marathons, 341n.72; name for torture, 44; national, 112; and political resistance, 346–47n.108; in scriptures, 202; in strategy writings, 71
Da Vinci, Leonardo, 81, 141, 284, 346n.102
Davis, Jack, 354n.178
Death: body's loss of reference, 118–19; cited to certify both victory and defeat, 117; designated "useless" after war, 73; in Dickens, 292–93; and George Eliot, 31; in Homer, 123; nonsentience perceived as invincibility, 126; pain as sensory equivalent for, 31, 61; Sartre's Pablo Ibbieta, 31–32; Shakespeare's Cordelia, 330–31n.15; and torture, 49
Dedication: "For my country," 73, 121–24, 131; objects made "for anyone," 292
Defense, 65, 139, 337n.21
Defoe, Daniel, *Robinson Crusoe*, 319–20
Delbrück, Hans, 102
Denial: of injury in war, 64–69, 136; of Jesus by Peter, 214; of pain in torture, 56–57
Desire, 161–62, 166, 168, 262, 284